## About Island Press

Island Press is the only nonprofit organization in the United States whose principal purpose is the publication of books on environmental issues and natural resource management. We provide solutions-oriented information to professionals, public officials, business and community leaders, and concerned citizens who are shaping responses to environmental problems.

In 1994, Island Press celebrated its tenth anniversary as the leading provider of timely and practical books that take a multidisciplinary approach to critical environmental concerns. Our growing list of titles reflects our commitment to bringing the best of an expanding body of literature to the environmental community throughout North America and the world.

Support for Island Press is provided by Apple Computer, Inc., The Bullitt Foundation, The Geraldine R. Dodge Foundation, The Energy Foundation, The Ford Foundation, The W. Alton Jones Foundation, The Lyndhurst Foundation, The John D. and Catherine T. MacArthur Foundation, The Andrew W. Mellon Foundation, The Joyce Mertz-Gilmore Foundation, The National Fish and Wildlife Foundation, The Pew Charitable Trusts, The Pew Global Stewardship Initiative, The Rockefeller Philanthropic Collaborative, and individual donors.

# Metapopulations and
# Wildlife Conservation

# Metapopulations and Wildlife Conservation

Edited by

Dale R. McCullough

**ISLAND PRESS**

Washington, D.C.   Covelo, California

The authors are grateful for permission to include the following previously copyrighted material: Figure 4.3 is reprinted from L. Hansson, L. Fahrig, and G. Merriam, eds., 1995. *Mosaic Landscapes and Ecological Processes.* Published by Chapman and Hall, London. Figure 12.2 is reprinted from T. R. Loughlin et al., 1992. *Marine Mammal Science,* vol 8. Published by The Society for Marine Mammalogy.

No copyright claim is made in chapters 7, 12, and 16, work produced by employees of the U. S. government.

**Library of Congress Cataloging in Publication Data**

Metapopulations and wildlife conservation / edited by Dale R.
  McCullough
        p.   cm.
     "Based on a symposium . . . held at the First Annual Meeting of the
  Wildlife Society on 22 September 1994 in Albuquerque, New Mexico"—
  Pref.
     Includes bibliographical references and index.
     ISBN 1-55963-457-X (cloth). — ISBN 1-55963-458-8 (paper)
     1. Mammal populations—Congresses.   2. Bird populations—
  Congresses.   3. Wildlife conservation—Congresses.   4. Wildlife
  management—Congresses.   I. McCullough, Dale R., 1933–   .
  II. Wildlife Society.   Meeting (1st : 1994 : Albuquerque, (N.M.))
  QL708.6.M48   1996
  599.052′48—dc20                                      96-11225
                                                           CIP

Printed on recycled, acid-free paper ⊗ ✿

Manufactured in the United States of America

10   9   8   7   6   5   4   3   2   1

# Contents

# Preface

This book is based on a symposium on Metapopulations and Wildlife Conservation held at the first annual meeting of The Wildlife Society on 22 September 1994 in Albuquerque, New Mexico. When I first proposed the symposium, I knew that metapopulation dynamics was an area of growing interest in the wildlife field, but I would not have dared to hope for the overwhelming response the symposium elicited.

On the day of the symposium, the initial room assigned, a large one, filled quickly, so the conference organizers folded back a partition. When that space filled, another partition (the last one available) was opened. The aisles, sides, and back of the room were filled with people sitting on the floor and standing in every space; the doorways were filled with heads, and people milled about in the hallway trying to hear the presentations. About 700 people, an astounding crush of bodies, crowded in (who knows how many others gave up and went away), and most stayed until the session ended at 10:00 P.M. Phil Hedrick, one of the speakers, commented wryly to me that he hadn't spoken to so many people since he last taught Biology 1A.

Although I had intended to publish the symposium from the start, this audience erased any doubts about the demand for a book on this topic. Not only is the subject timely, but the authors are well recognized in the field and interest is high in the species in question, most of which have been landmark problems in wildlife conservation.

This book is not just a proceedings of the symposium. Some papers given at the symposium have not been included, and other chapters not presented were invited later. Every chapter was subjected to anonymous peer review by at least two experts in the field and underwent significant alteration and refinement during the review and editing process. My goal has been to produce a book not only technically correct but also an integrated and coherent whole.

The symposium and the preparation of this volume were funded by the A. Starker Leopold Chair in Wildlife Biology at the University of California,

Berkeley, of which I am the holder. I am indebted to the authors who took time in their busy schedules to contribute their expertise and insights to this endeavor and am grateful for their patience with requests for additional material, revisions, just one more photo, and a myriad of other details. I wish to thank the reviewers of the various chapters for suggesting improvements and checking technical accuracy: Fred Allendorf, Mark Boyce, Terry Bowyer, Todd Fuller, Dan Goodman, Alan Hastings, Ed Heske, Walter Koenig, Paul Krausman, Tom Kucera, Bill Lidicker, Kevin McKelvey, James Patton, Kathy Ralls, John Sauer, Mark Shaffer, Tony Starfield, Tom Valone, Jerry Verner, Lisette Waits, Jeff Walters, and Bob Wayne.

At Island Press, Barbara Dean and Barbara Youngblood were most helpful and supportive of the project and facilitated the realization of this book. My greatest debt is to my assistants Margaret Jaeger and Lori Merkle. Margaret did a major share of checking accuracy and consistency and suggested many improvements in the writing. She also kept in contact with authors about the status of their manuscripts in relation to our requirements. Lori, who has long experience working with me, was her usual stalwart word processor, catcher of obscure errors, dealer with details, and rock of calmness in a sea of chaos.

# Metapopulations and Wildlife Conservation

# 1

# Introduction

*Dale R. McCullough*

In recent years, the word "metapopulation" has found its way into the discourse about wildlife conservation and management. The concept is not particularly new, but its relevance has grown with the changes wrought on natural environments by expansion of modern human civilization. This book considers the ideas concerning metapopulations and explores their usefulness to applied programs to conserve wildlife in a rapidly changing world.

Typically a population is thought of as an interacting collection of animals of the same species occupying a defined geographic area. The boundaries of this area might be set by a number of different criteria. In various cases, the area chosen may be local distributional limits, the entire range for narrowly distributed species, geographic units in which movement and interaction of the animals are greater within than between units, or simply a land scale amenable to administrative efficiency. In traditional usage, movements and interactions by individuals were relatively continuous over the population area, even though the habitat (the physical and biological environment that satisfies the species' prerequisites of life) may vary somewhat in overall quality from place to place. Often such populations were termed "panmictic," which means completely mixed in terms of genetics. Most populations of vertebrates have age, sex, and social structure, however, and do not mix freely. Breeding does not occur at random even in the case of very small populations. For the sake of consistency in this book, such traditional populations will be referred to as "continuous" populations.

A metapopulation, by contrast, is discontinuous in distribution. It is distributed over spatially disjunct patches of suitable habitat "patches" separated by intervening unsuitable habitat ("matrix") in which the animals cannot survive. Because of the risk of mortality in crossing hostile conditions of the matrix, movement of animals between the patches is not routine. Consequently, movement between patches ("dispersal") is restricted. Furthermore, because many of the habitat patches may be small, consequently supporting small population sizes, extinction in given patches ("local extinction") may be a common event compared to extinction of continuous populations. A metapopulation's persistence depends on the combined dynamics of extinction within given patches and recolonization among patches by dispersal. So long as the rate of recolonization exceeds the rate of extinction, the metapopulation can persist even though no given subpopulation in a patch may survive continuously over time.

Levins (1969, 1970) first coined the term "metapopulation"—that is, a population of populations. His interests related to the control of insect pests in a patchy environment. Prior work on subdivided populations had been done by people like Huffaker et al. (1963), who studied mite populations on food patches in a laboratory setting. In these experiments, food patches were arranged in uniform $x$ and $y$ arrays. Levins' (1970) model, sometimes called a "classical metapopulation," closely mimicked the circumstances of these laboratory experiments. In Levins' model, habitat patches separated by nonhabitat space were all of equal size, quality, and spacing, and organisms had equal probabilities of dispersing from one patch to any other patch in the array. The only matter of concern was the presence or absence of the individuals on each patch: if they were present, the assumption was that they would quickly increase to carrying capacity. Levins asked what rates of recolonization by random dispersal were necessary to balance random rates of extinction by patch so that the metapopulation would persist through time.

Levins' (1970) model, like Huffaker et al.'s (1963) experiments, had an elegant simplicity that illuminated the concept of how a constellation of partially isolated patches in space could yield overall stability to a system that was inherently chaotic at the level of the individual patch. This kernel of an idea, after lying largely dormant for nearly 20 years, has recently burgeoned into our current appreciation of metapopulation dynamics. Although some insects and plant populations conformed sufficiently to continue a small interest in the metapopulation idea, it suffered from being a solution in want of a problem. As pointed out by Harrison (1991, 1994) and others, the simplified assumptions of the Levins (1970) model do not apply very well to the complexity of most populations in nature. The problem that reawakened the metapopulation idea was the rapid and rampant fragmentation of once continuous habitats by the spread of human activity. As the remnant patches of habitat receded to a size too small to support populations without incurring local extinction, biologists quickly realized that hope for future perpetuation of many species rested upon maintaining many habitat patches and having animals disperse among them. The problem of fragmentation demanded a solution, and metapopulation ideas floating on the fringes of biological thinking were brought into central focus.

Given that few populations fit the assumptions of the Levins (1970) model—landscapes and fragmentation seldom possess such symmetry—just what is a metapopulation? Why do we use the term in a way that clearly violates the intent of the originator? The first and simplest response is, who cares? We are facing a crisis of loss of habitat, loss of species, and loss of biodiversity unprecedented in human experience in its degree and global scale. It may rival the great upheavals of paleontological history in that the rules of the game for survival are being irreversibly changed. Biologists hoping to arrest, or at least slow, this process want fixes, and they want them now. Metapopulation dynamics, as broadly and loosely conceived, offer one such fix. Whether it will work, ultimately, only time will tell.

A second answer is found in the manner of the concept's growth. In the traditional course of academic development, each step in the progression of an idea is clearly demarcated. Thus Smith's original model is elaborated by Jones' model, and additional permutations are addressed by Johnson's model, much like a biblical listing of who begat whom. But when the right idea arrives at the right time to fill a deeply felt need, it can take on a life of its own, evolving in a larger context outside the rules of priority and progression. It is as if a virus escaped from the laboratory and infected a multitude of people, mutating as it spread. Possession passes to the users, and the idea evolves according to their needs. To those in the arena, the nuances of an abstract model are the luxury of academicians.

In view of the continuing evolution of the metapopulation idea, it perhaps is neither desirable nor possible to give a rigorous definition to the term. For the purposes of the conservation and management audience this book addresses, a working definition will do. The definition must include populations that are already functioning as metapopulations, but it must remain cognizant also of the processes that are converting continuous populations to metapopulations. By definition a continuous population is not a metapopulation, nor is a spatially segregated population with high dispersal that eliminates the possibility of local patch extinction. Consequently, the two keys to the metapopulation idea are, first, a spatially discrete distribution and, second, a non-trivial probability of extinction in at least one or more of the local patches.

Purists may find this rather loose definition unsatisfactory, but it is the needs of applied ecologists that this book addresses. Some ambiguity is needed to admit the vagaries of nature across space and dynamic processes over time. For example, the fine distinction of a population as continuous versus metapopulation is of little importance if it can be seen that the processes operating on that population are inexorably shifting it from the former to the latter. For applied ecologists, it is the outcome that matters. A precise scheme that describes nicely but conserves nothing is of little help. Consequently, the definitions that will categorize populations as continuous or metapopulations must be contained within a broader delineation of the processes that shift populations to metapopulation status. The case histories in this book, therefore, include not only populations that clearly function as metapopulations, but others that illustrate the transition from continuous to metapopulation. Grizzly bear populations (Chapter 14 in this volume), for example, vary from natural metapopulations in the absence of anthropogenic influences, to extensive continuous populations, to metapopulations created by habitat fragmentation, depending on which part of their geographic distribution one chooses to examine.

Theoretical explorations of metapopulation dynamics, elucidated by computer models, are far ahead of empirical studies on the reality of metapopulation behavior in nature. Gilpin and Hanski's (1991) predominantly theoretical treatment of metapopulation concepts was extremely mathematical and abstract and not readily accessible to many of the wildlife conservators and land managers on the ground.

Some wildlife populations are natural metapopulations, although in the past they were not viewed in that manner. The metapopulation concept has given theoretical structure to the dynamics of such species, and technological developments have contributed greatly to our knowledge. Tools such as radiotelemetry to study dispersal, powerful DNA techniques to illuminate genetic structure, GPS instruments to locate accurately, and desktop computers and GIS software to handle voluminous quantities of data open up whole new avenues of inquiry. One set of natural metapopulations includes species adapted to a unique habitat disjunct in distribution. Examples include hyraxes (*Heterohyrax brucei*) and klipspringers (*Oreotragus oreotragus*) on rock outcrops in the Serengeti Plain; marmots (*Marmota* spp.) and pikas (*Ochotona princeps*) of mountain meadows and talus slopes in the western United States; an array of species occupying relict habitats on mountaintops in the Great Basin ranges; and gulls, terns, and other nesting birds on islands. A second set of metapopulations includes secondary-successional species occupying disturbed areas in forests or other extensive habitats. In fact, naturally occurring metapopulations are rather common. We have long recognized the biology unique to these species but lacked a unifying concept with which to view them collectively. The recent adaptation of the metapopulation concept has given that unity.

— The greatest concern, however, is not with naturally occurring metapopulations but with the rapid conversion of what were originally continuous populations to metapopulations by fragmentation of vast expanses of natural habitats through human activity. The northern spotted owl (*Strix occidentalis caurina*) is a celebrated case. Initially it occupied the undisturbed old-growth forests of the Pacific Northwest (Figure 1.1). When clear-cut logging first began in the H. J. Andrews Experimental Forest (Figure 1.2), on the western slope of the Cascade Range in Oregon in the 1950s, no one worried about the fate of spotted owls. Indeed, such cuttings were seen as diversifying the landscape and increasing biodiversity—which, in fact, they did. It was only years later, after logging had removed much of the original old-growth forest, that the plight of the spotted owl and associated species was fully realized (Thomas et al. 1990; Chapters 7 and 8 in this volume).

It is curious that over the years we have made a mental transposition of the matrix and patches. Originally the matrix was viewed as the extensive climax habitats found by Europeans on arrival on the North American continent. The patches were the disturbed areas caused by fire, storm, and flood. As humans have increased disturbance through logging, cultivation, drainage, planting, and spraying, the amount of climax habitat has declined

Figure 1.1. View eastward from Carpenter Mountain to Three Sisters Peak at the Cascade Mountains crest in western Oregon in 1958. Note the complete absence of clear-cut logging: the only disturbed area (in the hill in the middle foreground) is a fire scar.

Figure 1.2. View of the H. J. Andrews Experimental Forest on the western slope of the Cascade Mountains in Oregon showing recent clear-cut logging blocks in 1958.

(Figure 1.3). Thus the remaining undisturbed habitat has become the patches and the extensive disturbed areas are now the matrix. In fact, these categorizations do not adequately express the complexity of either the original (which included aboriginal human effects) or the modern human-altered landscapes. Both are better viewed as complex mosaics.

Nonetheless, abstractions that simplify (as in Figure 1.3) clarify the paradigm of metapopulation dynamics. Why did the wildlife field not think of the original landscape in metapopulation terms? Certainly disturbed patches were recognized as such, and a cardinal principle was the importance of creating edge to favor game species by producing disturbed patches in climax habitats. Fragmentation, for most of the history of the wildlife field, was viewed as good, not bad. But like the mineral, vitamin, or medicine that in small doses is beneficial but in large amounts is poisonous, too much fragmentation creates a different set of problems.

Probably the main reason disturbance-adapted animals were not thought of in terms of metapopulation ideas was their high capacity for dispersal—the consequence of a long evolutionary history of being adapted to finding and colonizing a shifting distribution of disturbed areas caused almost randomly across the landscape by meteorological phenomena. And if dispersals are high enough, discontinuously distributed subpopulations function as a continuous population because the matrix does not constitute a significant barrier.

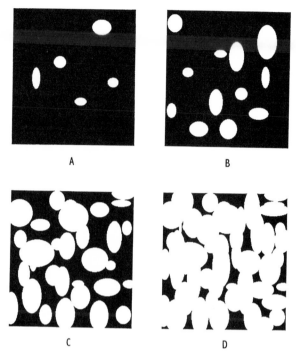

A

B

C

D

Figure 1.3. Hypothetical landscapes showing how the matrix and habitat patches in North America have been transposed over the years. Black indicates a mature habitat; white signifies disturbed habitats. **A.** The pre-European condition, the mature habitat, was the matrix in which scattered disturbed habitat patches (10 percent) were located. **B.** Disturbance by anthropogenic events (30 percent). **C.** Accelerated anthropogenic disturbance (50 percent). **D.** Advanced anthropogenic disturbance (80 percent) in which the mature habitat has become the patches and the disturbed area has become the matrix.

Although disturbance species were well adapted to dispersal across a matrix of climax habitat, rapid fragmentation of climax habitat is forcing a myriad of animal species that formerly were in continuously connected habitats to function as metapopulations in the currently fragmented habitats. Because subpopulations are small, they become more prone to extinction from both natural and anthropogenic causes. The problem is exacerbated by the array of predators and competitors that are favored in the altered habitat, and if the patches are small, their size is not sufficient to buffer the patch from intrusion. Thus, predators and competitors not only lower the likelihood of successful dispersal between habitat patches but also increase the likelihood of extinction within habitat patches. In the absence of successful dispersal, subpopulations will gradually become locally extinct patch by patch until the l

subpopulation, and the species, passes into extinction. Unfortunately, these "interior" species, for example, the denizens of the deep forest, are often not good dispersers because they avoid open areas, which they did not have to cross in the extensive pristine forest.

Although fragmentation is the most prevalent cause of continuous populations being reduced to metapopulations, other processes can create metapopulations. The second category includes deterioration of habitat quality without major conversion, which can have the same effect of isolating subpopulations to a set of the most suitable remaining patches of habitat in the degraded environment. These isolates, too, must function as metapopulations in order to persist.

A third category of human-induced metapopulation structure is caused by overexploitation, which extirpates a species population from much of its range. Reduction in population numbers commonly results in confinement of the species to isolated areas that are of higher quality (thus better able to re-plenish the local population), inaccessible, possessed of more escape cover, or otherwise protected.

A fourth category pertains to recovered species. Decline in population size and isolation to patches often leads to establishment of reserves to protect the wildlife species and, if that fails, implementation of captive breeding pro-grams. Such parks may function much like habitat "islands" in a sea of un-suitable habitat (or developed area) where dispersing animals are killed inten-tionally or accidentally. This situation has prompted the application of island biogeography theory to landscape planning. But island biogeography theory alone is not sufficient for planning and managing such situations. Because re-covered species often are reestablished by artificially transplanting them to widely isolated patches of suitable habitat or areas protected from exploita-tion, they become functional metapopulations. If habitat patches and trans-plant sites are isolated, humans may have to accomplish dispersal artificially in order for the species to persist.

The ultimate force driving this process is the inexorable increase in the human population, which shows no sign of abating. Planet Earth will con-tinue to be further fenced, plowed, grazed, logged, mined, and paved for the foreseeable future. It is the inevitability of this fate that has led to the devel-opment of the field of conservation biology, variously called a "crisis disci-pline" or a "rescue science" for salvaging biodiversity. Natural habitats in the future will progressively become reduced in area and more fragmented and isolated. An increasing number of animal populations will be altered from continuous populations to metapopulations.

By knowing in advance what a species needs to persist as a metapopula-tion, we are in a better position to advocate landscape designs that encom-

pass the size, number, and spatial distribution of habitat patches that will enhance the function and survival of metapopulations. Creating habitat corridors and influencing the character of the intervening matrix can foster successful dispersal between patches, the lifeblood of a metapopulation's survival.

Failures are inevitable in a complex world, and some species will slide toward extinction despite our best efforts. Knowledge of metapopulation dynamics, however, may allow us to artificially perform the function of dispersal and recolonization of locally extinct patches for some species of vertebrates. In this way we may manage to maintain a metapopulation in the wild and avoid the last resort: establishment of captive breeding colonies. Even if captive breeding should prove necessary, metapopulation thinking can guide recovery and restoration programs by advising on the placement of translocated animals, the numbers and sex and age composition used, and the genetic source of the stock.

Metapopulation thinking and analysis are critical to the modern practice of wildlife conservation and management. This book addresses the needs of an applied professional audience for comprehensible information to integrate into their practices. The anticipated readership ranges from professors and students in applied ecology, through natural resource agencies, to environmental consulting firms and local land-planning offices. Judging from the large audience at the symposium on this topic offered at The Wildlife Society's annual conference in September 1994, there is great interest indeed in metapopulations.

This book offers professionals a way of thinking about the environments, both present and future, that will have to serve the needs of wildlife conservation. It begins with four background chapters to set the principles: models, genetics, landscape configuration, and edges and corridors. Then follow two chapters that give detailed methods of analyzing metapopulation structure. Then follows a series of case histories on an array of vertebrate species in a variety of habitats to illustrate how this thinking is currently being applied in the real world. Examples include large and small vertebrates, marine and terrestrial environments, and different levels of plight with reference to human impacts. A brief concluding chapter sums up where we stand today with metapopulation thinking. In most cases this book does not furnish "how to" information, for each situation is unique. Instead, it presents examples of how metapopulation thinking has been applied in other situations and suggests the analysis required in a given case. It is less a road map than a set of guidelines on how to make decisions.

The book's goal is to explore how metapopulations function—whether metapopulation status is natural or induced by fragmentation—so that this knowledge can be applied to assure population persistence. One hopes, of course, that we can apply such knowledge before the fact; the ideal goal of wildlife conservation is to avoid anthropogenic creation of metapopulations. If habitat fragmentation is inevitable, we can direct land use changes along lines favorable for metapopulation persistence. Similarly, where successional changes will follow disturbance, careful landscape planning may promote a return to continuous population status.

Although the presentation emphasizes empiricism and biology, the reader will require some knowledge of mathematics (modeling and statistics) to follow all of the detail. Still, the treatment is such that the math-impaired reader should be able to follow the reasoning and consequences that underpin the larger messages the book intends to convey.

## REFERENCES

Gilpin, M. E., and I. Hanski, eds. 1991. Metapopulation dynamics: Empirical and theoretical investigations. *Biological Journal of the Linnean Society* 42:1–336.

Harrison, S. 1991. Local extinction in a metapopulation context: An empirical evaluation. *Biological Journal of the Linnean Society* 42:73–88.

———. 1994. Metapopulations and conservation. Pages 111–128 in P. J. Edwards, R. M. May, and N. R. Webb, eds., *Large-Scale Ecology and Conservation Biology*. Oxford: Blackwell.

Huffaker, C. B., K. P. Shea, and S. G. Herman. 1963. Experimental studies on predation: Complex dispersion and levels of food in an acarine predator-prey interaction. *Hilgardia* 34:305–330.

Levins, R. 1969. Some demographic and genetic consequences of environmental heterogeneity for biological control. *Bulletin of the Entomological Society of America* 15:237–240.

———. 1970. Extinction. Pages 77–107 in M. Gerstenhaber, ed., *Some Mathematical Questions in Biology*. Providence, R.I.: American Mathematical Society.

Thomas, J. W., E. D. Forsman, J. B. Lint, E. C. Meslow, B. R. Noon, and J. Verner. 1990. A conservation strategy for the northern spotted owl. Report of the Interagency Scientific Committee to address the conservation of the northern spotted owl. Portland: USDA Forest Service; USDI Bureau of Land Management/Fish and Wildlife Service/National Park Service.

# 2

# Metapopulations and Wildlife Conservation: Approaches to Modeling Spatial Structure

*Michael Gilpin*

This chapter is about the spatial structure exhibited by most species populations and, in particular, those populations fragmented by human modification of the landscape. I wish to carry out a general exploration with particular attention to the word "metapopulation," a term with a totally theoretical pedigree. My aim is to provide a conceptual framework that will help wildlife biologists and conservation biologists to understand the general class of spatially structured population-dynamic models—a group of models with sufficient resolution to address some of the decision problems they confront.

As conservationists and species managers, we are concerned with active manipulation and effective protection of biodiversity, which in large part we carry out at the level of single species. Faced with a deteriorating natural biological system, we comprehend as best we can the current state, including threats and possible remedies, and from this we attempt to project the system's future. Employing such projections and alternative scenarios, we engage the biological system, modify its current state, and, we hope, alter its future course of development to an end nearer our desires.

For amelioration, we add or subtract or rearrange individuals of target species; establish new colonies; build captive propagation facilities for sustained supplementation; modify or reconfigure habitat; and affect, positively or negatively, nontarget species that directly or indirectly have an impact on the target species. We do all this using our knowledge, views, and comprehension of the system on whose behalf we are acting and call this broadly interpreted understanding our "model" of the system. Simply put, then, we decide what management actions to take based on our model of the system. Good models are thus the key to good conservation management.

Much of what we do to preserve, conserve, restore, or manage depends on the spatial positions of the species and its members against their natural landscape and, as well, on the underlying spatial structure of the habitat that supports the system. Stated differently, our models of conservation and management are commonly map-based. Furthermore, our management recommendations are spatially explicit. Reserves are configured with boundaries drawn at *these* locations, not those. Harvesting is carried out with *these* quotas from *these* particular places. Individuals are translocated from here to there. Supplementation and restoration have spatial foci. Position makes a difference.

Having made this point, we must note that many of the population viability analyses (PVAs; Gilpin and Soulé 1986) and reserve design exercises currently being conducted and debated utilize zero-dimensional models and strategies. That is, they are based on demography, on birth–death branching-process models, and on simplified, graphic interpretations from island biogeography theory. I argue this approach to be inadequate, and in the remainder of the chapter I develop the rationale for a contrary view.

Today, the term "map-based" makes one think of geographic information systems (GIS). But a GIS is a passive data structure, rather like a spatial spreadsheet. A GIS holds an abstraction of raw, unprocessed, geo-referenced data. It is a wonderful platform, with a rich tool kit on which to carry out analyses, but this computer representation of data must be distinguished from the actual biological analysis or modeling, which is necessarily based on an ecological understanding of the system being portrayed.

To fix these ideas, consider the following hypothetical case study of a kangaroo rat population threatened by human habitat modification. Ideally we monitor the kangaroo rat's ecology, life history, and population dynamics. We plot densities onto a two-dimensional map of the landscape that also contains information on resources and competitors. We abstract these possibly continuous distributions of densities to discrete local units. Based on possibly inadequate data that parameterized movements and population fluctuations within and between these local units, we search for a possibly disconnected subset of the system that has minimal viability (say, 98 percent survival probability for 100 years), and through this analysis we decide where best to locate a system of public reserves. The progression is:

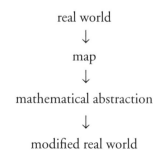

real world
↓
map
↓
mathematical abstraction
↓
modified real world

Important decisions take place at each step in the process. In this chapter we are concerned with the set of decisions involved in the map → mathematical-abstraction transformation.

At the outset, I distinguish four categories of metapopulation models with spatial structure: spatially implicit models, spatially explicit models, grid-based (cellular automata) models, and individual-based models. After comparing these models, I will suggest the conditions under which they may effectively be employed.

## Spatial Structure: The End Points

The category "spatially structured models" covers models in which positions of populations or their constituent individuals are located, or referenced, against a spatial background. I limit my discussion in this chapter to two-dimensional systems. I base my distinctions on stylized habitat and species density maps contrived for purposes of illustration. In speaking of the term "metapopulation," we must distinguish foreground from background. Following Hanski and Gilpin (1991), the metapopulation is defined at the species level: it is a population of populations. Under this strict usage, the underlying habitat structure is ignored, though this is not something one ever would do in conservation planning. To illustrate spatial structuring, I will consider two extreme types of spatial configuration on opposite sides of the scale: the panmictic population and the Levins metapopulation (Figure 2.1).

Many ecological models are without spatial extension, that is, zero-dimensional. Consider the well-known logistic growth model:

$$\frac{\mathrm{d}n}{\mathrm{d}t} = rn\left(1 - \frac{n}{K}\right)$$

The animal population size is characterized by a single number, a scalar variable, $n$. There is no indication of where these $n$ individuals are, over what range they extend, or whether there is internal structure to this population. Logically this number should be interpreted as total population size, but it is frequently interpreted as a density, which does imply extension in space. Many multispecies models are structured and interpreted the same way.

Although the issue is normally ducked, it is reasonable to assume, when scalar population state variables are interpreted as densities, that this density is uniform through some range and that, with change in time, the density change is the same throughout the unaltered range of the population (see Figure 2.1). This assumes that the population is extremely well mixed, or panmictic, which implies that the animals in the system can move anywhere in the range of the system within the time step (even if infinitesimal) of the

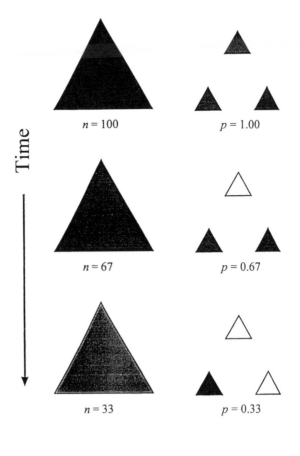

Figure 2.1. The panmictic population is on the left; the three local populations of the Levins metapopulation are on the right. Time is assumed to run down the figure. In each case, a regionwide loss of population is being illustrated. Darker shading corresponds to higher density. Note that the panmictic population maintains spatially uniform density throughout the process. With the Levins metapopulation the internal densities of colonized local populations remain constant, but more go extinct with time.

abstracted process—subject, of course, to the constraint that the density distribution remains uniform. Additionally, this seems to suggest that the habitat quality and other environmental features are the same throughout the domain of this local population.

At the other end of the spectrum of spatial structure is the Levins (1970) metapopulation. I will not discuss the motivations and theoretical background of this system. (The reader is referred to Hanski and Gilpin 1991.) Rather, I want to illustrate the features of the system and compare them graphically to the panmictic system. The Levins metapopulation is composed of local populations (demes), all of which are equal in all aspects. The size (or density) of each local population can be either zero or $K$, that is, extinct or at carrying capacity, with each local population being internally panmictic. The actual distinction is between absence and occupancy of a patch; consequently, this kind of model is sometimes called an "occupancy model."

Under the Levins metapopulation model, local populations go extinct with a constant probability and realized extinction events are independent from patch to patch. This is tantamount to viewing the environmental stochasticity over the region of the local populations as being independent or uncorrelated. Locally extinct populations may be recolonized by dispersal from extant populations. Recolonization is proportional to the fraction of extant populations in the system, regardless of their position. Dispersal and patch recolonization events are normally assumed to be infrequent. The state of the Levins metapopulation is commonly given by the scalar variable $p$, which describes the frequency of extant populations.

The most mathematical and idealized version of the Levins metapopulation model is not in fact map-based but an implicit spatial model. That is, the locations and sizes of the patches are ignored. Under these assumptions, the dynamic equation that governs $p$ is

$$\frac{dp}{dt} = cp(1 - p) - ep$$

where $c$ is the parameter for colonization and $e$ is the per-patch parameter for extinction, with an equilibrium value (for $c > e$) of $p$ given by

$$p^{eq} = \frac{(c - e)}{e}$$

Some investigators (for example, Lande 1987; Hanski 1991) have extended this spatially implicit model to give it more realism and more applicability to real-world management.

Later I deal with a spatially explicit, or map-based, version of the Levins metapopulation model in which equal-sized patches are arrayed on a landscape. If dispersal leading to colonization is independent of distance between patches, this spatially explicit form can be approximated by the preceding equations. But it should be recognized that colonization is much more likely to be from nearby occupied patches in such systems.

Figure 2.1 compares the changes of state of the panmictic population with the equivalent change for a Levins metapopulation. For illustration and contrast, both systems show a decline in the total number of animals with time. The panmictic population does this through a uniform decrease of density; the Levins metapopulation does it with a decrease in the number of extant populations. At equilibrium, the panmictic population would have a constant density and the Levins metapopulation would have a stable fraction of occupied patches, all occupied patches having the same internal density. Due to the requirement for ongoing extinction and recolonization in a metapopulation, it is impossible that, at equilibrium, the actual fraction of occupied patches in a finite, spatially explicit Levins metapopulation would be perfectly constant, for the extinction and recolonization events are assumed to be independent. The best that could be anticipated would be a steady-state behavior with a stochastic fluctuation about a constant expectation for the fraction $p$.

No real-world population or system of local populations fits either of these two extremes. Both models are too stylized and too simplified to be useful for map-based conservation management. We must move to a representation that renders both the landscape and the behavior of animals more faithfully. We must begin by detailing some of the important factors and considerations that govern spatially extended populations. Only then can we characterize which subset of the possible systems are "metapopulational" and which others are better characterized by one of the two other modeling approaches to space—grid or individual—though I must caution at this point that the same real-world system could often, with profit, be modeled with more than one of these alternative approaches.

Here is a partial list of considerations that affect spatially explicit models—that is, realistic features that may need to be accounted for to make accurate future projections:

- Most populations, local or regional, show density variation throughout their range.
- Local populations have densities that can vary over time between zero and some upper limit. This is sometimes called "structure."
- Virtually all fragmentation or patchy population structure exhibits variation in both patch size and patch spacing.
- Most systems have partially correlated environmental variation even within a continuous range.
- Long-term anthropogenic changes can modify underlying habitat structure such as forest succession and spotted owl habitat (Thomas et al. 1990; Smith and Gilpin 1996; Chapter 7 in this volume).
- The edge between the good habitat and the bad habitat, probably corresponding to the spatial limit of the local population, can be of different severities, possibly contributing an "edge effect" (Chapter 10 in this volume).

### Distribution, Abundance, and Movement

Figure 2.2 sketches some of the spatial elements we will consider in the analysis to follow. The basic idea is to represent the continuities and discontinuities of spatial variation in density and simultaneously to consider the effect of animal movement on density.

At the map level for population density, we can consider some purely geometrical factors for spacing and patch dimensions as illustrated in Figure 2.3. All four of the configurations lie between panmixia and the Levins metapopulation presented in Figure 2.1. As we will see, these different geometries will have differing interactions with other aspects of spatial structure and individual movement.

Ignoring demographic stochasticity (MacArthur and Wilson 1967; Gilpin 1992)—which, by definition, is independent among local populations—the environmentally driven population dynamics (that is, environmental stochasticity) of the independent units may have different degrees of correlation (see Gilpin 1988, 1990; Harrison and Quinn 1989). Some patterns of correlated environmental stochasticity are illustrated in Figure 2.4. In Figure 2.4a, the fluctuations of density (population size) in the three polygonal units are

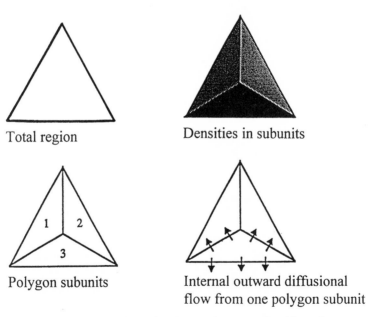

Figure 2.2. The total region is arbitrarily taken to be triangular. The subunits are taken as equally sized smaller triangles. Densities may vary in subunits. The bottom figure shows the localized "diffusional" movement of animals out of one of the cells. The movement of animals out of the system along the lower edge may contribute to an "edge effect."

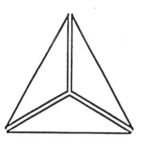

A.  Minor fragmentation          B.  Unequal separations

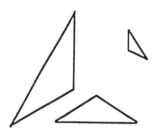

C.  Unequal sizes                D.  Unequal sizes and separations

Figure 2.3. Some geometrical forms between the panmictic and the Levins metapopulation. These forms may or may not be viewed as metapopulation systems, depending on the correlation of the environment and the degree of animal movement between patches.

identical. In Figure 2.4*b*, the correlation is perfect, but the amplitudes are different in the local populations. In Figure 2.4*c*, the three subunits show uncorrelated dynamics. In Figure 2.4*d*, not only are the fluctuations uncorrelated, but patches 1 and 2 at times fluctuate to extinction; presumably their recolonization is mediated from the other patches. The system in Figure 2.4*a*, although spatially disjoint, seems to mimic panmixia. The situation in Figure 2.4*d* behaves more like the Levins metapopulation. The others have an intermediate behavior.

Diffusion of individuals (migration, dispersal, and the like) can modify the environmentally driven dynamics on the separate patches. Typically, a local population with a temporarily or permanently high density may act as a source (or mainland) to a nearby population that is at low density (which

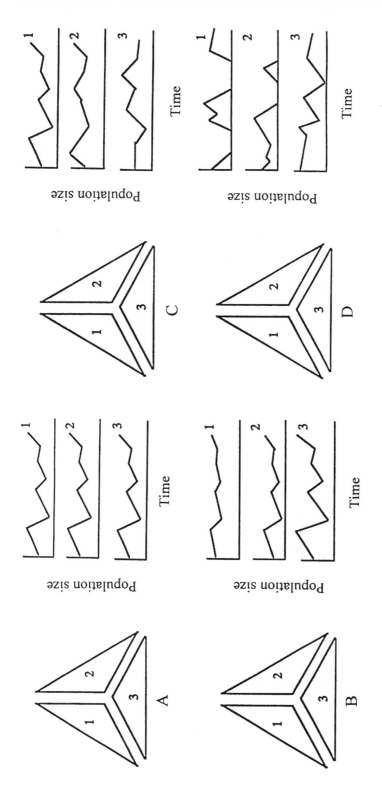

Figure 2.4. Patterns of population change in the three polygonal subunits of a mildly fragmented system. Patterns **A** and **B** show perfect correlation. Pattern **C** shows uncorrelated population changes. Pattern **D** shows uncorrelated change with extinctions. (The recolonization of subunits 1 and 2 is due to immigration of individuals, which is not shown.)

could be called a sink population). Figure 2.5 illustrates some influences of diffusional exchange of individuals. In Figure 2.5*a*, the environment is correlated between the patches and the symmetric diffusion of individuals between these units alters nothing. In Figure 2.5*b*, diffusion partially smooths out the uncorrelated environment. Extreme diffusion could cancel out all irregularities of the environmental stochasticity. Thus, even with fragmentation, the system could most closely resemble a single panmictic population. Figure 2.5*c* illustrates the "rescue effect" first identified by Brown and Kodric-Brown (1977). In this case, the dynamics would drive local populations extinct, but diffusion introduces enough individuals to preclude this extinction. Figure 2.5*d* illustrates source/sink dynamics. Polygon 1 (the sink) has dynamics that would drive it extinct—negative population growth—but the movement of individuals from the stable (source) populations in polygons 2 and 3 rescues this sink population from its fate. Observe that the system depicted in Figure 2.5*b* could also be called source/sink dynamics, but over the time shown, the net movement of animals between pairs of patches shifts back and forth. That is, each patch can act as both source and sink.

Diffusional flow, by strengthening patch-to-patch correlation, can defeat or preclude metapopulation analysis. In the Levins metapopulation, it is uncorrelated dynamics between patches that lead to sudden extinctions, independent from patch to patch. As the distances between local populations become greater, the movement of animals between them lessens and is less likely to produce an effect that mimics panmixia. Rescue and source/sink patterns will also be lessened or eliminated. That is, all else being equal, the greater the separation between patches, the more like a spatially explicit Levins metapopulation the system becomes.

Figure 2.6 shows some situations that have further complications, probably due to more structure in the underlying habitat. In Figure 2.6*a* the population is actually continuous. It is at lower density everywhere other than the three triangular regions, but it is nonetheless connected. In Figure 2.6*b* there are "stepping stones" between the three larger local populations. While either of these cases may be perfectly representative of many natural cases, they are somewhat more difficult to model, especially with a metapopulation approach.

### Nonequilibrium Spatially Structured Populations

In a system analysis, one must distinguish between the initial conditions, the transient behavior, and the equilibrium, whether deterministic or stochastic. Just as a panmictic population can exhibit deterministic negative growth, a metapopulation or other spatially structured system can be in collapse mode:

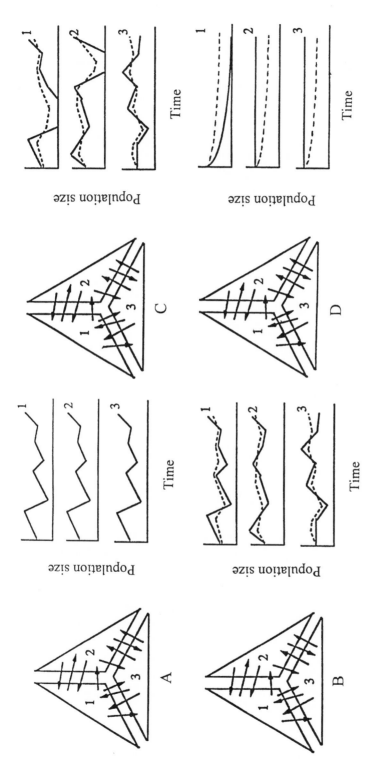

Figure 2.5. The influence of short-distance migration on population dynamics. For each of the three subunits, the density change with time is plotted at the right. The solid lines are the population densities expected without the effects of diffusion; the dashed lines represent the effect of including the immigration due to diffusion. See the text for discussion.

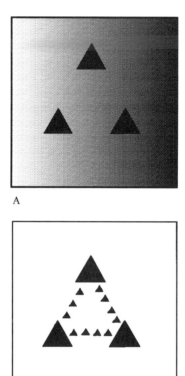

A

B

Figure 2.6. Two spatial variations from the stylized representations considered in the preceding figures. **A** depicts a continuous population with areas of varying population density. **B** shows three large local populations linked by smaller populations.

each local population is subject to some probability of extinction but with a trivial probability of recolonization. That is, the metapopulation system would exhibit little or no turnover of any local patch. From an evolutionary perspective, such systems are uninteresting. They will not exist in the future, and any such systems that existed in the past will not have made it to the present. From a conservation standpoint, however, they represent something that may too often be the case—seriously fragmented species populations cut off from dispersal and recolonization, systems doomed to extinction. Clearly, with such fragmented systems some management of the environment and dispersal must be instituted to reverse the monotonic trends that doom the population to ultimate extinction. A metapopulation model may well suggest fruitful approaches.

## Time Scales and Spatial Structure

Time and space are intertwined in the analysis of population dynamics. A decision about the scale of one may limit or preclude certain behavior in the other. If one wanted to talk about human population density changes in the United States following the passage of some piece of trade legislation, for example, one might focus on job creation and ignore demography. That is, the time scale might be short enough to neglect birth and death events. Similarly, some planning horizons may be short enough that metapopulation dynamics—that is, extinction and recolonization—could be ignored to a first approximation. But to the extent that an action will permanently disturb an equilibrium configuration, the spatial implications must be examined. Different species and different landscapes will require different minimum time spans for the emergence of spatially structured population dynamics.

## Metapopulation Models

All of the real-world features described above can be incorporated into a general and spatially explicit metapopulation model in which the state variables are the presences and absences of populations on a patch; that is, a vector of 1s and 0s. A more detailed approach would specify the population size on each patch. Such inclusive modeling is well beyond analytical mathematics. Analysis and investigation are possible only through computer-based simulation—repeated simulations that allow characterizations of central tendencies. The idea is that each patch is individually parameterized with a growth rate, a carrying capacity, some features of edge, area and shape, and some function, possibly density dependent, for emigration. A partially or fully correlated environment affects local growth, as does immigration from the set of other occupied patches in the system. A number of such models have been created (Gilpin 1989; Thomas et al. 1990; LaHaye et al. 1994), and many of the chapters in this book employ versions of generalized metapopulation models. The ultimate utility of the metapopulation approach will be determined by the success of these approaches.

# Grid-Based and Individual-Based Models

There are two other classes of spatially explicit models that must be compared to the spatially explicit metapopulation model. These are alternative representations of the same underlying biological reality. In a grid-based or cellular automata approach, all spatial units are squares of the same size. Each cell has a state—typically the number of individuals of the target species in that cell. The state values in the cells change with time in a manner that typically depends on inputs and interactions with closely neighboring cells. Depending on the cell size, a patch in a metapopulation model could be composed of multiple grid cells, all of which are contiguous.

In a geo-referenced individual-based model, all individuals are followed spatially and in demography and behavior over the course of their lives. The spatial indexing of animals is based on continuous $x$ and $y$ coordinates. Thus an animal inscribes a trajectory over its landscape during the course of its life. These trajectories are necessarily stochastic. But as animals move and make random encounters, the outcomes of the encounters may depend deterministically on behavioral and demographic states of the two animals. These three spatially explicit approaches are contrasted in Figure 2.7.

I must emphasize that, with the exception of the spatially implicit metapopulation models, all three of these approaches require computers to simulate the dynamics of the system. The grid-based and the individual-based approaches, however, have many more state variables and may require supercomputers or parallel computers for effective implementation. All of these spatially explicit approaches are "data hungry," but each in a different way.

The metapopulation approach is parameterized best with long time series of population sizes on multiple patches. The cell-based grid approach requires high spatial resolution of the population dynamics, but it may permit a more even-handed application of rules to each cell. With the individual-based approach, the population dynamics are a consequence of the parameterization of the individual's demography and various choices of behavior. Thus one could chose between these models solely on the basis of available data and on an a priori understanding of the behavior of the species.

It is beyond the scope of this introduction to go more deeply into the mathematical structure of these three approaches. But it should be noted that each of these different approaches, and also zero-dimensional approaches, could be applied to any system—that is, to any map-based management problem. Different approaches would have different strengths. The metapopulation approach requires the least data and is useful for population viability analysis (Gilpin and Soulé 1986). The grid-based approach might work best for designing reserves (Gilpin 1993; Chapter 10 in this volume). The individual-based approach might be best for situations where environmental mitigation is required throughout the stable range of a species population.

The choice of a modeling strategy initially depends on the character and quality of the data. Typically, in conservation planning for threatened and endangered species, the data are scant and were collected for reasons only tangentially connected to the issue of predicting extinctions or designing reserves. Indeed, the data may be inadequate for any of the spatially explicit

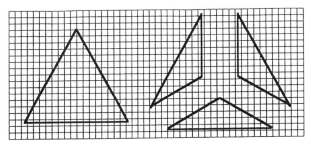

A. Underlying structure

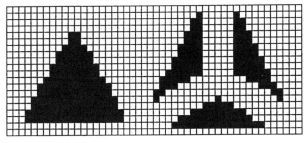

B. Grid model

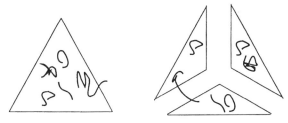

C. Individual-based model

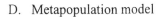

D. Metapopulation model

Figure 2.7. Part **A** shows the map structure of the density of the species. In **B** the structure is decomposed into densities in a number of small, equally sized spatial cells. In **C** the approach is individual-based: the exact movement of each individual is tracked over the landscape. Some individuals remain in their local population throughout their lifetime; others are shown moving between the subunits. **D** is the metapopulation approach: the subunits are abstracted to dimensionless points on a two-dimensional surface.

approaches. Yet real-world conservation and wildlife management are on-going processes in which each element of the effort informs other elements. The initial modeling should guide the second round of data collection. From this point on, there is no excuse not to geo-reference the data. Thus, at some point in the process, the modeling should begin to contain spatial elements that are modeled in one of the fashions outlined above.

## REFERENCES

Brown, J. H., and A. Kodric-Brown. 1977. Turnover rates in insular bio-geography: Effect of immigration on extinction. *Ecology* 58:445–449.

Gilpin, M. 1988. Comment on Quinn and Hastings: Extinction in subdi-vided habitats. *Conservation Biology* 2:290–292.

————. 1989. Population viability analysis. *Endangered Species Update* 6:15–18.

————. 1990. Extinction of finite metapopulations in correlated environ-ments. Pages 177–186 in B. Shorrocks and I. R. Swingland, eds., *Living in a Patchy Environment*. Oxford: Oxford University Press.

————. 1992. Demographic stochasticity: A Markovian approach. *Journal of Theoretical Biology* 154:8–18.

————. 1993. A viability model for Stephens' kangaroo rat in western River-side County. Report no. 12, vol. II: Technical reports. Draft habitat con-servation plan for the Stephens' kangaroo rat in western Riverside County, California. Riverside: Riverside County Habitat Conservation Agency.

Gilpin, M. E., and M. E. Soulé. 1986. Minimum viable populations: Processes of species extinction. Pages 19–34 in M. E. Soulé, ed., *Conser-vation Biology: The Science of Scarcity and Diversity*. Sunderland, Mass.: Sinauer Associates.

Hanski, I. 1991. Single species metapopulation dynamics: Concepts, models and observations. *Biological Journal of the Linnean Society* 42:17–38.

Hanski, I., and M. Gilpin. 1991. Metapopulation dynamics: Brief history and conceptual domain. *Biological Journal of the Linnean Society* 42:3–16.

Harrison, S., and J. F. Quinn. 1989. Correlated environments and the persis-tence of metapopulations. *Oikos* 56:293–298.

LaHaye, W. S., R. J. Gutiérrez, and H. R. Akçakaya. 1994. Spotted owl metapopulation dynamics in southern California. *Journal of Animal Ecology* 63:775–785.

Lande, R. 1987. Extinction thresholds in demographic models of territorial populations. *American Naturalist* 130:624–635.

Levins, R. 1970. Extinction. Pages 77–107 in M. Gerstenhaber, ed., *Some Mathematical Questions in Biology.* Providence, R.I.: American Mathematical Society.

MacArthur, R. H., and E. O. Wilson. 1967. *The Theory of Island Biogeography.* Princeton, N.J.: Princeton University Press.

Smith, A. T., and M. Gilpin. 1996. Correlated dynamics in a pika metapopulation. In I. Hanski and M. Gilpin, eds., *Metapopulation Dynamics: Ecology, Genetics and Evolution.* New York: Academic Press.

Thomas, J. W., E. D. Forsman, J. B. Lint, E. C. Meslow, B. R. Noon, and J. Verner. 1990. A conservation strategy for the northern spotted owl. Report of the Interagency Scientific Committee to address the conservation of the northern spotted owl. Portland: USDA Forest Service; USDI Bureau of Land Management/Fish and Wildlife Service/National Park Service.

# 3

# Genetics of Metapopulations: Aspects of a Comprehensive Perspective

*Philip W. Hedrick*

Conservation biology has been the major focus of extensive research and interest in recent years. Much of the reason for this emphasis is the critical situation of many endangered species and an effort to understand the factors that lead to their extinction. In documenting the factors that can result in extinction, Shaffer (1981, 1987) suggested that they can be divided into four major categories. (See Lande 1993 for a theoretical update of the nongenetic factors.) Two of these factors are extrinsic to the species: environmental uncertainty (such as variation in the influence of other species like pathogens or predators) and catastrophe (such as the effects of floods, fires, or droughts). The other two, demographic stochasticity and genetic deterioration, are intrinsic to the species.

The approach to determine the impact of the two intrinsic factors was termed the "small population paradigm" in conservation biology by Caughley (1994), who contrasted it to the "declining population paradigm" generally used by wildlife biologists. (See Hedrick et al. 1996 for a perspective.) According to Caughley, one of the weaknesses of the small population approach is that it is not obvious how often intrinsic factors have been involved in the extinction of a species. Indeed, he finds it much easier to document that extrinsic factors, such as introduced predators, may be important in causing the extinction of particular species.

For basic scientific researchers, the application of genetic or evolutionary principles to conservation biology, as in the small population paradigm, is particularly appealing because a number of theoretically based approaches have been developed. Even so, their application is somewhat controversial and their apparent exactitude may be misleading. (See the discussion in

Lande and Barrowclough 1987 and Hedrick and Miller 1996.) For example, the general suggestion that a population should have an effective population size of 500 to maintain standing genetic variation has recently been revised to approximately 5000 based on data demonstrating that many mutant variants that have an effect of increasing the genetic variance also have an overall detrimental effect on fitness (Lande 1995). Another important concept, based on ecological and evolutionary principles, is that of metapopulation analysis. This approach assumes that a population is divided into patches and that these subpopulations may themselves go extinct or become recolonized—dynamics that influence the probability of extinction of the whole metapopulation. It remains to be seen how important metapopulation dynamics are to the genetic constitution of endangered species, but it is presently a topic of great research interest. (See Hastings and Harrison 1994 and the references cited there for a comprehensive coverage.)

In leading up to an analysis of some of the genetic implications of metapopulation dynamics, I will first give some examples that concern the genetic influence on the long-term survival of a small population. These and other recent studies add significantly to our understanding of the small population paradigm in conservation biology. I begin by briefly discussing types of genetic variation and the association (or lack thereof) of allozyme heterozygosity and fitness in Scots pine. I then summarize the factors that can influence the extent of genetic variation with emphasis on genetic bottlenecks, give an example of the estimation of the effective population size in winter-run chinook salmon, and discuss an experiment in which the detailed effects of inbreeding depression were estimated in *Drosophila melanogaster*. Finally, I consider some of the effects that metapopulation dynamics may have on effective population size and, consequently, maintenance of genetic variation. Because the genetic implications of metapopulation dynamics are quite complicated, a thorough comprehension of the knowledge (or lack of it) on the population genetics of small populations is an essential perspective. In the following pages, I discuss a variety of topics, mainly from recent research on topics related to the small population paradigm, that are fundamental to a full understanding of genetics of metapopulations.

## Measuring Genetic Variation

A great deal of information has been accumulated about the extent of genetic variation within a number of endangered species for certain molecular markers, including allozymes and mitochondrial DNA, and more recently DNA fingerprints, microsatellites, and nuclear DNA sequences (for example,

Hedrick and Miller 1992; Avise 1994; Smith and Wayne 1996). (Statistical techniques for analysis of molecular data are well developed (as in Weir 1990) and software to analyze these data is readily available.)

Molecular studies have often given invaluable insights into the population structure or phylogenetic relationships of endangered species. For example, molecular research has shown that although the Florida panther (*Felis concolor coryi*) has low genetic variation, part of the population has genetic variation that appears to be from an introduction that included South American ancestry (O'Brien et al. 1990; Roelke et al. 1993). A microsatellite survey of the certified Mexican wolf population (*Canis lupus baileyi*) and a comparison to two other putative Mexican wolf captive lineages has demonstrated that none of these three groups appears to have ancestry from dogs or coyotes and that they are close to each other genetically (García-Moreno et al. 1996; Hedrick 1995*a*).

Such molecular studies are making important contributions to the understanding of the genetics of endangered species. I should caution, however, that most of this molecular variation is probably not adaptive—although it is valuable for determining relationships between groups, past gene flow, and so on or may become adaptive in the future—and that other loci that may or may not show similar patterns are the ones of adaptive significance. One set of loci that may be of important adaptive significance is that involved in the major histocompatibility complex, which is thought to be essential for pathogen resistance. (For a review see Hedrick 1994.) Furthermore, the genes determining most adaptively important quantitative traits are not known, and it would be probably be a rare situation if a molecular marker was such a gene or was associated with an adaptive gene. For example, there are a number of cases from forestry provenance trials in which samples that show little or no differentiation among populations for allozymes have large genetic differences in adaptively important quantitative traits. Some of these differences are so dramatic that, for example, samples of Scots pine (*Pinus sylvestris*) transplanted to other areas cannot survive or reproduce in the transplant area (Muona 1990 and the references cited there).

### *Association of Heterozygosity and Fitness in Scots Pine*

At one time it was hoped that a simple measure of overall molecular genetic variation, such as allozyme heterozygosity, would be useful in predicting the fitness of a population or individuals in a population—that is, high-fitness individuals would have high allozyme heterozygosity (for example, Mitton and Grant 1984). These optimistic suggestions have been cooled by a realistic evaluation of the situation (for example, Hedrick et al. 1986), however, and by the fact that a number of studies do not report such a positive association.

Of course, this lack of an association is not surprising when one realizes that most vertebrates have 20,000 or more genes and that the typical allozyme study surveys 50 or fewer loci, many of which are not polymorphic in a given species (Chakraborty 1981). It appears that with larger numbers of highly variable molecular markers, as from microsatellites, that a molecular approach can allow, for example, identification of highly inbred individuals in a generally random-mating population. On a population level, there are several examples of species that have rebounded from very low numbers and appear to have good fitness levels—such as the northern elephant seal (*Mirounga angustirostrus;* Hoelzel et al. 1993) and the beaver (*Castor fiber;* Ellegren et al. 1992) in Sweden—but have very low genetic variation as measured by molecular techniques.

In a detailed study of Scots pine in Finland, Savolainen and Hedrick (1995) have measured the genetic variation at 12 polymorphic allozyme loci and six different quantitative traits related to fitness in three different populations. One of these populations was from above the Arctic Circle, a harsh climate even for Scots pine, and the other two were from southern Finland grown as clones from a limited number of genotypes, making the estimate of the actual genotype effect influencing the quantitative traits more accurate than in a natural population. Scots pine has extensive gene flow, as measured both by genetic and direct indicators, and an extremely large population size. As a result, if an association between heterozygotes and fitness was found, it would likely be due to an intrinsic heterozygote advantage effect at the loci being examined and not from other factors such as a statistical association with other genes resulting from a small population size or inbreeding.

Of 156 comparisons of heterozygotes and homozygotes possible in our study, approximately 8 percent were significant at the 5 percent level, but only half of these showed a positive association while the other half showed a negative association. A useful way to illustrate these results is given in Figure 3.1, where the level of significance is given for the three populations and 12 different loci using ANOVA. If there were a tendency for a statistically significant association, then the values would pile up on the left near or below the broken line that indicates 5 percent statistical significance. Obviously there is no pattern for these results (except a more or less random one), suggesting that there is no advantage for heterozygotes overall or at particular loci.

As another approach to examine the association of heterozygosity and the various quantitative traits related to fitness, the association of the number of heterozygous loci in individual trees and value of the six different quantitative traits in the three populations was examined using multiple regression. Out of 16 combinations of traits and populations, none was significant at the 5 percent level. The combination of highest significance was diameter size in

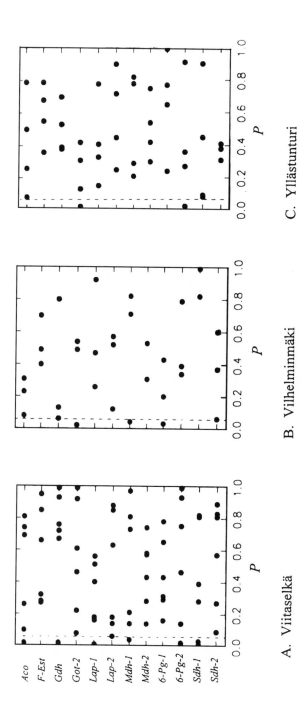

A. Viitaselkä          B. Vilhelminmäki          C. Yllästunturi

Figure 3.1. The probability of significance level ($P$) for a difference between heterozygotes and homozygotes from ANOVA for the 12 polymorphic loci in three different populations of Scots pine. The individual points are the probability level for a given quantitative trait and allozyme locus. From Savolainen and Hedrick (1995).

one of the southern populations, which explained about 7 percent of the variance. In this case, however, there was a negative association of the trait and heterozygosity—that is, the most heterozygous individuals had the lowest values for the trait. As a graphic way to illustrate these results, Figure 3.2 gives for different numbers of heterozygous loci the mean values of three traits: pollen production, cone production, and height for the three different populations. Obviously there is no apparent positive association in which one would expect low values for these traits to have low heterozygosity and high values to have high heterozygosity. In fact, there appears to be a general lack of pattern between these traits (and the other traits that are not shown here) and individual heterozygosity.

A number of previous studies have searched for a relationship between individual heterozygosity and traits related to fitness. (For reviews see Mitton and Grant 1984; Houle 1989; and Pogson and Zouros 1994.) Of the published reports, a reasonable number have not shown such an association, a rather surprising finding given the difficulty in getting negative results published. In addition, many of the positive findings have been published by a few research groups, suggesting that there may be some bias from them in reporting results that only show a positive association between heterozygosity and traits related to fitness. Overall, it seems that one should not depend on a measure such as allozyme heterozygosity to determine fitness. More direct indicators of fitness should be used.

## Factors Influencing Genetic Variation

The factors that influence the extent and pattern of genetic variation in a population are traditionally defined as selection, gene flow (population structure), mutation, inbreeding, and genetic drift (for example, Hedrick 1985; Hartl and Clark 1989). Particularly important in endangered species is genetic drift, which may have an influence on the extent of genetic variation because of a long-term small population size, bottlenecks (constrictions of population size in time), or small founding numbers for a population (for example, Hedrick and Miller 1992; Chapter 16 in this volume). All of these factors may be important in metapopulations, but the major impact is probably that of the small effective population size, through either founding events or small numbers of individuals in the different patches. Of course, when there is a small population size, the rate of inbreeding will be higher because of the lack of mates that are not relatives. Also important in metapopulations are the rate and pattern of movement between patches as well as the rate and cause of extinction within the patches.

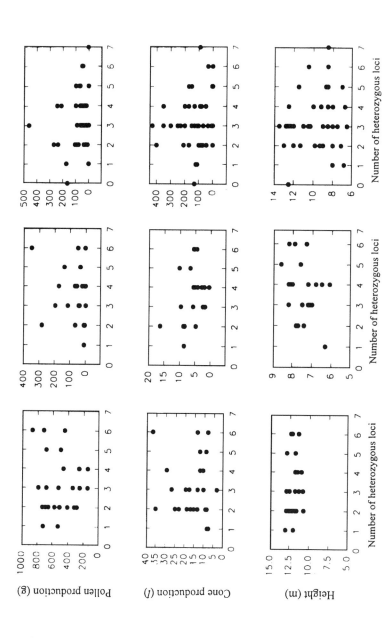

Figure 3.2. The phenotypic values of three quantitative traits—height, cone production, and pollen production—for three populations of Scots pine (left to right, Viitaselkä, Vihelminmäki, and Yllästunturi) with different numbers of heterozygous loci. From Savolainen and Hedrick (1995).

Often several different factors may simultaneously influence the extent of genetic variation or the level of fitness in a population, and the guidelines that have been generally recommended for avoiding extinction in endangered species are based on such principles. For example, the guideline that at least 50 individuals should be used to form the initial basis of a population and that 50 is the minimum effective number to avoid inbreeding depression is the result of practical experience with livestock. (See Franklin 1980; but see Lande and Barrowclough 1987 and Hedrick and Miller 1992 for discussions.) The guideline of 500 to maintain genetic variation in a finite population is the result of the combination of the input of mutation and the loss of variation due to genetic drift. (See Lande 1995 for a revision of this number.) The guideline of one migrant per generation ($Nm = 1$; for example, Lacy 1987) is based on the combination of the rate of gene flow, $m$, and the effective population size, $N$. Although these guidelines make general sense in a population genetics context, they should be used judiciously, case by case, and other information about the habitat, population ecology, or other factors should be given strong consideration.

## Population Bottlenecks

Population bottlenecks are common occurrences in endangered species, even if the species are able to rebound in number and not go extinct. The population bottleneck in the northern elephant seal, for example, was documented to some extent because these animals were hunted to near extinction in the late nineteenth century (for example, Hoelzel et al. 1993). After they were protected, a remnant population of a very few individuals was discovered on Guadalupe Island off the coast of Baja California in 1892. (Eight were found, and seven of these were killed.) From the few other animals that were not discovered, the population expanded to over 120,000 in 1980. This bottleneck and the low population numbers in those generations (and the immediate generations following it when the population began increasing) appear to have resulted in a great reduction in genetic variation—that is, much less mitochondrial DNA (mtDNA) and allozyme variation than the related southern elephant seal. Using either a detailed demographic model (Hoelzel et al. 1993) or a simple population genetics model (Hedrick 1995*b*), the low observed mtDNA variation is explainable by a bottleneck of approximately one generation with an effective population size of 10 to 20. But the complete lack of allozyme variation at 55 loci (Hoelzel et al. 1993) is not explainable by such a bottleneck (Hedrick 1995*b*), suggesting either that the northern elephant seal already had less variation than the southern elephant seal before the human-induced bottleneck or that some other unknown factor was important in reducing genetic variation.

Presently, the Florida panther, a subspecies of the mountain lion, is undergoing a bottleneck and exists in only a small area of southern Florida with an estimated population size between 30 and 50. This population appears to have lower genetic variation than other mountain lion subspecies (O'Brien et al. 1990; Roelke et al. 1993) and appears to have low fitness (Roelke et al. 1993; Barone et al. 1994) because of its long isolation from other mountain lions and its small effective population size. As a result, a panel of scientists and panther workers has recommended that mountain lions from Texas be translocated to Florida (for example, Seal 1994). The level of gene flow recommended, 0.2 in the first generation and about 0.02 to 0.04 per generation thereafter, will in theory result in a restoration in fitness, restoration of genetic variation, and retention of adaptive Florida panther traits (Hedrick 1995c).

Bryant et al. (1986) carried out some laboratory bottleneck experiments on the housefly and from their results suggested that bottlenecks may in fact increase the amount of additive genetic variation in a population and, therefore, the potential for future adaptation. But in their experiments and those of Lopez-Fanjul and Villaverde (1989), the value of traits related to fitness greatly declined as the result of the bottlenecks. In a later study, Bryant et al. (1990) found that over a series of bottlenecks, the initial fitness loss was ameliorated and the fitness returned to its prebottleneck level (see also Miller and Hedrick 1996). (I should note that the probability of extinction from demographic factors under such a series of bottlenecks would be quite high for most endangered species, which of course do not have the reproductive potential of houseflies.) Both Bryant et al. (1986) and Goodnight (1987) gave explanations of this observation based on the generation of additive genetic variation from epistatic variation, but Willis and Orr (1993) have shown that detrimental variants not having epistatic effects might be responsible.

O'Brien et al. (1983, 1985, 1987) have suggested that cheetahs underwent one or more bottlenecks that resulted in a low amount of genetic variation and low fitness. Pimm et al. (1989) and Gilpin (1991), however, have pointed out that the probability of extinction, given a bottleneck that would greatly reduce the amount of genetic variation, would be quite high. As an alternative explanation, Pimm et al. (1989) suggest that a metapopulation structure could result in the extensive loss of genetic variation apparently observed in cheetahs and also a low probability of extinction.

### Effective Population Size in Winter-Run Chinook Salmon

Winter-run chinook salmon from the Sacramento River, California, are listed as a federally endangered species (Hedrick 1994). The estimated annual run dropped from an average of 86,000 in the period from 1967 to 1969 to low numbers in the late 1980s and early 1990s with an extreme estimated low of

191 spawners in 1991. As a result, a program was started to capture adults, artificially spawn them, raise the young in Coleman National Fish Hatchery, and then directly release these progeny to augment the natural population.

There was concern, however, that these released hatchery fish could overwhelm the natural population and that the overall effective population size would be a reflection of the few number of breeders used at the hatchery. As a result, a breeding protocol was instituted as an attempt to equalize the contributions from the captured spawners and make the effective population size from the hatchery as large as possible (Hedrick et al. 1995). In this protocol, eggs from each female are divided into two lots and, when possible, fertilized with gametes from two different males. Furthermore, the gametes of each male are used to fertilize at least two different females. In this way, the contributions of fertile individuals are retained even if one of their mates is not fertile. Furthermore, this protocol will tend to equalize the contributions from different individuals, a factor that results in a larger effective population size (for example, Lande and Barrowclough 1987; Caballero 1994; Hedrick et al. 1995).

Hedrick et al. (1995) evaluated this program over its first 3 years. In the first year, 1991, there were only six females and nine males that were successfully combined in 12 matings. To have exactly equal contributions, each female should contribute 0.167 of the female gametes and each male 0.111 of the male gametes. However, the breeding protocol was not yet in place; and as a result of this and uneven maturation in the spawners, a high proportion of the gametes were contributed by one 4-year-old male (0.614 of the male gametes) and one 4-year-old female (0.411 of the female gametes). As a result of these unequal contributions, the effective population size from the hatchery was estimated to be only 7.02 individuals. (See Hedrick et al. 1995 for the estimation approach.)

In 1992, more spawners were captured and 13 males and 13 females contributed to 22 matings. Table 3.1 gives the number of progeny contributed from these parents and their proportionate contributions. For equal contributions from each parent, they should contribute 0.077 each. Notice that the largest value for any female was 0.138 and the largest for any male was 0.197, much closer to the equality goal than in the 1991 matings. The estimated effective population size from the hatchery in 1992 was significantly higher at 19.02.

In 1993, only 18 spawners were caught and there were only 12 matings among the surviving nine females and three males that produced smolts. Even with a number of problems encountered with these spawners, the distribution of progeny from these individuals was fairly even with the largest female contribution at 0.254 and largest male contribution at 0.509. (The

TABLE 3.1.

Number of progeny or gametes from the different
female and male spawning winter-run chinook salmon
in the 1992 brood year

| Female (Age) | No. of progeny (proportion) | Male (Age) | No. of progeny (proportion) |
|---|---|---|---|
| A (3) | 2890 (0.105) | 1 (4) | 2778 (0.101) |
| B (4) | 3440 (0.125) | 2 (3) | 2220 (0.080) |
| C (4) | 3429 (0.124) | 3 (3) | 5441 (0.197) |
| D (3) | 2763 (0.100) | 4 (3) | 2992 (0.108) |
| F (3) | 3897 (0.138) | 5 (3) | 2272 (0.082) |
| G (3) | 290 (0.010) | 7 (3) | 1452 (0.053) |
| H (?) | 1962 (0.071) | 8 (3) | 153 (0.006) |
| I (3) | 1581 (0.057) | 9 (3) | 137 (0.005) |
| J (4) | 268 (0.010) | 10 (3) | 867 (0.031) |
| K (3) | 609 (0.022) | 11 (3) | 3479 (0.126) |
| L (3) | 3167 (0.115) | 13 (4) | 1076 (0.039) |
| M (3) | 867 (0.031) | 14 (3) | 3853 (0.140) |
| N (3) | 2524 (0.092) | 15 (3) | 877 (0.032) |
| Mean | 2123 (0.077) | Mean | 2123 (0.077) |

*Source:* Hedrick et al. (1995).

goals for females and males were 0.111 and 0.333, respectively.) The effective population size was only 7.74 in 1993, but this low number is primarily due to the unequal numbers of the two sexes (only 3 males out of 12 spawners) and not because of unequal contribution of different individuals within the two sexes.

The bounds of the effective population size of the natural run in the 3 years were also estimated based on a lower bound of 0.1 and an upper bound of 0.333 of the run estimate (after Bartley et al. 1992; R. Waples, National Marine Fisheries Service, personal communication). In 1992, for example, the run was fairly strong and these lower and upper bounds were 115.1 and 383.6, respectively. Ryman and Laikre (1991) have shown how the hatchery and natural run estimates of the effective population size can be combined to give the overall effective population size. Figure 3.3 gives these estimates as a function of the contribution that the captive (hatchery) progeny makes to the population. At one extreme, if the captive proportion is 1.0 and there is no survival from the natural run, then the overall effective population size is 19.02. The estimated contribution from the hatchery in 1992 was 0.061, so that the bounds on the overall effective population size using this value were

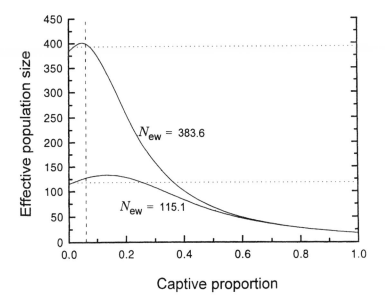

Figure 3.3. The estimated effective population size for the 1992 run of the winter-run chinook salmon using the estimated effective population size from the hatchery stock and a high and a low estimate of the effective population size of the natural run ($N_{ew}$). The vertical broken line indicates the estimated captive proportion; the horizontal dotted lines indicate the estimated population sizes if no salmon were removed for artificial spawning. From Hedrick et al. (1995).

between 127.3 and 401.0. In other words, the hatchery program does not seem to have reduced the overall effective population size in 1992 and may have slightly increased it. (In the other years, there was a slight increase at the lower bound and a slight decrease at the upper bound; see Hedrick et al. 1995.)

Winter-run chinook salmon, like other anadromous salmon, exist in year classes that tend to mate within the year class, making a given run a metapopulation existing in time rather than space. One of the main concerns is that if one of the year classes goes extinct, then recolonization from other year classes may be difficult. If year classes do go extinct and are subsequently recolonized, then such dynamics may greatly reduce the overall effective population size. (See the discussion of spatial metapopulations in the next section.)

## Inbreeding Depression in Drosophila melanogaster

In an effort to document the overall effects of inbreeding on fitness, we have carried out a series of laboratory experiments in wild-caught *Drosophila melanogaster* (Miller and Hedrick 1993; Miller et al. 1993; Miller 1994). There is a long history of using *Drosophila* as an experimental organism to investigate various aspects of evolutionary genetics (for example, Hedrick and Murray 1983) and recently in conservation genetics (for example, Frankham 1995) because of its short generation time, ease of culture, and our vast knowledge of its genetics. It is important to note, however, that some factors influencing extinction under the small population paradigm are not addressed in experimental studies using *Drosophila* because of its high reproductive value compared to most endangered species.

We determined in detail the effects, on the main components of fitness, survival, female fecundity, and male mating success, of a single chromosome that constitutes about 40 percent of the *Drosophila melanogaster* genome when it was made completely homozygous in a single generation (Miller and Hedrick 1993). The impact of inbreeding on the different fitness components for six individual wild chromosomes is given in Table 3.2. Male mating ability and female fecundity were not evaluated for line 7 because of the very low homozygote viability in that line. Only two of the six chromosomes showed significantly lower viability as homozygotes (lines 7 and 16), and only one of the lines showed significantly lower female fecundity (line 17). On the other hand, all five of the lines when made homozygous had a greatly reduced ability for male mating success. In line 17, the worst mating line, only 2 out

TABLE 3.2.

Fitness component estimates for six experimental
lines of *Drosophila melanogaster*

| Line | Viability | Male mating ability | Female fecundity |
|------|-----------|---------------------|------------------|
| 4 | 1.260 (0.053)* | 0.061 (0.133)* | 1.116 (0.202) |
| 7 | 0.198 (0.012)* | — | — |
| 8 | 1.063 (0.043) | 0.297 (0.103)* | 1.202 (0.202) |
| 16 | 0.800 (0.084) | 0.212 (0.062)* | 1.325 (0.225) |
| 17 | 0.999 (0.051) | 0.023 (0.017)* | 0.554 (0.110)* |
| 21 | 1.041 (0.040) | 0.237 (0.071)* | 0.878 (0.179) |
| Mean | 0.894 (0.371) | 0.275 (0.212)* | 1.015 (0.305) |

*Source:* Miller and Hedrick (1993).

*Note:* Values indicate the fitness component of the chromosomal homozygote relative to that of the heterozygote.

* Significantly different from 1.

of 87 stock females were inseminated by homozygous males. (See Miller et al. 1993 for more details on the cause of the decrease in male mating ability.)

When considering the effect of inbreeding depression, juvenile viability is usually measured because data for this component of fitness are most readily available (Ralls et al. 1988). But our results demonstrate that there may be as much or more inbreeding depression in the other components of fitness and that the effect on different components may not be correlated. Interestingly, in cheetahs, Florida panthers, and lions, male aspects of fitness such as sperm quality also appear to suffer most from what appears to be either inbreeding depression or chance fixation of detrimental alleles in a small population size, a result concordant with ours in *Drosophila*.

Using a model that incorporates all three fitness components, we were able to predict quite closely the change of chromosome frequency over ten generations in a separate experiment and the eventual equilibrium frequency (the other homozygote has a lower fitness because of a mutant marker), indicating that we were able to estimate quite accurately the total amount of selection (inbreeding depression) occurring on these chromosomes (Miller and Hedrick 1993). Interestingly, the extent of selection—as observed by a lower equilibrium frequency in these lines in a stress environment of increased temperature and/or lead in the media—is somewhat greater (Miller 1994). This suggests that inbreeding depression may be greater in natural (or stressful) environments where the environment is not near the optimum conditions of our laboratory experiments or in captivity in general. (See also Dudash 1990; Chen 1993; Wolfe 1993; Jiménez et al. 1994.)

## Metapopulation Dynamics and Effective Population Size

In recent years it has become recognized that endangered species often exist in isolated populations among which exchange of individuals may be quite infrequent. This stage may be a step on the way to loss of these populations and eventual extinction of the species. If further degradation of the habitat can be stopped, however, such species may form a metapopulation in which there is extinction of local populations and recolonization from other populations. In fact, human translocation of plants or animals between isolated populations may cause some species to have dynamics similar to those of natural metapopulations (see Chapter 16 in this volume). Of course, some species that naturally exist in habitat patches or in social groups may have metapopulation dynamics. (See Hastings and Harrison 1994 for a review of some aspects of genetics and metapopulations.)

In a series of widely publicized studies, O'Brien and his colleagues have demonstrated that cheetahs have a low amount of genetic variation compared to most other large cat species (for example, O'Brien et al. 1983, 1985, 1987; Yukhi and O'Brien 1990). (It should be noted that a number of other species also have low amounts of genetic variation as noted by Caughley 1994 and that cheetahs do have genetic variation at two allozyme loci, for minisatellites, and for mitochondrial DNA; see Menotti-Raymond and O'Brien 1993). As a result, O'Brien et al. (1983) have hypothesized that cheetahs went through a bottleneck about 10,000 years ago and at that time lost most of their genetic variation. But Pimm et al. (1989) suggest that a bottleneck small enough and long enough to result in this loss of genetic variation would have also resulted in a very high probability of extinction due to demographic considerations (see Hedrick 1996). As an alternative explanation, they suggest that a metapopulation structure and the resulting dynamics may have resulted in loss of genetic variation without a high probability of extinction. (Menotti-Raymond and O'Brien acknowledge indirectly that the suggestion of Pimm et al. is a valid alternative explanation.)

Gilpin (1991) has illustrated how the total population (census) number could be fairly large but the loss of genetic variation because of a small effective population size could still be substantial. Figure 3.4 gives a diagrammatic representation of this theoretical situation in which there are three patches in a metapopulation. Initially all patches or subpopulations have high heterozygosity, and each has an effective (and census) population size of 500. The important sequence of events starts in generation 48 when patch 2 goes extinct and is subsequently recolonized from patch 3 with a consequent reduction in heterozygosity. Next patch 1 is colonized from patch 2 with a founder population having no genetic variation. When patch 2 goes extinct in generation 71, the metapopulation has no variation, although there are still 500 individuals in patch 1. All of these individuals can be traced back to some individuals in patch 3 before generation 51. In other words, there is, in Gilpin's terminology, a rapid coalescence of ancestry to a few individuals in the metapopulation.

The general effect of extinction and recolonization dynamics on the effective population size can be estimated by measuring the rate of loss of heterozygosity. Hedrick and Gilpin (1996) have developed a computer simulation to determine the influence of the different parameters in a metapopulation model on the effective population size. The impact of metapopulations dynamics is illustrated in Table 3.3, for example, where there are ten patches of size $K$ and the per-generation rate of colonization is 0.2 (when all patches are occupied) and the per-generation rate of extinction is 0.05. In this case, the probability of metapopulation extinction is very low, about 0.2 percent over the first 50 generations. The estimated overall effective population size of the metapopulation ($N_{e(T)}$) is given in the third column: it is apparent that it does not increase very much as the

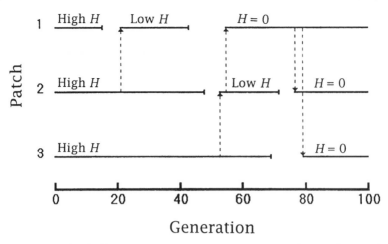

Figure 3.4. The level of heterozygosity ($H$) over time in a simulation of a population existing in three patches. The short vertical bars on the right-hand end of horizontal lines indicate extinctions in a patch; the arrows indicate recolonization. After Gilpin (1991).

number of individuals in a patch ($K$) increases or the actual observed (census) number of the metapopulation ($N$) increases. As a result, the ratio of the effective population number to the observed number declines to a very low value as $K$ increases. The effective size within a patch ($N_{e(S)}$) given in column 2 also remains small but becomes a larger proportion of the overall effective size as the value of the carrying capacity increases.

The primary basis for the low effective population size of the metapopulation is the assumption that each recolonization event is by one fertilized female, or the genomes of two founder individuals. Because it is assumed that the patch increases to its $K$ value in one generation, the small founder number is the factor that greatly reduces the heterozygosity in the patch. With a series of extinction and recolonization events, the whole metapopulation becomes genetically connected, the metapopulation heterozygosity is markedly reduced, and this results in the low effective population for the metapopulation.

The impact of other parameters varies from the highly predictable to the surprising (Hedrick and Gilpin 1996). Increasing the number of founders, for example, increases the effective population size as expected, and increasing the turnover rate, the rate of extinction and subsequent recolonization, greatly reduces the effective population size. But adding gene flow ($m$) to each generation among the patches increases the effective population size (Table 3.4)—a finding that is somewhat counterintuitive given the previous findings on substructured populations (for example, Lacy 1987; Lande and Barrowclough 1987). In other population-structure models, the lowest amount of gene flow leads to high differentiation among patches (as it does here as indicated by the high value of $F_{ST}$, a statistic to measure the level of between-patch diversity) and therefore retention of overall genetic variation. In the basic metapopulation model presented earlier, there is no gene flow between patches: thus a patch that once had no genetic variation can have restored nonzero heterozygosity only if there was an extinction and a recolonization. Gene flow allows restoration of heterozygosity without such a series of events. Because the effective population size is measured by loss of heterozygosity, adding gene flow results in a higher effective population size.

In discussing the impact of population structure on the retention of genetic variation, it has been suggested that more genetic variation in the total population could be retained with a low amount of gene flow because different subpopulations could become fixed for different alleles (for ex-

TABLE 3.3.

Estimated effective population sizes for different local carrying capacities ($K$) when the colonization and extinction rates are 0.2 and 0.05 per generation, respectively, and there are ten patches

| $K$ | $N_{e(S)}$ | $N_{e(T)}$ | $N$ | $N_{e(T)} / N$ |
|---|---|---|---|---|
| 25 | 16.4 | 38.2 | 174.4 | 0.219 |
| 50 | 22.4 | 46.6 | 345.9 | 0.134 |
| 100 | 30.4 | 64.2 | 690.2 | 0.093 |
| 200 | 36.0 | 66.7 | 1392.9 | 0.048 |
| $\infty$ | 40.7 | 76.6 | $\infty$ | 0.000 |

*Source:* Hedrick and Gilpin (1996).

TABLE 3.4.

Estimated effective population size for different levels of gene flow between patches when the number in a patch is infinite, the colonization and extinction rates are 0.2 and 0.05 per generation, respectively, and there are ten patches

| $m$ | $N_{e(S)}$ | $N_{e(T)}$ | $N_{e(S)} / N_{e(T)}$ | $F_{ST}$ |
|---|---|---|---|---|
| 0.00 | 40.7 | 76.6 | 0.531 | 0.312 |
| 0.00125 | 66.5 | 95.1 | 0.699 | 0.224 |
| 0.0025 | 88.9 | 115.7 | 0.769 | 0.167 |
| 0.005 | 125.8 | 140.1 | 0.898 | 0.114 |
| 0.01 | 156.7 | 172.4 | 0.909 | 0.069 |
| 0.02 | 217.7 | 219.7 | 0.991 | 0.040 |

*Source:* Hedrick and Gilpin (1996).

ample, Lacy 1987; Lande and Barrowclough 1987). But this retention of overall variation occurs only if there is no metapopulation dynamics—that is, there is no population extinction and recolonization from other patches that causes the coalescence of ancestry demonstrated above.

## Lessons

Many endangered species appear to have some of the characteristics of a metapopulation—that is, subdivision into small isolated populations and subsequent extinction of local populations. Because of extreme isolation, however, the recolonization of extinct patches may not readily occur and may have to rely on human translocations. To understand the effects of metapopulation dynamics on the extent and pattern of genetic variation entails understanding and documenting a number of basic population genetic parameters, such as the effective population size, level of inbreeding depression, and type of selection. As Caughley (1994) suggests, there is a distinct lack of information on many of these factors, but the examples given here as well as those from other recent studies (see the citations in Frankham 1995) are adding to our collective knowledge of the small population paradigm's parameters. The population structure of a metapopulation adds reality and complexity to the general picture, of course, but the fundamental aspects still need to be understood before we can determine the additional influence of metapopulation dynamics.

## Acknowledgments

This research was supported by grants from the National Science Foundation, the California Department of Water Resources, and the Finnish Research Council for Agriculture and Forestry.

## REFERENCES

Avise, J. C. 1994. *Molecular Markers, Natural History, and Evolution.* New York: Chapman & Hall.
Barone, M. A., M. E. Roelke, J. G. Howard, J. L. Brown, A. E. Anderson, and D. E. Wildt. 1994. Reproductive characteristics of male Florida pan-

thers: Comparative studies from Florida, Texas, Colorado, Latin America, and North American zoos. *Journal of Mammalogy* 57:150–172.

Bartley, D., M. Bagley, G. Gall, and B. Bentley. 1992. Use of linkage disequilibrium data to estimate effective size of hatchery and natural fish populations. *Conservation Biology* 6:365–375.

Bryant, E. H., S. A. McCommas, and L. M. Combs. 1986. The effect of an experimental bottleneck upon quantitative genetic variation in the housefly. *Genetics* 114:1191–1211.

Bryant, E. H., L. M. Meffert, and S. A. McCommas. 1990. Fitness rebound in serially bottlenecked populations of the house fly. *American Naturalist* 136:542–549.

Caballero, A. 1994. Review article: Developments in the prediction of effective population size. *Heredity* 73:657–679.

Caughley, G. 1994. Directions in conservation biology. *Journal of Animal Ecology* 63:215–244.

Chakraborty, R. 1981. The distribution of the number of heterozygous loci in an individual in natural populations. *Genetics* 98:461–466.

Chen, X. 1993. Comparison of inbreeding and outbreeding in hermaphroditic *Arianta arbustorum* (L.) (land snail). *Heredity* 71:456–461.

Dudash, M. R. 1990. Relative fitness of selfed and outcrossed progeny in a self-compatible protandrous species, *Sabatia angularis* L. (Gentianaceae): A comparison in three environments. *Evolution* 44:1129–1139.

Ellegren, H., G. Hartman, M. Johansson, and L. Andersson. 1992. Major histocompatibility complex monomorphism and low levels of DNA fingerprinting variability in a reintroduced and rapidly expanding population of beavers. *Proceedings of the National Academy of Sciences* (U.S.A.) 90:8150–8153.

Frankham, R. 1995. Conservation genetics. *Annual Review of Genetics* 29:305–327.

Franklin, I. R. 1980. Evolutionary changes in small populations. Pages 135–149 in M. E. Soulé and B. A. Wilcox, eds., *Conservation Biology: An Evolutionary-Ecological Perspective*. Sunderland, Mass.: Sinauer Associates.

García-Moreno, J., M. S. Roy, E. Geffen, and R. K. Wayne. 1996. Relationships and genetic purity of the endangered Mexican wolf based on analysis of microsatellite loci. *Conservation Biology* 10:376–379.

Gilpin, M. 1991. The genetic effective size of a metapopulation. *Biological Journal of the Linnean Society* 42:165–175.

Goodnight, C. J. 1987. On the effect of founder events on the additive genetic variance. *Evolution* 41:80–91.

Hartl, D. L., and A. G. Clark. 1989. *Principles of Population Genetics.* Sunderland, Mass.: Sinauer Associates.

Hastings, A., and S. Harrison. 1994. Metapopulation dynamics and genetics. *Annual Review of Ecology and Systematics* 25: 167–188.

Hedrick, P. W. 1985. *Genetics of Population.* Boston: Jones & Bartlett.

———. 1994. Conservation biology of endangered Pacific salmonids. *Conservation Biology* 8:863–894.

———. 1995*a*. Genetic evaluation of the three captive Mexican wolf lineages and consequent recommendations. Report of the Genetics Committee of the Mexican Wolf Recovery Team, U.S. Fish and Wildlife Service, Albuquerque, N.M.

———. 1995*b*. Elephant seals and the estimation of a population bottleneck. *Journal of Heredity* 86:232–235.

———. 1995*c*. Gene flow and genetic restoration: The Florida panther as a case study. *Conservation Biology* 9:996–1007.

———. 1996. Bottleneck(s) or metapopulations in cheetahs. *Conservation Biology:* in press.

Hedrick, P. W., and M. Gilpin. 1996. Metapopulation genetics: Effective population size. In I. Hanski and M. Gilpin, eds., *Metapopulation Dynamics: Ecology, Genetics, and Evolution.* New York: Academic Press.

Hedrick, P. W., and P. S. Miller. 1992. Conservation genetics: Techniques and fundamentals. *Ecological Applications* 2:30–46.

Hedrick, P. W., and E. Murray. 1983. Selection and measures of fitness. Pages 61–104 in M. Ashburner, H. Carson, and J. Thompson, eds., *The Genetics and Biology of* Drosophila. Vol. 3. New York: Academic Press.

Hedrick, P. W., D. Hedgecock, and S. Hamelberg. 1995. Effective population size in winter-run chinook salmon. *Conservation Biology* 9: 615–624.

Hedrick, P. W., R. Lacy, F. Allendorf, and M. Soulé. 1996. Directions in conservation biology: Comments on Caughley. *Conservation Biology:* in press.

Hedrick, P. W., P. F. Brussard, F. W. Allendorf, J. A. Beardmore, and S. Orzack. 1986. Protein variation, fitness and captive propagation. *Zoo Biology* 5:91–99.

Hoelzel, A. R., J. Halley, S. J. O'Brien, C. Campagna, T. Arnbom, B. Le Boeuf, K. Ralls, and G. A. Dover. 1993. Elephant seal genetic variation and the use of simulation models to investigate historical population bottlenecks. *Journal of Heredity* 84:443–449.

Houle, D. 1989. Allozyme-associated heterosis in *Drosophila melanogaster.* *Genetics* 123:789–801.

Jiménez, J. A., K. A. Hughes, G. Alaks, L. Graham, and R. C. Lacy. 1994. An experimental study of inbreeding depression in a natural habitat. *Science* 265:271–274.

Lacy, R. C. 1987. Loss of genetic diversity from managed populations: Interacting effects of drift, mutation, immigration, selection, and population subdivision. *Conservation Biology* 1:143–158.

Lande, R. 1993. Risks of population extinction from demographic and environmental stochasticity and random catastrophes. *American Naturalist* 142:911–927.

————. 1995. Mutation and conservation. *Conservation Biology* 9:782–791.

Lande, R., and G. F. Barrowclough. 1987. Effective population size, genetic variation, and their use in population management. Pages 87–123 in M. E. Soulé, ed., *Viable Populations for Conservation.* Cambridge: Cambridge University Press.

Lopez-Fanjul, C., and A. Villaverde. 1989. Inbreeding increases genetic variance for viability in *Drosophila melanogaster. Evolution* 43:1800–1804.

Maruyama, T., and M. Kimura. 1980. Genetic variation and effective population size when local extinction and recolonization of subpopulations are frequent. *Proceedings of the National Academy of Sciences* (U.S.A.) 77:6710–6714.

Menotti-Raymond, M., and S. J. O'Brien. 1993. Dating the genetic bottleneck of the African cheetah. *Proceedings of the National Academy of Sciences* (U.S.A.) 90:3172–3176.

Miller, P. S. 1994. Is inbreeding depression more severe in a stressful environment? *Zoo Biology* 13:195–208.

Miller, P. S., and P. W. Hedrick. 1993. Inbreeding and fitness in captive populations: Lessons from *Drosophila. Zoo Biology* 12:333–351.

————. 1996. Fitness in bottlenecked populations of *Drosophila melanogaster:* Is inbreeding depression purged? Unpublished manuscript.

Miller, P. S., J. Glasner, and P. W. Hedrick. 1993. Inbreeding depression and male-mating behavior in *Drosophila melanogaster. Genetica* 88:29–36.

Mitton, J. B., and M. C. Grant. 1984. Associations among protein heterozygosity, growth rate, and developmental homeostasis. *Annual Review of Ecology and Systematics* 15:79–500.

Muona, O. 1990. Population genetics in forest tree improvement. Pages 282–298 in A. H. D. Brown, M. T. Clegg, A. L. Kahler, and B. S. Weir, eds., *Plant Population Genetics, Breeding, and Genetic Resources.* Sunderland, Mass.: Sinauer Associates.

O'Brien, S. J., M. E. Roelke, N. Yuhki, K. W. Richards, W. E. Johnson, W. L. Franklin, A. E. Anderson, O. L. Bass, Jr., R. C. Belden, and J. S. Martenson. 1990. Genetic introgression within the Florida panther *Felis concolor coryi. National Geographic Research* 6:485–494.

O'Brien, S. J., M. E. Roelke, L. Marker, A. Newman, C. A. Winkler, D. Metzler, L. Colly, J. F. Everman, M. Bush, and D. E. Wildt. 1985. Genetic basis for species vulnerability in the cheetah. *Science* 227:1428–1434.

O'Brien, S. J., D. E. Wildt, D. Goldman, C. R. Meril, and M. Bush. 1983. The cheetah is depauperate in genetic variation. *Science* 221:459–461.

O'Brien, S. J., D. E. Wildt, M. Bush, T. M. Caro, C. FitzGibbon, I. Aggundey, and R. E. Leakey. 1987. East African cheetahs: Evidence for two population bottlenecks. *Proceedings of the National Academy of Sciences* (U.S.A.) 84:508–511.

Pimm, S. L., J. L. Gittleman, G. F. McCracken, and M. Gilpin. 1989. Plausible alternatives to bottlenecks to explain reduced genetic diversity. *Trends in Ecology and Evolution* 4:46–48.

Pogson, G. H., and E. Zouros. 1994. Allozyme and RFLP heterozygosities as correlates of growth rate in the scallop *Placopecten magellanicus:* A test of the associative overdominance hypothesis. *Genetics* 137:221–231.

Ralls, K., J. D. Ballou, and A. Templeton. 1988. Estimates of lethal equivalents and the cost of inbreeding in mammals. *Conservation Biology* 2:185–193.

Roelke, M. E., J. S. Martenson, and S. O'Brien. 1993. The consequences of demographic reduction and genetic depletion in the endangered Florida panther. *Current Biology* 3:344–350.

Ryman, N., and L. Laikre. 1991. Effects of supportive breeding on the genetically effective population size. *Conservation Biology* 5:325–329.

Savolainen, O., and P. W. Hedrick. 1995. Heterozygosity and fitness: No association in Scots pine. *Genetics* 140:755–766.

Seal, U. S. 1994. A plan for genetic restoration and management of the Florida panther (*Felis concolor coryi*). Report to the U.S. Fish and Wildlife Service, Conservation Breeding Specialist Group, SSC/IUCN, Apple Valley, Minn.

Shaffer, M. L. 1981. Minimum population sizes for species conservation. *BioScience* 31:131–134.

———. 1987. Minimum viable populations: Coping with uncertainty. Pages 69–86 in M. E. Soulé, ed., *Viable Populations for Conservation.* Cambridge: Cambridge University Press.

Smith, T. B., and R. K. Wayne. 1996. *Molecular Genetic Approaches in Conservation Biology.* Oxford: Oxford University Press.

Weir, B. S. 1990. *Genetic Data Analysis.* Sunderland, Mass.: Sinauer Associates.

Willis, J. H., and H. A. Orr. 1993. Increased heritable variation following population bottlenecks: The role of dominance. *Evolution* 47:949–957.

Wolfe, L. M. 1993. Inbreeding depression in *Hydrophyllum appendiculatum:* Role of maternal effects, crowding, and parental mating history. *Evolution* 47:374–386.

Yukhi, N., and S. J. O'Brien. 1990. DNA variation of the mammalian major histocompatibility complex reflects genomic diversity and population history. *Proceedings of the National Academy of Sciences* (U.S.A.) 87: 836–840.

# 4

# Wildlife in Patchy Environments: Metapopulations, Mosaics, and Management

*John A. Wiens*

Natural environments are patchy. Resource levels, predation risks, physiological stresses, and a host of other factors that influence individuals and populations vary in space. Land use by humans often intensifies this patchiness (see Meyer and Turner 1994), especially through the loss and fragmentation of "natural" habitat. Indeed, habitat fragmentation is widely regarded as a— if not *the*—central issue in conservation biology.

When a habitat such as forest or grassland undergoes fragmentation, remnant patches of the habitat are increasingly isolated in a matrix of altered and often heavily used lands (clear-cuts, agriculture). To predict the fate of wildlife populations or communities occupying these remnants, ecologists have sought insights from theory. As the analogy between habitat fragments and true islands was not hard to make, island biogeography theory (MacArthur and Wilson 1967) was quickly embraced as the guiding framework for thinking about fragmentation (Bennett 1990; Wiens 1995a). Island biogeography theory, by postulating a simple relationship between island (or fragment) isolation, area, and species number, offers a way to anticipate the magnitude of species loss as habitat patches become smaller and increasingly isolated by fragmentation. These relationships also led to the formulation of several principles of reserve design (Wilson and Willis 1975), some of which have been implemented in practice (for example, Diamond 1986).

The island analogy, however, is superficial at best. Patches of habitat in a terrestrial landscape are not true islands. Their edges are often gradients rather than crisp boundaries, and the matrix is not an ecologically hostile "sea" but offers benefits (such as food supplies) as well as risks (such as predator or competitor populations) to individuals within the patch. The matrix also permits (or enhances) interpatch movements, so fragment isolation is not simply a matter of distance between patches. The island model is simple and general, to be sure, but it leaves out so much of what goes on in the dynamics of spatially fragmented habitats and populations that its value in conservation and management has turned out to be severely limited (see Simberloff and Abele 1982; Zimmerman and Bierregaard 1986; Angelstam 1992; Soberón 1992; Wiens 1995a). As Saunders et al. (1991) have observed, the species-area equation "may give a manager a rough idea of how many species will be maintained on a remnant of a given area, but will yield absolutely no information on the practical issue of which habitats contribute most to species richness or on which species are most likely to be lost from the remnant."

Metapopulation models have largely replaced island biogeography as the theoretical framework for thinking about fragmentation issues. Metapopulation models (at least those following in the tradition of Levins 1970; see Harrison 1991) deal with the dynamics of regional (meta)populations that are subdivided into small subpopulations linked by dispersal. Local extinctions of subpopulations may occur, but under the right conditions colonization from other subpopulations will reestablish populations in those patches before all of the local subpopulations suffer extinction. As a consequence, the persistence of the regional population is increased over that of a similar population that is not spatially subdivided (Hanski 1991). (See Gilpin and Hanski 1991 and Hanski and Gilpin 1996 for a full elaboration of metapopulation theory and its empirical support.)

Wildlife and conservation biologists have been attracted to metapopulation theory for at least two reasons. First, most habitats are fragmented or at least patchy, and the pattern of spatial subdivision of populations portrayed in the models seems to coincide with what we observe in reality. Second, the prediction that a metapopulation structure may enhance overall population persistence is satisfying, since it offers the hope that extinctions of populations within local patches or fragments need not inevitably lead to global extinctions.

The development of management strategies for spotted owls (*Strix occidentalis caurina*) in the Pacific Northwest of the United States (Thomas et al. 1990; Chapters 7 and 8 in this volume) provides a good example of the use of metapopulation thinking founded on both of these reasons. The distribution of the old-growth forest habitat required by the owls is well defined and highly fragmented. As a consequence, the regional population is subdivided and local breeding populations are small. These conditions accord well with the spatial framework of metapopulation theory. Because dispersal among local subpopulations is thought to be limited, recolonization of available habitat patches from which local subpopulations have disappeared due to environmental and demographic stochasticity may be infrequent. Continuing removal of old-growth forest by logging will only increase the probability of local extinctions and decrease the likelihood that empty patches will be recolonized. Based on a combination of field studies and modeling of metapopulation dynamics, Thomas et al. (1990) recommended a management framework that focused on a minimum size and maximum spacing of old-growth habitat patches that might have a reasonable likelihood of ensuring metapopulation persistence. Further elaborations of the models (for example, Noon and McKelvey 1992; McKelvey et al. 1993; Chapter 7 in this volume) have retained the emphasis on metapopulation structure and dynamics.

All theories contain assumptions and conditions, and metapopulation theory is no exception. Before embracing the theory and its predictions too enthusiastically, wildlife biologists and conservationists would do well to consider these assumptions and conditions carefully (Doak and Mills 1994). For metapopulation theory to apply, is it sufficient that the populations of interest be spatially subdivided, or must other conditions also be satisfied? My objective in this chapter is to provide a context for thinking about the applicability of metapopulation thinking to conservation issues. I focus particularly on three aspects of this relationship: spatial scale, movement and dispersal, and landscape structure. Before doing so, however, it is necessary to review the critical assumptions and conditions of metapopulation theory.

## Assumptions of Metapopulation Theory

Like many current ideas in ecology, the general features of metapopulation thinking were outlined by Andrewartha and Birch in 1954 (Caughley 1994). The development of formal metapopulation theory and the coining of the term, however, are generally credited to Levins (1970). Levins' model addressed the generation of population stability or persistence through a balance between local extinctions and recolonizations of vacant, but suitable, patches. His model contained three critical elements (Hanski 1991): density dependence in local population dynamics, spatial asynchrony in local population dynamics, and limited dispersal linking the local populations. In addition, the model made several simplifying assumptions: all patches were considered to be similar in size and quality; because the spatial arrangement of patches was ignored (there was no spatial correlation in the state of patches), all patches were assumed to be equally accessible to dispersers; the number of patches was very large; local population dynamics were not affected by dispersal; and patches were modeled as either occupied or unoccupied (that is, abundance within patches was not considered or, alternatively, carrying capacity was considered to be constant among patches). Subsequent modeling (for example, Hastings 1991; Gyllenberg and Hanski 1992; Verboom et al. 1993; Hanski and Gyllenberg 1993; Hanski 1994; Hanski and Thomas 1994; Chapter 2 in this volume) has relaxed many of these simplifying assumptions. At the same time, definitions of "metapopulation" have broadened to the point of referring to any population that has spatial structure (for example, Opdam et al. 1993; Harrison 1994).

Metapopulation theory is relevant to conservation and wildlife management because of its predictions about the *dynamics* of subdivided populations. Thus, although it may be convenient to use "metapopulation" as a label for any spatially subdivided population, our attention should be focused on the attributes of such populations that produce distinctive metapopulation dynamics, especially population persistence in the face of local extinctions. Hanski and his colleagues (Hanski et al. 1995, 1996, Hanski and Kuussaari 1995) have stipulated four necessary conditions for metapopulation persistence that follow from theory and modeling:

1. Local breeding populations occur in "relatively discrete" habitat patches (that is, demography and population interactions are spatially structured).
2. No local population is so large that its expected lifetime is long relative to that of the metapopulation as a whole (that is, there is no large "mainland" population).
3. The dynamics of local populations are sufficiently asynchronous to make simultaneous extinction of all local populations quite unlikely.
4. Habitat patches are not so isolated that recolonization is prevented.

Biologists contemplating the application of metapopulation theory to management or conservation problems must consider whether these conditions are likely to be met in a particular situation. This brings us to the matters of scale, movement, and landscapes.

## The Effects of Spatial Scale

Issues of scale have become a major concern of ecology (Wiens 1989; Levin 1992, 1993; Malmer and Enckell 1994). Ecologists have come to recognize that many of the patterns and processes they study are scale-dependent—what happens at one scale of observation does not necessarily translate into the same thing at other scales. Our perceptions of how a population is spatially subdivided and our notions of extinction and colonization dynamics, for example, depend on the scale at which the population is viewed. Observation scale involves two components: *grain* (the minimum level of resolution—for example, the sample area) and *extent* (the broadest scale of resolution—for example, the size of a study area) (Wiens 1989, 1990). Changes in grain or extent may have differing effects on what we observe. Thus, the likelihood of documenting a local patch-level extinction for a given species will decrease as the grain size of observation is enlarged because we are less likely to distinguish between individual patches. We therefore miss the absence of the species from particular patches and instead observe the average dynamics across several patches. The extinction rate of species in a community, however, may increase with increasing extent of observation because the broader scale contains more rare species that have high extinction probabilities. Glenn and Collins (1992) offer an example of this effect for prairie plants.

### Spatial Patterns of Habitats and Populations

Our perceptions of habitat fragmentation or population subdivision (the first necessary condition of a metapopulation) depend on the scale. Simply stated, "fragmentation" refers to a disruption of continuity (Lord and Norton 1990), but there is a continuous gradient in the degree to which continuity may be disrupted (Wiens 1995a). Moreover, such gradients may be expressed on many scales. So in an absolute (map) sense, whether or not a habitat is "fragmented" depends both on the scale of our observation of the pattern and on what patterns on a gradient we choose to regard as fragmented. The effects of fragmentation may be strongly nonlinear, however: loss of some area from a continuous habitat may initially have relatively little effect on patch isolation; but at some threshold, further habitat loss may break landscape connectivity and result in rapidly accelerating patch isolation. At this threshold, populations occupying the habitat shift into a subdivided structure from a more-or-less continuous distribution (Opdam et al. 1993).

What sorts of patterns we regard as fragmented may also depend on the question asked, which is often closely related to the level of organization studied. What may be a patchy, fragmented habitat to an individual selecting foraging patches within its home range may not be fragmented at the scale of a population of the same species. For example, individual capercaillie (*Tetrao urogallus*) cocks require 20 to 50 ha of old-growth forest for a territory in a breeding lek, whereas a local population of birds at a lek requires 200 to 500 ha of forest; some 10,000 ha of multi-aged forest landscape is needed to satisfy the needs of a local population through the breeding cycle (Rolstad and Wegge 1987; Rolstad et al. 1988; Angelstam 1992). Fragmentation that is expressed at any one of these scales may not necessarily have effects at other scales.

Beyond this, different species respond to environmental patterns on quite different scales (Kotliar and Wiens 1990; Wiens et al. 1993). Grazing elk (*Cervus elaphus*) and bison (*Bison bison*) in Yellowstone National Park, for example, responded more strongly to habitat heterogeneity at broader scales (81 and 255 ha) than on a per-hectare scale (Pearson et al. 1995). Grazing microtines in the same areas would probably be more strongly affected by fine-scale variation at scales of square meters to hectares. In a comparison of seed-eating birds and mammals in small forest fragments versus larger forests in Spain, Santos and Tellería (1994) reported that wood mouse (*Apodemus sylvaticus*) density was much greater in the smaller fragments, whereas thrush (*Turdus* spp.) densities were much greater in the large forests. Clearly, different species respond differently to fragmentation at a given scale, and what is fragmented to one species (say, humans) may not be to another (say, wildlife species) (Nilsson and Ericson 1992; Wiens 1995*a*). Thus the scale on which we evaluate whether a population occurs in "relatively discrete" patches depends on the grain and extent of our observations, which must be adjusted to the appropriate level of resolution (local populations rather than individual home ranges), the scale of the processes in which we are interested (local extinctions and colonizations), and the scale at which certain organisms perceive environmental grain or the spatial extent of their activities (Wiens 1989; Cale and Hobbs 1994).

Some wide-ranging species may operate on extents far greater than are encompassed by traditional management efforts. Peregrine falcons (*Falco peregrinus*) banded as nestlings on the Yukon River in the Yukon, for example, have been recorded as breeding adults near Calgary, Alberta, some 1500 km away (P. Kennedy, Colorado State University, personal communication). Although allometric relationships with body mass may provide insight into how organisms scale their environments (Milne et al. 1992; Holling 1992), the peregrine example indicates that allometry alone may be an insufficient index of scale. These differences in the scales on which different species respond to habitat patchiness also cast doubt on the prospects for using a single species as an "umbrella species" to ensure adequate protection for other species occupying the same general habitats. Establishment of reserves to protect a target species, for example, may have only limited value in simultaneously meeting the needs of other species (Pyle and Franklin 1992; Launer and Murphy 1994). Species that operate at finer scales than the target species may differ not only in extent but also in the ways in which they respond to environmental patchiness (grain); landscape-mosaic effects, as we shall see, may further complicate the relationship.

## Local Population Dynamics

Scale may also affect the third necessary condition for metapopulation dynamics—that local populations have asynchronous dynamics. Local populations will be likely to develop asynchronous dynamics if they are small and only loosely linked by dispersal (as metapopulation theory suggests) *and* if their dynamics are controlled by factors acting at a local scale. Tip O'Neill, former Speaker of the U.S. House of Representatives, once remarked that "all politics is local"; ecologists traditionally have studied their systems as if all ecology is local. Yet there are important ecological patterns that emerge only at broad scales (Brown 1995) and, moreover, many patterns at the local scale are strongly influenced by factors acting over much broader regional or biogeographic scales (for example, Mönkkönen and Welsh 1994; Medel and Vásquez 1994; Bestelmeyer and Wiens 1996; Ricklefs and Schluter 1993). Episodic, broad-scale events such as insect outbreaks, epizootics, El Niño Southern Oscillations, or droughts may affect populations over wide areas, subjecting most or all of the subpopulations in a metapopulation to the same effects and synchronizing their dynamics. The importance of such events in determining the dynamics and persistence of populations has not been fully recognized (Mangel and Tier 1994).

## Patch Isolation

Scale considerations also affect the fourth feature of metapopulations: patch isolation. Whether or not patches are sufficiently isolated to promote the independence of local population dynamics, the underpinning of metapopulation persistence, depends on two factors: how organisms perceive environmental patchiness at a given scale ("functional heterogeneity"; Kolasa and Rollo 1991) and their movement distances and rates. Patch perception is related both to the grain and extent on which organisms scale environmental variation and to how they respond behaviorally to patch boundaries (Stamps et al. 1987; Wiens 1992; Chapter 5 in this volume). In other words, a pattern of patchiness in the spatial distribution of a habitat of interest may or may not be relevant to a given organism, depending on whether the pattern falls within its perceptual range and whether it recognizes the same patch boundaries that we do. Dispersal characteristics affect the probability that habitat fragments separated by a certain distance will be recognized by individuals as habitat patches or can be recolonized if local extinctions occur. For species that move rapidly over relatively large areas (such as the pine marten, *Martes martes*), fragmentation that occurs at fine scales may fall within the home-range movements of individuals and be effectively irrelevant to metapopulation considerations. A less vagile species occupying the same environment (such as the tundra vole, *Microtus oeconomus*) might respond to the same spatial pattern at a population level (Wiens 1985; Wiens et al. 1993). Simulation models (Fahrig 1988; Fahrig and Paloheimo 1988) have shown that the spatial arrangement of habitat patches is important when dispersal distances are small relative to the average interpatch distance or when the distance over which moving individuals can detect new patches is intermediate. If detection distances are large, patch isolation is effectively destroyed because the probability of movement among patches is high, whereas small patch-detection distances may lead to complete patch isolation, reducing the probability of patch recolonization following local extinctions.

## Movement and Metapopulation Dynamics

Because species respond differently to fragmentation and the establishment of subpopulations, understanding how particular species or communities will respond to a particular form of fragmentation or a certain habitat-management design requires knowledge of the details of the species' behavior, especially dispersal (Kareiva 1987; Merriam 1991; Saunders et al. 1991; Wiens et al. 1993; Wiens 1995a). This brings us to a consideration of movement in metapopulations.

### Dispersal

Once a population has become spatially subdivided, movement is the key to determining whether metapopulation models may apply. Too much dispersal among patches will homogenize population dynamics and erode the asynchrony that supports metapopulation persistence. Too little dispersal will make it unlikely that habitat patches in which local extinctions occur will be recolonized promptly, destroying the balance between extinctions and recolonizations that promotes persistence. Equilibrium persistence solutions of metapopulation models require that dispersal lie somewhere between these extremes. As Nunney and Campbell (1993) have noted, the amount of movement required to counteract inbreeding and preserve genetic variation in subpopulations may be greater than that required to recolonize empty patches. How much dispersal is "just right" depends on whether one is interested in population genetics or demography.

One reason for thinking that metapopulation theory might be relevant to spotted owl populations is the observation that although movements among local populations are rare, they do occur. Much of the recent effort in both field studies and modeling of spotted owl populations has focused on documenting natural dispersal (Thomas et al. 1990) and exploring the effects of varying dispersal patterns in relation to interpatch distances (Noon and

McKelvey 1992). As Verner (1992) has noted, setting a distance between blocks of suitable habitat is one of the most critical parts of a conservation strategy, and this decision depends on having a reliable knowledge of the distribution of dispersal distances.

Despite the importance of a knowledge of movement patterns to any sort of spatially-related ecology (May and Southwood 1990; Opdam et al. 1993; Wiens et al. 1993; Ims 1995; Wiens 1995b) and the existence of book-length treatments of dispersal (for example, Bunce and Howard 1990; Stenseth and Lidicker 1992), we know relatively little about the dispersal patterns and dynamics of most species (see Chapter 5 in this volume). Some examples of how some animals move in fragmented landscapes, however, may be instructive. In New Zealand, for example, brown kiwis (*Apteryx australis*) are now often restricted to small patches of forest and scrub separated by large tracts of pasture. Because the birds are flightless, their movements among habitat fragments can be determined with considerable accuracy by radio tracking. In one area, Potter (1990) found that all forest remnants isolated by less than 80 m were occupied, more isolated remnants were used only if they were relatively large, and fragments isolated by more than 330 m were not used (Figure 4.1). However, movements of up to 1.2 km from a large forest reserve were made by birds using small remnants as "stepping stones."

In the kiwi study, individual movements were not directly related to population distributions. In another study, Matthysen et al. (1995) compared dispersal of nuthatches (*Sitta europaea*) in a highly fragmented landscape in

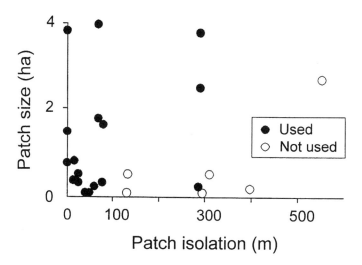

Figure 4.1. Occupancy of forest remnants by brown kiwis in an agricultural landscape in New Zealand in relation to patch size and isolation. After Potter (1990).

Belgium with that reported in populations occupying larger forests. Although local recruitment (the proportion of nestlings establishing territories relatively close to their birth site) did not differ among the populations, mean dispersal distances were considerably greater in the population occupying the fragmented landscape. In this species, then, dispersal distance varies with the degree of habitat fragmentation. Matthysen et al. also found that young nuthatches in the forest fragments were less likely to move again once they had established territories, so subpopulations in this landscape were potentially more isolated than those in the larger forests.

Finally, Hanski and his colleagues (Hanski and Thomas 1994; Hanski et al. 1994, 1995, Hanski and Kuussaari 1995) have conducted extensive studies of the population structure of the Glanville fritillary (*Melitaea cinxia*) in the Åland archipelago of Finland. The distribution of the butterfly is highly subdivided, and local populations appear to meet the conditions necessary for metapopulation-level persistence. In a mark–recapture study in one area, Hanski et al. (1994) found that 9 percent of the recaptures were of individuals originally marked in a different patch. The mean distance moved by migrants among the patches was 590 m (maximum 3050 m); because the mean nearest-neighbor distance among habitat patches was only 240 m, recolonization of empty patches was likely (although by no means certain). In another study in the same area, the number of marked butterflies immigrating to empty habitat patches increased with increasing patch area and butterfly density and decreased with increasing isolation from the release point (Kuussaari et al. 1996). On the Finnish mainland, where *Melitaea cinxia* went extinct in the late 1970s, suitable habitat patches still remain. As a result of agricultural modernization, however, such patches are so widely separated that interpatch dispersal and recolonization following local subpopulation extinctions do not occur (Hanski et al. 1995).

These examples illustrate the importance of dispersal and its relation to the patchiness of an environment. Although the distribution of dispersal distances in a population is commonly skewed toward short distances, there is some evidence that the tail of the distribution function may be inordinately important in population dynamics. On the basis of simulation models, Goldwasser et al. (1994) have suggested that the presence of even a few rapidly dispersing individuals in a population can increase the rate of population spread. If the arrangement of patches of habitat in the environment is irregular, however, the most isolated patches may be relatively unimportant in overall population dynamics, especially if dispersal entails distance-dependent costs (Adler and Nuernberger 1994). There is little field evidence to support either expectation, however, and information on long-distance dispersal and dispersal costs is notoriously difficult to obtain.

*Other Influences on Patch Recolonization*

Dispersal is not the only behavior that affects the probability of patch recolonization following local extinctions. If individuals disperse preferentially to patches occupied by conspecifics, the random-dispersal assumption of most metapopulation models will be violated (Ray et al. 1991). Weddell (1991), for example, reported that Columbian ground squirrels (*Spermophilus columbianus*) readily moved between habitat patches, but colonization of new (and apparently suitable) habitats did not occur because emigrants settled near other squirrels. Such social attraction may reduce patch-recolonization probabilities and make populations more sensitive to habitat fragmentation. Habitat or patch selection may also affect recolonization probabilities. Habitat selection may be an ongoing process during dispersal, and the degree of flexibility of habitat choice will influence how far dispersing individuals may move and whether they find suitable patches in which to settle. One might expect that natural selection would favor a broadening of habitat choice in populations occupying fragmented environments, but the rate of adaptive response in populations may not be able to match the rate of fragmentation of many habitats.

Competitive relations among species may also interact with dispersal properties to affect patterns of patch occupancy and extinctions. Tilman et al. (1994) suggest that because dominant competitors are often poor dispersers, even a small amount of fragmentation and habitat destruction can lead to the extinction of the dominant competitors in remnant patches. Because the competitive dominance of such species may lead them to be relatively abundant in undisturbed habitats, however, such extinctions may take generations to develop, creating an "extinction debt." This counterintuitive result has important management implications, but it depends strongly on the assumption that competitive dominance and dispersal are inversely related, which is not obviously so. Again, the importance of a detailed knowledge of movement patterns is evident.

*Incidence Functions and Diffusion*

Whatever the underlying mechanisms, the distributional responses of populations to habitat fragmentation are often portrayed as "incidence functions" that plot the proportion of fragments that are occupied by a species as a function of patch area (for example, Opdam et al. 1985). Such plots indicate zones in which metapopulation dynamics (at least those characterized by extinction–colonization turnover) are likely to occur (Figure 4.2*a*). Although incidence functions are often thought to illustrate the area requirements of a species, the relationship between patch area and patch isolation in affecting patch occupancy (Figure 4.1) suggests that the shape of an incidence function

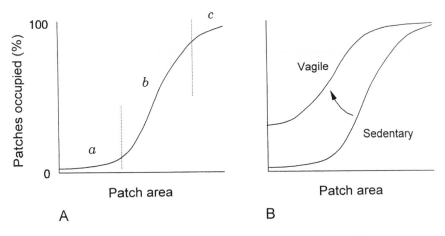

Figure 4.2. Hypothetical example of an incidence function portraying the proportion of patches in an area that are occupied by a species in relation to patch area. **A:** $a$ = zone in which metapopulation structure is unlikely because too few of the available patches are occupied to maintain the metapopulation; $b$ = zone in which strict metapopulation dynamics are possible; $c$ = zone in which strict metapopulation dynamics (as gauged by turnover of subpopulations) is unlikely because all available patches are continuously occupied. **B:** the shift in incidence function that may be associated with increased movement (dispersal) by individuals.

will also be sensitive to dispersal characteristics. If individuals are relatively sedentary, patches of a given size in a landscape with a certain level of fragmentation or patch isolation will be less likely to be occupied than if individuals are more vagile (Figure 4.2*b*). Thus, if the frequency distributions of dispersal distances vary within a species in response to changes in patch distribution (as in the nuthatches studied by Matthysen et al. 1995), the incidence functions for the species will depend on the structure of the landscape, not just on patch areas.

Although movement is a property of individuals, ecologists have been fond of thinking of movement in terms of functions that aggregate the behavior of many individuals in a population. Modeling movement as a diffusion process (for example, Okubo 1980; Turchin 1989; Johnson et al. 1992*a*) has been especially popular—perhaps because of an expectation that this approach might lead to something resembling the gas laws of physics. The spread of muskrats (*Ondatra zibethicus*) following their introduction at several places in Europe, for example, shows a reasonable fit to the predictions of diffusion models, as does the range expansion of several other invading species (Hengeveld 1989; Andow et al. 1990).

Despite the attractiveness of the idea that dispersal of individuals among habitat patches might follow a diffusion-based distance-decay function, reality is more complex. Not only do patch boundaries alter diffusion patterns (Wiens 1992), but the matrix between the patches may be differentially permeable to movement. This brings us to a consideration of landscape effects.

## Landscape Effects on Metapopulations

Metapopulation theory assumes that the matrix separating subpopulations is homogeneous and featureless. That is, the only "cost" involved in traveling through the matrix between patches is associated with interpatch distance and the decreasing probability of reaching (or finding) empty patches. In reality, of course, the matrix is a complex mosaic of habitats. Subpopulations in a metapopulation occur in a landscape. I have considered the implications of landscapes for metapopulation theory elsewhere (Wiens 1996); here I will comment only on a few aspects of landscapes that seem particularly relevant to thinking about wildlife populations and their management.

### Spatial Configuration

First, it is obvious that the spatial configuration of patches in a landscape has important consequences, regardless of the complexity of the matrix. A simple clumping of patches, for example, increases the probability of metapopulation persistence by reducing extinction probability in the clump (Adler and Nuernberger 1994), and increasing the variance among patches in patch size may further enhance the probability of persistence (source/sink metapopulations; Pulliam 1988, Danielson 1991). Thus, there is a spectrum of spatial patterns of patches in a landscape that promotes increasing probability of metapopulation persistence (Harrison and Fahrig 1995; Figure 4.3).

### Patch Context

The presence of a species in a patch may be a consequence not only of patch size and isolation, but also of the structure and composition of the surrounding landscape (Merriam 1988; Andrén 1994; Lidicker 1994). In Pearson's (1993) analysis of habitat occupancy by birds on the Georgia piedmont, for example, characteristics of the surrounding matrix explained as much as 65 percent of the variance for some species (such as the rufous-sided towhee, *Pipilo erythropthalmus,* and several *Parus* species), but little or none of the variation for others (song sparrow, *Melospiza melodia,* and white-throated sparrow, *Zonotrichia albicollis*). Thus, although the elements of the landscape surroundings may not be ideal habitat for a species, individuals may be able to live and even reproduce there. Habitat generalists, in particular, may be

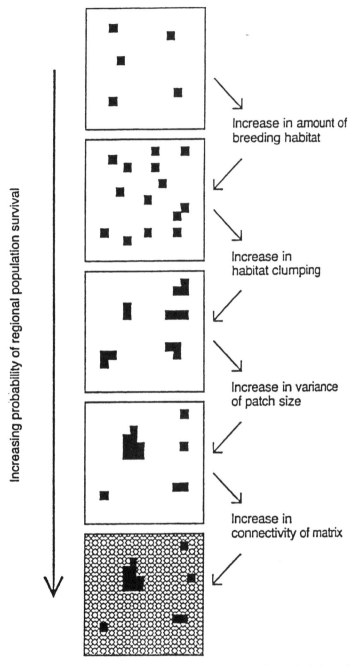

Figure 4.3. The effects of changes in landscape structure on the probability of survival (persistence) of a regional metapopulation. The expected changes are derived from simulation modeling studies. From Harrison and Fahrig (1995).

able to utilize nearby habitats. Instead of striking out in a desperate rush across an inhospitable "sea," dispersing individuals of such species may instead be able to wander in a more leisurely fashion through the mosaic, sampling habitats (and perhaps settling) as they go. Some landscape elements may be quite unsuitable or contain great risks, of course, acting to block or redirect dispersers. As a consequence of landscape structure, then, dispersal pathways between habitat patches may be much longer than the straight-line distances separating them. "Patch isolation" takes on an entirely different meaning, and measures of dispersal distances gathered from mark–recapture studies may give a misleading indication of the total distance traveled.

Among the various features of landscape structure and patch context, patch edges or ecotones have attracted particular attention from ecologists and conservation biologists (for example, Holland et al. 1991; Gosz 1993; Murcia 1995; Risser 1995). Because they are the intersections between different habitats in a landscape mosaic, ecotones may be important areas of population interactions, enhanced biodiversity, or abrupt changes in the flows of individuals or materials (Wiens et al. 1985; Hansen and di Castri 1992; Chapter 5 in this volume). Although the evidence is somewhat equivocal, predation risks are widely regarded to be greater at ecotones than in the interior of habitat patches, especially in forest (for example, Angelstam 1992; Andrén 1992, 1994). Because habitat fragmentation increases the amount of edge per unit area of habitat, edge-related predation has become an issue of central concern in conservation biology. Wilcove's study (1985), however, shows clearly that the magnitude of predation at ecotones is a function of the composition of the surrounding habitat (see also Andrén 1992), so once again patch context matters.

Scale enters into the picture as well. Whether or not an individual can respond to an ecotone depends on the relationship between the scale of landscape structure and the perceptual range (grain to extent) of the species. On Vancouver Island, for example, black-tailed deer (*Odocoileus hemionus*) did not respond to edges between forest, clear-cuts, and second-growth habitats when the landscape occurred as a fine-scale mosaic of these elements, but they showed clear habitat responses when the mosaic was expressed on a broader scale (Kremsater and Bunnell 1992).

### Connectivity and Corridors

A third feature of landscapes that is particularly relevant to conservation and management issues is *connectivity*—the pattern of interconnectedness or "networking" of a landscape that, along with movement behavior, determines how individuals move among landscape elements (Taylor et al. 1993). Landscape connectivity is most clearly manifested as corridors: areas of habitat that link similar habitat patches in a landscape. Because corridors may direct and facilitate the movement of animals over the landscape—thereby reducing the effects of patch isolation and small within-patch population sizes and increasing the probability of recolonization of empty patches—they have received considerable attention from conservation biologists (for example, Bennett 1990, 1991; Saunders and Hobbs 1991; Hobbs 1992; Merriam and Saunders 1993; Chapter 5 in this volume). Simulation models (for example, Lefkovitch and Fahrig 1985; Henein and Merriam 1990) demonstrate clearly that populations in interconnected patches have a greater survival probability (that is, persistence) than those in isolated patches and, moreover, that survival probability in connected patches increases with the degree of clustering among patches and with corridor quality. Clearly some species do orient their movements along habitat corridors—such as cockatoos and honeyeaters in Australia (Bennett 1991), blue jays (*Cyanocitta cristata*) or harvester ants (*Pogonomyrmex occidentalis*) in North America (Johnson and Adkisson 1985; DeMers 1993), white-footed mice (*Peromyscus leucopus*) in Canada (Merriam and Lanoue 1990), and passerine birds in Poland (Dmowski and Kozakiewicz 1990). Faunal diversity may also be enhanced in corridors (Osborne 1984; van Dorp and Opdam 1987), perhaps from a combination of edge effects and the presence of a variety of species using the corridors in their movements between landscape patches.

Despite the logical appeal of corridors as a potential solution to fragmentation problems, it is not clear that corridors in fact function as we imagine (or wish) them to. The presumed value of corridors is in facilitating movement, yet there is very little clear empirical evidence that corridors serve this function for very many species (Hobbs 1992). Some ecologists have argued that corridors may confer costs as well as benefits to populations, through facilitating the spread of disease, predators, or disturbances (such as fire) or acting as population "sinks" that draw dispersing individuals to edge-dominated habitats where predation may be high (Simberloff and Cox 1987; Hobbs 1992; Merriam and Saunders 1993). We need to know how animals move and respond to corridor configuration and how corridors influence the spatial distribution or dynamics of factors that affect species to predict how or whether corridors might enhance patch linkage and counteract fragmentation effects. Our lack of information places severe constraints on the development of ecologically based corridor designs (Harrison 1992; Chapter 5 in this volume).

Corridors are only one aspect of landscape connectivity. Indeed, a preoccupation with corridors may lead to a neglect of the broad relationships among landscape structure, population subdivision, movement, and metapopulation dynamics. Just as patches do not exist in a featureless matrix, neither do corridors. Knowledge of the suitability of corridor habitats, the home ranges and movements of the species of interest, and the effects of corridor width (Harrison 1992) may not be sufficient to predict the use or occupancy of corridors by species; information on the surrounding landscape may also be required (Lindenmayer and Nix 1993). Recognizing this, habitat management plans for spotted owls (Thomas et al. 1990) have recommended that 50 percent of the forest area surrounding population reserves be in stands averaging 11 inches diameter at breast height and at least 40 percent canopy cover (the "50–11–40 rule"). Whether or not this particular management scheme is appropriate, it does recognize that the landscape surroundings of habitat patches may be more important in promoting movements and subpopulation linkages than narrow corridors of a particular habitat, such as old-growth forest. Landscape connectivity is the important factor affecting inter-patch movement in metapopulations, and connectivity is a function of the varying suitabilities and "permeabilities" of *all* the elements of a landscape mosaic and their spatial configuration, not just corridors (Taylor et al. 1993; Wiens 1995*a*, 1996).

There is one other aspect of landscape connectivity that deserves mention. As an area of relatively continuous habitat becomes fragmented, habitat that is suitable for some species of interest is replaced by less suitable habitat and the remaining patches of suitable habitat are increasingly isolated. Fragmentation thus involves both habitat loss and patch isolation (Wiens 1995*a;* Harrison and Fahrig 1995). The relative importance of loss and isolation, however, does not change continuously as fragmentation proceeds. Rather, habitat loss may be most important when landscapes contain a high proportion of the suitable habitat; but at a certain threshold of habitat loss, patch isolation may quickly come to dominate population dynamics (Andrén 1994).

Gardner and his colleagues (Gardner et al. 1989, 1992; Lavorel et al. 1993) have used percolation theory to explore such thresholds in landscape connectivity. In simple percolation models, cells in a two-phase landscape are assigned to one or the other habitat type at random: particles (individual organisms) move among cells at random in the four cardinal directions, and cells of one of the habitat types are impermeable to movement (cannot be entered). Under these conditions, individuals have a high probability of moving from one side of the cell grid to another, or "percolating" through the landscape, when the coverage of the permeable habitat type is high. At a certain coverage threshold, however, the probability of moving across the landscape suddenly diminishes to very low values. The landscape has become disconnected. Under the assumptions given earlier, this critical threshold occurs on average at a coverage of the permeable habitat of 0.59.

These assumptions, of course, are restrictive and not very realistic biologically. When the assumptions are relaxed to permit individuals to move to diagonal as well as adjacent cells or to "leapfrog" cells (equivalent to viewing the grain of the landscape at a different scale) or to consider various nonrandom configurations of habitat distributions (as in Figure 4.3), the critical percolation threshold shifts to much lower values (Dale et al. 1994; Pearson et al. 1996; With et al. in press). While such percolation models are still simplified abstractions of the complexity of real landscapes, they do indicate that there may be strong nonlinearities in the effects of changing landscape structure on connectivity. Assessing the roles of dispersal and the spatial distribution of patches in maintaining metapopulation dynamics in real-world (or even abstract) landscapes is therefore a much thornier problem than portrayed in traditional metapopulation theory.

## Realistic Metapopulations and Reserve Management

Scale, movement, and landscape effects collectively complicate the predictions of metapopulation theory and limit its direct applicability to real-world situations. Despite this, metapopulation thinking is useful to reserve design and management by drawing attention to spatially explicit relationship and the importance of dispersal and of landscape structures.

### *Metapopulation Theory and Ecological Reality*

If we are to expect the theoretical predictions of metapopulation persistence to apply to real-world populations, the underlying conditions of the theory should be satisfied. For there to be a balance between local extinctions and patch recolonizations, metapopulation theory asks that regional populations be spatially structured, that none of the subpopulations be very large, that the dynamics of local populations be asynchronous, and that dispersal be sufficient to ensure recolonization of empty habitat patches but not so great as to erode local asynchrony. Clearly, many populations are spatially subdivided, at least at some scales. Beyond this, however, it is not obvious that the remaining conditions are often satisfied. Hanski et al. (1995) have documented that Glanville fritillaries in the Åland Islands appear to meet these requirements, as do pool frogs (*Rana lessonae*) at the northern limit of their distribution in Sweden (Sjögren 1991; Chapter 6 in this volume). Harrison (1991, 1994) and Simberloff (1995), however, suggest that few populations exhibit metapopulation structure, at least in the sense of Levins' (1970) model and its variations. Instead, real-world populations seem most often to have a large "core" distribution, a source/sink structure, or simply to be patchily distributed without clear evidence of any equilibrium in local turnover. Moreover, metapopulation models implicitly assume habitat stability: patches that become vacant through local (stochastic) extinction remain suitable for subsequent recolonization. In many cases, habitat change, through either natural succession or anthropogenic influences, contributes to local extinction and leaves patches unsuitable. In such situations, achieving a regional balance between local extinctions and patch recolonizations may not be possible unless the habitats recover (Thomas 1994). This situation may characterize many declining species.

Does this mean that metapopulation theory is useless? Of course not. Metapopulation theory, like any theory, simplifies reality in an effort to produce tractable models that generate interesting predictions. Just because nature does not match the theory precisely does not mean that the theory should be abandoned. The challenge is to establish the domain of situations to which metapopulation theory may offer insights, if not testable predictions. The importance of landscape configuration and connectivity in metapopulation considerations suggests that current metapopulation theory could be enhanced by adopting spatially explicit formulations and incorporating ideas from landscape ecology (Wiens 1996). Progress in developing spatially explicit approaches is now rapid (Lidicker 1994; Turner et al. 1995). Although landscape ecologists have developed methods for measuring and analyzing landscape patterns at multiple scales (for example, Turner and Gardner 1991), the growth of landscape theory that could lend insights to metapopulation thinking has been slow (Merriam 1988; Wiens 1995b).

The empirical foundation for evaluating the applicability of metapopulation theory is also weak. We particularly lack information on movement and dispersal and how they are affected by the structure of the landscape (Wiens et al. 1993; Hanski and Thomas 1994; Dunning et al. 1995; Chapter 5 in this volume). Such information is critical to determining the effects of fragmentation, the efficacy of corridors, the form of landscape connectivity, and the potential dynamics of metapopulations—all issues of central concern in management and conservation biology.

The vogue in ecology is to deal with such issues experimentally, but experimentation at the scales of regional landscapes or metapopulations is difficult if not impossible. One approach is to use small-scale systems in which experiments can be performed as "experimental model systems" (EMS; see Johnson et al. 1992b; Ims et al. 1993; Wiens et al. 1993) that may serve as analogs of systems operating at broader scales. Robinson et al. (1992), La Polla and Barrett (1993), and Ims and his colleagues (Ims et al. 1993; Wiens et al. 1993) have explored the effects of differing patterns of habitat fragmentation by experimentally modifying meadow habitats and monitoring the responses of insects and small mammals, for example, and Robinson and Quinn (1988) have employed a similar approach in their studies of fragmentation effects on plant populations. My colleagues and I have charted the movements of tenebrionid beetles (*Eleodes* spp.) through microlandscape mosaics in semiarid grasslands to gain insights into how landscape structure affects movement patterns (Wiens and Milne 1989; Crist et al. 1992; Johnson et al. 1992*a*).

Although the EMS approach is a central component of research in other disciplines (physics, engineering, biomedicine, population genetics), it has not been widely used in ecology, perhaps because of the perception that the results cannot be extrapolated to other systems or scales. Many ecological patterns and processes are scale-dependent (Wiens 1989; Levin 1992), and the problems in translating findings from one scale to another may be formidable (King 1991; Rastetter et al. 1992). Simulation models may provide one way to explore this extrapolation problem. For example, With and Crist (1995) have used spatially explicit models to ask whether the broad-scale dispersion patterns of grasshopper populations can be predicted from an understanding of the fine-scale movement patterns of individuals through different landscape components. Their models provide some hope, but they also indicate that additional information on how landscapes influence movements (how individuals behave at patch boundaries, for example) might improve the extrapolation functions. Experimental model systems, combined with spatially explicit modeling, may offer one way to develop a better understanding of the relationships between dispersal, landscape structure, and metapopulation dynamics, but they cannot entirely replace detailed information on the movement dynamics and habitat responses of the species of interest.

### Reserve Management in Landscape Mosaics

Traditionally, the management of wildlife populations, especially of rare or endangered species, has relied heavily on the establishment of nature reserves or refuges. Some area of suitable habitat for a target species or community is set aside with the expectation or hope that it will be sufficient to ensure the persistence of the populations. If several reserves are established, these may be located where habitat availability, economics, and politics rather than population biology dictate (Figure 4.4*a*). This approach can easily lead to a "reserve mentality" (Brussard et al. 1992): politicians and the public believe that once reserves are established, the problem is solved and the lands outside the reserves can be subjected to any sort of use. Indeed, in some situations, fragmentation creates mosaics in which discrete patches of suitable habitat are separated by large expanses of totally unsuitable habitat. This is the context in which island biogeography theory may provide useful guidelines for reserve design.

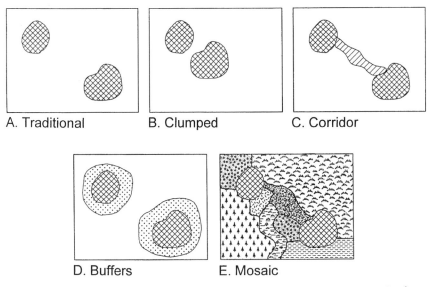

A. Traditional        B. Clumped        C. Corridor

D. Buffers        E. Mosaic

Figure 4.4. Hypothetical examples of various reserve-management strategies. In these examples, two reserves (hatched areas) are maintained. See the text for a description of the different strategies.

These conditions may be met for some species, such as forest specialists, in some environments, such as forest patches in a matrix of clear-cuts or intensive agriculture. Most species are not so ecologically restricted, however, and most environments are not so clearly separated into discrete habitable and uninhabitable patches. Although traditional reserve management may stop at the reserve boundary, fluxes of water, particulates, and organisms do not (Saunders et al. 1991). As a result, the dynamics of populations and their long-term persistence may be as much a result of what goes on in the broader landscape as of events within the reserves (Hanski and Thomas 1994). Reserve designs that are founded on island biogeography theory are therefore totally inappropriate. The "single large or several small" (SLOSS) debate that has continued for so long (Soulé and Simberloff 1986; Burkey 1989) can be seen to be largely irrelevant.

Recognition that reserves do not exist in a vacuum can lead to several management adjustments. If concerns about isolation of local populations are important, for example, reserves may be clustered to facilitate movements among them (Figure 4.4*b*) or habitat corridors may be included in the design (Harrison 1992; Figure 4.4*c*). Such designs are often justified in terms of metapopulation theory. If populations within the reserves are sensitive to external influences—predators, brood parasites, human activities, or the spread of disturbances such as fires—a buffer zone around the reserve may be created (Figure 4.4*d*; Baker 1992).

The arguments I have made in this chapter suggest that reserve design should involve more than corridors or buffers, however, especially if the populations of interest are spatially subdivided and dispersal is important. What is needed instead is management that focuses on both the most suitable habitat patches and the entire surrounding landscape mosaic (Figure 4.4e). Such "integrated landscape management" (Saunders et al. 1991; Hobbs et al. 1993) requires that attention be given to both reserved and nonreserved areas (Harris 1984; Woinarski et al. 1992; Barrett et al. 1994; Turner et al. 1995) and implies that varying levels of regulation of human land use in areas outside of reserves may be necessary. It requires recognition that not all suitable habitat areas may be occupied by a species at a particular time and that unoccupied areas may have substantial conservation value. And it requires an empirical assessment of how landscape pattern and individual movements interact to produce a level of landscape connectivity that will promote metapopulation dynamics. Because much of this information is species-specific, "ecosystem management" will have to retain a substantial autecological component. Modeling can provide conceptual guidance in these endeavors, and experimental model systems may provide helpful results. But as Hobbs (1992) has observed, "answers to practical conservation questions can come only from studies of real populations in real landscapes."

## Acknowledgments

Dale McCullough invited me to participate in the symposium on metapopulations at The Wildlife Society conference in Albuquerque, and then kept after me to put my thoughts on paper. I appreciate his persistence. Pat Kennedy, Bill Lidicker, and an anonymous reviewer offered comments on the manuscript. My work on populations in patchy environments has been supported by the National Science Foundation (DEB-9207010). I dedicate this chapter to my adviser and friend, John T. Emlen, who encouraged me to think across the boundary between basic ecology and wildlife management.

## REFERENCES

Adler, F. R., and B. Nuernberger. 1994. Persistence in patchy irregular landscapes. *Theoretical Population Biology* 45:41–75.

Andow, D. A., P. M. Kareiva, S. A. Levin, and A. Okubo. 1990. Spread of invading organisms. *Landscape Ecology* 4:177–188.

Andrén, H. 1992. Corvid density and nest predation in relation to forest fragmentation: A landscape perspective. *Ecology* 73:794–804.

———. 1994. Effects of habitat fragmentation on birds and mammals in landscapes with different proportions of suitable habitat: A review. *Oikos* 71:355–366.

Andrewartha, H. G., and L. C. Birch. 1954. *The Distribution and Abundance of Animals.* Chicago: University of Chicago Press.

Angelstam, P. 1992. Conservation of communities—the importance of edges, surroundings and landscape mosaic structure. Pages 9–70 in L. Hansson, ed., *Ecological Principles of Nature Conservation.* London: Elsevier Applied Science.

Baker, W. L. 1992. The landscape ecology of large disturbances in the design and management of nature reserves. *Landscape Ecology* 7:181–194.

Barrett, G. W., H. A. Ford, and H. F. Recher. 1994. Conservation of woodland birds in a fragmented rural landscape. *Pacific Conservation Biology* 1:245–256.

Bennett, A. F. 1990. Habitat corridors: Their role in wildlife management and conservation. Department of Conservation and Environment, Melbourne, Australia.

———. 1991. Roads, roadsides and wildlife conservation: A review. Pages 99–118 in D. A. Saunders and R. J. Hobbs, eds., *Nature Conservation* 2: *The Role of Corridors.* Chipping Norton, NSW: Surrey Beatty & Sons.

Bestelmeyer, B. T., and J. A. Wiens. 1996. The effects of land use on the structure of ground-foraging ant communities in the Argentine Chaco. *Ecological Applications:* in press.

Brown, J. H. 1995. *Macroecology.* Chicago: University of Chicago Press.

Brussard, P. F., D. D. Murphy, and R. F. Noss. 1992. Strategy and tactics for conserving biological diversity in the United States. *Conservation Biology* 6:157–159.

Bunce, R. G. H., and D. C. Howard, eds. 1990. *Species Dispersal in Agricultural Habitats.* London: Belhaven Press.

Burkey, T. V. 1989. Extinction in nature reserves: The effect of fragmentation and the importance of migration between reserve fragments. *Oikos* 55:75–81.

Cale, P. G., and R. J. Hobbs. 1994. Landscape heterogeneity indices: Problems of scale and applicability, with particular reference to animal habitat description. *Pacific Conservation Biology* 1:183–193.

Caughley, G. 1994. Directions in conservation biology. *Journal of Animal Ecology* 63:215–244.

Crist, T. O., D. S. Guertin, J. A. Wiens, and B. T. Milne. 1992. Animal movement in heterogeneous landscapes: An experiment with *Eleodes* beetles in shortgrass prairie. *Functional Ecology* 6:536–544.

Dale, V. H., S. M. Pearson, H. L. Offerman, and R. V. O'Neill. 1994. Relating patterns of land-use change to faunal biodiversity in the central Amazon. *Conservation Biology* 8:1027–1036.

Danielson, B. J. 1991. Communities in a landscape: The influence of habitat heterogeneity on the interactions between species. *American Naturalist* 138:1105–1120.

DeMers, M. N. 1993. Roadside ditches as corridors for range expansion of the western harvester ant (*Pogonomyrmex occidentalis* Cresson). *Landscape Ecology* 8:93–102.

Diamond, J. M. 1986. The design of a nature reserve system for Indonesian New Guinea. Pages 485–503 in M. E. Soulé, ed., *Conservation Biology: The Science of Scarcity and Diversity.* Sunderland, Mass: Sinauer Associates.

Dmowski, K., and M. Kozakiewicz. 1990. Influence of a shrub corridor on movements of passerine birds to a lake littoral zone. *Landscape Ecology* 4:99–108.

Doak, D. F., and L. S. Mills. 1994. A useful role for theory in conservation. *Ecology* 75:615–626.

Dunning, J. B., D. J. Stewart, B. J. Danielson, B. R. Noon, T. L. Root, R. H. Lamberson, and E. E. Stevens. 1995. Spatially explicit population models: Current forms and future uses. *Ecological Applications* 5:3–11.

Fahrig, L. 1988. A general model of populations in patchy habitats. *Applied Mathematics and Computation* 27:53–66.

Fahrig, L., and J. Paloheimo. 1988. Determinants of local population size in patchy habitats. *Theoretical Population Biology* 34:194–212.

Gardner, R. H., R. V. O'Neill, M. G. Turner, and V. H. Dale. 1989. Quantifying scale-dependent effects of animal movement with simple percolation models. *Landscape Ecology* 3:217–227.

Gardner, R. H., M. G. Turner, V. H. Dale, and R. V. O'Neill. 1992. A percolation model of ecological flows. Pages 259–269 in A. J. Hansen and F. di Castri, eds., *Landscape Boundaries: Consequences for Biotic Diversity and Ecological Flows.* New York: Springer-Verlag.

Gilpin, M., and I. Hanski, eds. 1991. Metapopulation dynamics: Empirical and theoretical investigations. *Biological Journal of the Linnean Society* 42:1–336.

Glenn, S. M., and S. L. Collins. 1992. Effects of scale and disturbance on rates of immigration and extinction of species in prairies. *Oikos* 63:273–280.

Goldwasser, L., J. Cook, and E. D. Silverman. 1994. The effects of variability on metapopulation dynamics and rates of invasion. *Ecology* 75:40–47.

Gosz, J. R. 1993. Ecotone hierarchies. *Ecological Applications* 3:369–376.

Gyllenberg, M., and I. Hanski. 1992. Single-species metapopulation dynamics: A structured model. *Theoretical Population Biology* 42:35–61.

Hansen, A. J., and F. di Castri, eds. 1992. *Landscape Boundaries: Consequences for Biotic Diversity and Ecological Flows.* New York: Springer-Verlag.

Hanski, I. 1991. Single-species metapopulation dynamics: Concepts, models and observations. *Biological Journal of the Linnean Society* 42:17–38.

———. 1994. A practical model of metapopulation dynamics. *Journal of Animal Ecology* 63:151–162.

Hanski, I., and M. Gilpin, eds. 1996. *Metapopulation Dynamics: Ecology, Genetics, and Evolution.* New York: Academic Press.

Hanski, I., and M. Gyllenberg. 1993. Two general metapopulation models and the core-satellite species hypothesis. *American Naturalist* 142: 17–41.

Hanski, I., and M. Kuussaari. 1995. Butterfly metapopulation dynamics. Pages 149–171 in N. Cappuccino and P. W. Price, eds., *Population Dynamics: New Approaches and Synthesis.* San Diego: Academic Press.

Hanski, I., and C. D. Thomas. 1994. Metapopulation dynamics and conservation: A spatially explicit model applied to butterflies. *Biological Conservation* 68:167–180.

Hanski, I., M. Kuussaari, and M. Nieminen. 1994. Metapopulation structure and migration in the butterfly *Melitaea cinxia. Ecology* 75:747–762.

Hanski, I., T. Pakkala, M. Kuussaari, and G. Lei. 1995. Metapopulation persistence of an endangered butterfly in a fragmented landscape. *Oikos* 72:21–28.

Hanski, I., A. Moilanen, T. Pakkala, and M. Kuussaari. 1996. The quantitative incidence function model and persistence in an endangered butterfly metapopulation. *Conservation Biology* 10: 578–590.

Harris, L. D. 1984. *The Fragmented Forest.* Chicago: University of Chicago Press.

Harrison, R. L. 1992. Toward a theory of inter-refuge corridor design. *Conservation Biology* 6:293–295.

Harrison, S. 1991. Local extinction in a metapopulation context: An empirical evaluation. *Biological Journal of the Linnean Society* 42:73–88.

———. 1994. Metapopulations and conservation. Pages 111–128 in P. J. Edwards, R. M. May, and N. R. Webb, eds., *Large-Scale Ecology and Conservation Biology.* Oxford: Blackwell.

Harrison, S., and L. Fahrig. 1995. Landscape pattern and population conservation. Pages 293–308 in L. Hansson, L. Fahrig, and G. Merriam, eds., *Mosaic Landscapes and Ecological Processes.* London: Chapman & Hall.

Hastings, A. 1991. Structured models of metapopulation dynamics. *Biological Journal of the Linnean Society* 42:57–71.

Henein, K., and G. Merriam. 1990. The elements of connectivity where corridor quality is variable. *Landscape Ecology* 4:157–170.

Hengeveld, R. 1989. *Dynamics of Biological Invasions.* London: Chapman & Hall.

Hobbs, R. J. 1992. The role of corridors in conservation: Solution or bandwagon? *Trends in Ecology and Evolution* 7:389–392.

Hobbs, R. J., D. A. Saunders, and A. R. Main. 1993. Conservation management in fragmented systems. Pages 279–296 in R. J. Hobbs and D. A. Saunders, eds., *Reintegrating Fragmented Landscapes.* New York: Springer-Verlag.

Holland, M. M., P. G. Risser, and R. J. Naiman, eds. 1991. *Ecotones: The Role of Landscape Boundaries in the Management and Restoration of Changing Environments.* New York: Chapman & Hall.

Holling, C. S. 1992. Cross-scale morphology, geometry, and dynamics of ecosystems. *Ecological Monographs* 62:447–502.

Ims, R. A. 1995. Movement patterns related to spatial structures. Pages 85–109 in L. Hansson, L. Fahrig, and G. Merriam, eds., *Mosaic Landscapes and Ecological Processes.* London: Chapman & Hall.

Ims, R. A., J. Rolstad, and P. Wegge. 1993. Predicting space use responses to habitat fragmentation: Can voles *Microtus oeconomus* serve as an experimental model system (EMS) for capercaillie grouse in boreal forest? *Biological Conservation* 63:261–268.

Johnson, A. R., B. T. Milne, and J. A. Wiens. 1992*a*. Diffusion in fractal landscapes: Simulations and experimental studies of tenebrionid beetle movements. *Ecology* 73:1968–1983.

Johnson, A. R., J. A. Wiens, B. T. Milne, and T. O. Crist. 1992*b*. Animal movements and population dynamics in heterogeneous landscapes. *Landscape Ecology* 7:63–75.

Johnson, W. C., and C. S. Adkisson. 1985. Dispersal of beech nuts by blue jays in fragmented landscapes. *American Midland Naturalist* 113:319–324.

Kareiva, P. 1987. Habitat fragmentation and the stability of predator-prey interactions. *Nature* 326:388–390.

King, A. W. 1991. Translating models across scales in the landscape. Pages 479–517 in M. G. Turner and R. H. Gardner, eds., *Quantitative Methods in Landscape Ecology.* New York: Springer-Verlag.

Kolasa, J., and C. D. Rollo. 1991. Introduction: The heterogeneity of heterogeneity: A glossary. Pages 1–23 in J. Kolasa and S. T. A. Pickett, eds., *Ecological Heterogeneity.* New York: Springer-Verlag.

Kotliar, N. B., and J. A. Wiens. 1990. Multiple scales of patchiness and patch structure: A hierarchical framework for the study of heterogeneity. *Oikos* 59:253–260.

Kremsater, L. L., and F. L. Bunnell. 1992. Testing responses to forest edges: The example of black-tailed deer. *Canadian Journal of Zoology* 70:2426–2435.

Kuussaari, M., M. Nieminen, and I. Hanski. 1996. An experimental study of migration in the Glanville fritillary butterfly *Melitaea cinxia. Journal of Animal Ecology:* in press.

La Polla, V. N., and G. W. Barrett. 1993. Effects of corridor width and presence on the population dynamics of the meadow vole (*Microtus pennsylvanicus*). *Landscape Ecology* 8:25–37.

Launer, A. E., and D. D. Murphy. 1994. Umbrella species and the conservation of habitat fragments: A case of a threatened butterfly and a vanishing grassland ecosystem. *Biological Conservation* 69:145–153.

Lavorel, S., R. H. Gardner, and R. V. O'Neill. 1993. Analysis of patterns in hierarchically structured landscapes. *Oikos* 67:521–528.

Lefkovitch, L. P., and L. Fahrig. 1985. Spatial characteristics of habitat patches and population survival. *Ecological Modelling* 30:297–308.

Levin, S. A. 1992. The problem of pattern and scale in ecology. *Ecology* 73:1943–1967.

———. 1993. Concepts of scale at the local level. Pages 7–19 in J. R. Ehleringer and C. B. Field, eds., *Scaling Physiological Processes: Leaf to Globe.* New York: Academic Press.

Levins, R. 1970. Extinction. Pages 77–107 in M. Gerstenhaber, ed. *Some Mathematical Questions in Biology.* Providence, R.I.: American Mathematical Society.

Lidicker, W. Z., Jr. 1994. A spatially explicit approach to vole population processes. *Polish Ecological Studies* 20:215–225.

Lindenmayer, D. B., and H. A. Nix. 1993. Ecological principles for the design of wildlife corridors. *Conservation Biology* 7:627–630.

Lord, J. M., and D. A. Norton. 1990. Scale and the spatial concept of fragmentation. *Conservation Biology* 4:197–202.

MacArthur, R. H., and E. O. Wilson. 1967. *The Theory of Island Biogeography.* Princeton: Princeton University Press.

Malmer, N., and P. H. Enckell. 1994. Ecological research at the beginning of the next century. *Oikos* 71:171–176.

Mangel, M., and C. Tier. 1994. Four facts every conservation biologist should know about persistence. *Ecology* 75:607–614.

Matthysen, E., F. Adriaensen, and A. A. Dhondt. 1995. Dispersal distances of nuthatches, *Sitta europaea,* in a highly fragmented forest habitat. *Oikos* 72:375–381.

May, R. M., and T. R. E. Southwood. 1990. Introduction. Pages 1–22 in B. Shorrocks and I. R. Swingland, eds., *Living in a Patchy Environment.* Oxford: Oxford University Press.

McKelvey, K., B. R. Noon, and R. H. Lamberson. 1993. Conservation planning for species occupying fragmented landscapes: The case of the northern spotted owl. Pages 424–450 in P. M. Kareiva, J. G. Kingsolver, and R. B. Huey, eds., *Biotic Interactions and Global Change.* Sunderland, Mass.: Sinauer Associates.

Medel, R. G., and R. A. Vásquez. 1994. Comparative analysis of harvester ant assemblages of Argentinean and Chilean arid zones. *Journal of Arid Environments* 26:363–371.

Merriam, G. 1988. Landscape dynamics in farmland. *Trends in Ecology and Evolution* 3:16–20.

———. 1991. Corridors and connectivity: Animal populations in heterogeneous environments. Pages 133–142 in D. A. Saunders and R. J. Hobbs, eds., *Nature Conservation 2: The Role of Corridors.* Chipping Norton, NSW: Surrey Beatty & Sons.

Merriam, G., and A. Lanoue. 1990. Corridor use by small mammals: field measurements for three experimental types of *Peromyscus leucopus. Landscape Ecology* 4:123–131.

Merriam, G., and D. A. Saunders. 1993. Corridors in restoration of fragmented landscapes. Pages 71–87 in D. A. Saunders, R. J. Hobbs, and P. R. Ehrlich, eds., *Nature Conservation 3: Reconstruction of Fragmented Ecosystems.* Chipping Norton, NSW: Surrey Beatty & Sons.

Meyer, W. B., and B. L. Turner II, eds. 1994. Changes in land use and land cover: A global perspective. Cambridge: Cambridge University Press.

Milne, B. T., M. G. Turner, J. A. Wiens, and A. R. Johnson. 1992. Interactions between the fractal geometry of landscapes and allometric herbivory. *Theoretical Population Biology* 41:337–353.

Mönkkönen, M., and D. A. Welsh. 1994. A biogeographical hypothesis on the effects of human caused landscape changes on the forest bird communities of Europe and North America. *Annales Zoologica Fennici* 31:61–70.

Murcia, C. 1995. Edge effects in fragmented forests: Implications for conservation. *Trends in Ecology and Evolution* 10:58–62.

Nilsson, S. G., and L. Ericson. 1992. Conservation of plant and animal populations in theory and practice. Pages 71-112 in L. Hansson, ed., *Ecological Principles of Nature Conservation.* London: Elsevier Applied Science.

Noon, B. R., and K. S. McKelvey. 1992. Stability properties of the spotted owl metapopulation in southern California. Pages 187–206 in J. Verner et al., The California spotted owl: A technical assessment of its current status. General Technical Report PSW-GTR-133. Albany, Calif.: USDA Forest Service Pacific Southwest Research Station.

Nunney, L., and K. A. Campbell. 1993. Assessing minimum viable population size: Demography meets population genetics. *Trends in Ecology and Evolution* 8:234–239.

Okubo, A. 1980. *Diffusion and Ecological Problems: Mathematical Models.* New York: Springer-Verlag.

Opdam, P., G. Rijsdijk, and F. Hustings. 1985. Bird communities in small woods in an agricultural landscape: Effects of area and isolation. *Biological Conservation* 34:333–352.

Opdam, P., R. van Apeldoorn, A. Schotman, and J. Kalkhoven. 1993. Population responses to landscape fragmentation. Pages 147–171 in C. C. Vos and P. Opdam, eds., *Landscape Ecology of a Stressed Environment.* London: Chapman & Hall.

Osborne, P. 1984. Bird numbers and habitat characteristics in farmland hedgerows. *Journal of Applied Ecology* 21:63–82.

Pearson, S. M. 1993. The spatial extent and relative influence of landscape-level factors on wintering bird populations. *Landscape Ecology* 8:3–18.

Pearson, S. M., M. G. Turner, R. H. Gardner, and R. V. O'Neill. 1996. An organism-based perspective of habitat fragmentation. In R. C. Szaro and D.W. Johnston, eds., *Biodiversity in Managed Landscapes: Theory and Practice.* Oxford: Oxford University Press.

Pearson, S. M., M. G. Turner, L. L. Wallace, and W. H. Romme. 1995. Winter habitat use by large ungulates following fire in northern Yellowstone National Park. *Ecological Applications* 5:744–755.

Potter, M. A. 1990. Movement of North Island brown kiwi (*Apteryx australis mantelli*) between forest remnants. *New Zealand Journal of Ecology* 14:17–24.

Pulliam, H. R. 1988. Sources, sinks, and population regulation. *American Naturalist* 132:652–661.

Pyle, C., and J. F. Franklin. 1992. An examination of differences in the accessibility of habitat for four types of old-growth dependent organisms in fragmented landscapes. *Northwest Environmental Journal* 8:219–220.

Rastetter, E. B., A. W. King, B. J. Cosby, G. M. Hornberger, R. V. O'Neill, and J. E. Hobbie. 1992. Aggregating fine-scale ecological knowledge to model coarser-scale attributes of ecosystems. *Ecological Applications* 2:55–70.

Ray, C., M. Gilpin, and A. T. Smith. 1991. The effect of conspecific attraction on metapopulation dynamics. *Biological Journal of the Linnean Society* 42:123–134.

Ricklefs, R. E., and D. Schluter, eds. 1993. *Species Diversity in Ecological Communities.* Chicago: University of Chicago Press.

Risser, P. G. 1995. The status of the science examining ecotones. *BioScience* 45:318–325.

Robinson, G. R., and J. F. Quinn. 1988. Extinction, turnover and species diversity in an experimentally fragmented California annual grassland. *Oecologia* 76:71–82.

Robinson, G. R., R. D. Holt, M. S. Gaines, S. P. Hamburg, M. L. Johnson, H. S. Fitch, and E. A. Martinko. 1992. Diverse and conflicting effects of habitat fragmentation. *Science* 257:524–526.

Rolstad, J., and P. Wegge. 1987. Distribution and size of capercaillie leks in relation to old forest fragmentation. *Oecologia* 72:389–394.

Rolstad, J., P. Wegge, and B. B. Larsen. 1988. Spacing and habitat use of capercaillie during summer. *Canadian Journal of Zoology* 66:670–679.

Santos, T., and J. L. Tellería. 1994. Influence of forest fragmentation on seed consumption and dispersal of Spanish juniper. *Biological Conservation* 70:129–134.

Saunders, D. A., and R. J. Hobbs, eds. 1991. *Nature Conservation 2: The Role of Corridors*. Chipping Norton, NSW: Surrey Beatty & Sons.

Saunders, D. A., R. J. Hobbs, and C. R. Margules. 1991. Biological consequences of ecosystem fragmentation: A review. *Conservation Biology* 5:18–32.

Simberloff, D. 1995. Habitat fragmentation and population extinction of birds. *Ibis* 137:S105–S111.

Simberloff, D., and L. G. Abele. 1982. Refuge design and island biogeographic theory: Effects of fragmentation. *American Naturalist* 120:41–50.

Simberloff, D., and J. Cox. 1987. Consequences and costs of conservation corridors. *Conservation Biology* 1:63–71.

Sjögren, P. 1991. Extinction and isolation gradients in metapopulations: The case of the pool frog (*Rana lessonae*). *Biological Journal of the Linnean Society* 42:135–147.

Soberón, J. 1992. Island biogeography and conservation practice. *Conservation Biology* 6:161.

Soulé, M. E., and D. Simberloff. 1986. What do genetics and ecology tell us about the design of nature reserves? *Biological Conservation* 35:19–40.

Stamps, J. A., M. B. Buechner, and V. V. Krishnan. 1987. The effects of edge permeability and habitat geometry on emigration from patches of habitat. *American Naturalist* 129:533–552.

Stenseth, N. C., and W. Z. Lidicker, Jr., eds. 1992. *Animal Dispersal: Small Mammals as a Model*. London: Chapman & Hall.

Taylor, P. D., L. Fahrig, K. Henein, and G. Merriam. 1993. Connectivity is a vital element of landscape structure. *Oikos* 68:571–573.

Thomas, C. D. 1994. Extinction, colonization, and metapopulations: Environmental tracking by rare species. *Conservation Biology* 8:373–378.

Thomas, J. W., E. D. Forsman, J. B. Lint, E. C. Meslow, B. R. Noon, and J. Verner. 1990. A conservation strategy for the northern spotted owl: Report of the Interagency Scientific Committee to address the conservation of the northern spotted owl. Portland: USDA Forest Service: USDI Bureau of Land Management/Fish and Wildlife Service/National Park Service.

Tilman, D., R. M. May, C. L. Lehman, and M. A. Nowak. 1994. Habitat destruction and the extinction debt. *Nature* 371:65–66.

Turchin, P. 1989. Beyond simple diffusion: Models of not-so-simple movement in animals and cells. *Comments in Theoretical Biology* 1:65–83.

Turner, M. G., and R. H. Gardner, eds. 1991. *Quantitative Methods in Landscape Ecology.* New York: Springer-Verlag.

Turner, M. G., G. J. Arthaud, R. T. Engstrom, S. J. Hejl, J. Liu, S. Loeb, and K. McKelvey. 1995. Usefulness of spatially explicit population models in land management. *Ecological Applications* 5:12–16.

van Dorp, D., and P. Opdam. 1987. Effects of patch size, isolation and regional abundance on forest bird communities. *Landscape Ecology* 1:59–73.

Verboom, J., J. A. J. Metz, and E. Meelis. 1993. Metapopulation models for impact assessment of fragmentation. Pages 172–191 in C. C. Vos and P. Opdam, eds., *Landscape Ecology of a Stressed Environment.* London: Chapman & Hall.

Verner, J. 1992. Data needs for avian conservation biology: Have we avoided critical research? *Condor* 94:301–303.

Weddell, B. J. 1991. Distribution and movements of Columbian ground squirrels (*Spermophilus columbianus* (Ord)): Are habitat patches like islands? *Journal of Biogeography* 18:385–394.

Wiens, J. A. 1985. Vertebrate responses to environmental patchiness in arid and semiarid ecosystems. Pages 169–193 in S. T. A. Pickett and P. S. White, eds., *The Ecology of Natural Disturbance and Patch Dynamics.* New York: Academic Press.

———. 1989. Spatial scaling in ecology. *Functional Ecology* 3:385–397.

———. 1990. On the use of "grain" and "grain size" in ecology. *Functional Ecology* 4:720.

———. 1992. Ecological flows across landscape boundaries: A conceptual overview. Pages 217–235 in A. J. Hansen and F. di Castri, eds., *Landscape Boundaries: Consequences for Biotic Diversity and Ecological Flows.* New York: Springer-Verlag.

———. 1995a. Habitat fragmentation: Island v. landscape perspectives on bird conservation. *Ibis* 137:S97–S104.

———. 1995b. Landscape mosaics and ecological theory. Pages 1–26 in L.

Hansson, L. Fahrig, and G. Merriam, eds., *Mosaic Landscapes and Ecological Processes.* London: Chapman & Hall.

————. 1996. Metapopulation dynamics and landscape ecology. In I. Hanski and M. Gilpin, eds., *Metapopulation Dynamics: Ecology, Genetics, and Evolution.* New York: Academic Press.

Wiens, J. A., and B. T. Milne. 1989. Scaling of "landscapes" in landscape ecology, or, landscape ecology from a beetle's perspective. *Landscape Ecology* 3:87–96.

Wiens, J. A., C. S. Crawford, and J. R. Gosz. 1985. Boundary dynamics: A conceptual framework for studying landscape ecosystems. *Oikos* 45:421–427.

Wiens, J. A., N. C. Stenseth, B. Van Horne, and R. A. Ims. 1993. Ecological mechanisms and landscape ecology. *Oikos* 66:369–380.

Wilcove, D. S. 1985. Nest predation in forest tracts and the decline of migratory songbirds. *Ecology* 66:1211–1214.

Wilson, E. O., and E. O. Willis. 1975. Applied biogeography. Pages 522–534 in M. L. Cody and J. M. Diamond, eds., *Ecology and Evolution of Communities.* Cambridge, Mass: Harvard University Press.

With, K. A., and T. O. Crist. 1995. Critical thresholds in species responses to landscape structure. *Ecology* 76:2446–2459.

With, K. A., R. H. Gardner, and M. G. Turner. In press. Landscape connectivity and population distributions in heterogeneous environments. *Oikos.*

Woinarski, J. C. Z., P. J. Whitehead, D. M. J. S. Bowman, and J. Russell-Smith. 1992. Conservation of mobile species in a variable environment: The problem of reserve design in the Northern Territory, Australia. *Global Ecology and Biogeography Letters* 2:1–10.

Zimmerman, B. L., and R. O. Bierregaard. 1986. Relevance of the equilibrium theory of island biogeography and species-area relations to conservation with a case from Amazonia. *Journal of Biogeography* 13:133–143.

# 5

# Responses of Terrestrial Vertebrates to Habitat Edges and Corridors

*William Z. Lidicker, Jr., and Walter D. Koenig*

The fate of metapopulations depends on only three things: the persistence of local populations, the success of emigration and immigration, and movements in and out of the metapopulation as a whole. These processes are equivalent respectively to mortality, reproduction, and emigration/immigration in continuous populations. A key element in each is the dispersal of individuals, both within and among patches, and in some cases imports and exports relative to other metapopulations. It is critical that we understand what causes individuals to become dispersers, how they respond to habitat edges and various nonhabitat matrices, and what determines successful immigration and colonization of empty habitat patches. Paradoxically, these are all subjects about which we know relatively little (Stenseth and Lidicker 1992a).

Our lack of knowledge about dispersal poses an increasingly serious dilemma. As the Earth's biota becomes progressively fragmented through human activities, a metapopulation structure characterizes more and more species. Our efforts, therefore, to conserve as much biodiversity as possible depends greatly on our understanding of metapopulation dynamics. There are two primary reasons why dispersal has until recently received little attention: it is difficult to study, and investigators have traditionally avoided edges, instead focusing their attention on population processes within habitat-types. Now that metapopulations (Gilpin and Hanski 1991; Chapter 4 in this volume) and heterogeneous spatial arrays of community-types—that is, landscapes (Forman and Godron 1986; Lidicker 1995a)—have become popular subjects of investigation, our attention must be directed toward correcting this serious information gap.

In this chapter we assess what we know about how terrestrial vertebrates respond to habitat edges and to potential dispersal corridors between habitat patches. This is perhaps the least studied aspect of the dispersal process in metapopulations. We also consider whether there are life-history features that are correlated with the way organisms deal with edges and hence can be used as predictors of behavior in a fragmented landscape. Do individuals avoid edges or prefer edges? Are habitat generalists more inclined to cross habitat borders than are specialists? Do abundance, social system, body size, density variation, vagility, competitive ability, and community dominance influence a species' performance? Finally, we want to know if the role of these life-history parameters varies across vertebrate taxa.

## Some Theoretical Considerations

Before we consider vertebrate responses to various landscape elements, we should examine some definitions and our level of understanding of these elements.

### Edges

The concept "edge" is not easily defined. When one community-type changes abruptly into another, an objective edge or ecotone is formed (Figure 5.1). These edges may be anthropogenic (forest/wheat field or grassland/road) or natural (lake/forest or mangrove forest/intertidal mudflat). Often, however, the edge is more subtle, such as mature/secondary forest or perennial/annual grassland. Even more problematic is that the perception of edges is species-specific and hence scale-dependent. A beetle or a salamander may act as if the boundary of a decaying log is a habitat edge; a thrush might avoid a grassland patch even if it is surrounded by forest; a weasel might be deterred by a river or farmland. For metapopulation studies, it is therefore essential to know what discontinuities are important to the species being investigated. There are also scale-related effects that depend on whether community, landscape, or biome-level edges are being studied (Risser 1995), but these are not directly relevant to metapopulation analysis.

A second question concerns the width of edges (Yahner 1988). Species living on edges and those in the adjacent communities respond to edges in idiosyncratic ways. This makes it difficult to define a generalized edge width. Edges do have abiotic, direct biotic, and indirect biotic effects (Murcia 1995) that can be measured and characterized with respect to their importance to focal species. For forest edges, it has been found that abiotic effects penetrate up to 50 m into the forest. The invasion of exotic plants and penetration by

Figure 5.1. Bonanza Creek Forest, southeast of Fairbanks, Alaska, looking east toward the Tanana River. An organism traveling east would pass through natural
habitat patches of white spruce (*Picea glauca*) forest with aspen (*Populus tremuloides*)
groves, tamarack (*Larix laricina*) bog, meadow, willow (*Salix*) brush, black spruce
(*Picea mariana*) forest, and the river. Photo by W. Z. Lidicker, Jr., 24 July 1975.

predators and nest parasites, however, may extend to 500 m or more (Wilcove 1985). Similarly, interior species may respond to edges at some distance from the actual boundary; for example, ovenbirds (*Seiurus aurocapillus*) were found to decline in breeding density within 200 m of the edge of deciduous forests in Missouri (Wenny et al. 1993). Clearly, edge widths must be defined in terms relevant to particular species. In the case of ecotones between community-types, we can be confident that many species will perceive them as discontinuities just as we do.

### Corridors

Corridors are key components in metapopulation functioning, as well, but they are not consistently defined. Now that they are widely perceived as important management instruments, they are receiving a great deal of attention (Saunders and Hobbs 1991; Soulé and Gilpin 1991; La Polla and Barrett 1993), including spirited controversy over their effectiveness (Simberloff and Cox 1987). In the context of metapopulations, corridors can most usefully be defined as strips of habitat that facilitate movements of organisms between local populations (Figure 5.2). In this sense, they are not places that support breeding populations of the species, or they would then be considered small habitat patches. Thus, it is not necessary that they be the same habitat as the patches they connect or even be composed of native species. Of course, the more they resemble the patches they are connecting the more effective they are likely to be as conduits for the widest range of species.

Corridors work because many species are willing to traverse habitats that are not suitable for permanent residency to find a more suitable patch, to find mates, or to escape from their current patch. Some species will traverse corridors by phoresy—that is, they are carried as pollen, seeds, parasites, or as incidental baggage by other species. The greatest challenge for land managers and conservation biologists is species that are reluctant to venture out of their preferred habitats at any time and avoid edges as well. Since corridors tend to be heavily influenced by edge effects, they are not likely to be effective for such unadventuresome species. For these species, effective corridors must be wide enough to contain a band of habitat unscathed by edge effects that matter to the targeted species.

An interesting case of phoresy that is not accidental (Gehlbach and Baldridge 1987) involves screech owls (*Otus asio*) and blind snakes (*Leptotyphlops dulcis*). The owls carry the snakes to their nests alive whereupon the snakes take up residency in the nest, feeding on various invertebrates including ectoparasites of the owls. The young in such nests grow better and suffer less mortality than do nestlings unaccompanied by snakes. Phoresy oc-

Figure 5.2. Experimental landscapes including corridors (foreground) for study of space use by tundra voles (*Microtus oeconomus*), Evenstad, Østerdalen, Norway. Photo by W. Z. Lidicker, Jr., 3 August 1990.

curs when some snakes escape and others leave the nest after the owl nestlings fledge.

There are other important attributes of corridors that may affect their efficiency: length (for example, Yahner 1983), whether or not they have narrow places (bottlenecks), presence of gaps, or whether they harbor predators and aggressive competitors. It is generally assumed that corridors improve with increasing width. But if a corridor is truly only a transit route, then the more rapidly it is traversed the better (Soulé and Gilpin 1991). This is because corridors often sustain high predation pressures or may have marginal food and shelter provisions. In this connection, an ingenious experiment performed with tundra voles (*Microtus oeconomus*) has demonstrated that corridors narrower than 1 m inhibited use and those wider slowed transit significantly (Andreassen et al. 1996).

### Presaturation Dispersal

One life-history trait, presaturation dispersal (Lidicker 1975; Stenseth and Lidicker 1992*b*), we know can strongly influence both the amount and the timing of dispersal. Species with this trait produce dispersers over a wide range of densities and before the habitat becomes filled to the carrying capacity for that species. Moreover, presaturation dispersers are generally younger and in better condition than individuals moving under saturated or carrying capacity circumstances. They are also moving at times when resource conditions are relatively favorable. All of these features contribute to an increased likelihood of successful immigration or colonization.

We hypothesize that, other things being equal, species with presaturation dispersal are more likely to be successful in coping with a fragmented habitat. There are several ways such behavior could have evolved: adaptation to habitats already fragmented before human intervention; avoidance of inbreeding complications; and adaptation to ephemeral, early successional, or "weedy" habitats.

### Patch and Landscape Properties

Geometrical modeling by Stamps et al. (1987a) has shown that both the size and the shape of patches strongly influence the amount of dispersal (emigration) from those patches. Specifically, it is the edge-to-size ratio that is positively related to emigration, so long as the edges allow any crossings at all. Moreover, edge permeability is positively correlated with emigration. This factor is especially influential when permeabilities are low. With "softer" edges, the edge-to-size ratio dominates in effectiveness. Since permeability is clearly a function of the type of matrix community that borders patches and the presence or absence of suitable corridors, this is a landscape property influencing interpatch dispersal rates. Whether the models of Stamps et al. assumed that dispersers had random or straight trajectories did not change the qualitative nature of their conclusions, but straight-line dispersers had generally higher dispersal rates than did random wanderers. This is because straight-line movers encountered edges more often than did random dispersers.

The shape and size of patches as well as edge permeability may also affect the population dynamics within patches through their influence on both the level and the distribution of intruder pressure on territories within the patch (Stamps et al. 1987b). Intruder pressure influences how much time and energy an individual can expend on growth and reproduction and, hence, ultimately, its reproductive success and production of new dispersers.

More subtle landscape effects are increasingly being explored. For example, predation on a focal species can be affected by the nature of the matrix and patch sizes (Lidicker 1995b; Oksanen and Schneider 1995). Predators in turn can influence the population dynamics of their prey within patches as well as the success of dispersal between patches. Computer simulations have shown that the extent of the matrix or "sink" relative to patch size can also influence emigration rates from the patch (Buechner 1987). Thus it is evident that the study of metapopulation dynamics in a landscape context will be a major thrust of future research.

## Amphibians and Reptiles

Herpetologists seem hardly to have addressed the issues of edge and corridor use and the crossing of unfavorable matrix habitats by amphibians and reptiles. Several recent reviews of dispersal, activity, and space use in these taxa fail to mention these topics (Kiester 1985; Gibbons 1986; Gibbons and Semlitsch 1987; Clobert et al. 1994), although one paper does say that habitat edges may be important (Gregory et al. 1987). Nevertheless, most authors do assume that a sensitivity to habitat nuances generally characterizes these groups. Because these creatures are generally long-lived and have lower metabolic demands than endotherms, we might anticipate that habitat responses would be more restrained than in short-lived birds and mammals.

Movement among habitats on a seasonal basis is primitive in these vertebrates (Kiester 1985). Originally, of course, the pattern was to aggregate in aquatic sites for breeding and then subsequently to disperse into feeding and overwintering habitats. This remains the pattern for many amphibians. In three genera of iguanas, the nesting aggregations occur in clearings in the forest (Bock et al. 1985). Some species require three separate habitats on an annual basis: breeding, foraging, and hibernating. From this basic migratory pattern, there have evolved life histories in which mating has been separated spatially from egg laying (marine turtles, for example) and in which both mating and egg laying are dispersed. This last condition is found in most reptiles and terrestrial salamanders (Kiester 1985).

Considerable effort has been expended on documenting the timing and extent of movements among habitats (see the preceding reviews; Gibbons 1970; Gibbons and Bennett 1974), but very little is known about the actual behavior of dispersing individuals. Techniques used to track individuals have included toe and scute marks, radioisotope labeling (Gregory et al. 1987), radiotelemetry, and spools of thread (Marlow and Tollestrup 1982). An effective technique for capturing moving individuals has been the use of drift fences with pitfall traps. In addition to the regular seasonal movements that characterize so many species, dispersal in response to erratically occurring habitat deterioration has also been documented (Gibbons and Semlitsch 1987). This includes responses to drying of aquatic habitats, flooding of terrestrial habitats, and changes in prey abundance. Actual habitat choice has been demonstrated for a few species of snakes, and in one case (the timber rattlesnake, *Crotalus horridus*) choice of habitat depended on the color morph of the individual snakes (Gibbons and Semlitsch 1987).

In an excellent study of the European common lizard, *Lacerta vivipara,* Clobert et al. (1994) reported on many aspects of dispersal, including sex and age factors, response to density, and differing quality of habitat. An important discovery was that juveniles in better condition are more likely to disperse than those in poor condition. This result contradicts the often-stated assumption that dispersal is in response to habitat deterioration or is always in the direction of poor conditions to better. The questions addressed in this chapter are not addressed in that study, however, nor are they cited in the authors' lists of future research directions. Similarly, the impressive 19-year study of the box turtle, *Terrepene carolina,* by Schwartz et al. (1984) focused on seasonal and short-term movement patterns, distances, home ranges, and sex differences in these patterns, but not on the landscape processes that we now think are so important. Hopefully future studies of movements in amphibians and reptiles will start on the immense task of learning how these species respond to habitat discontinuities, how they fare in crossing unfavorable areas, and the extent to which they can utilize corridors connecting habitat patches.

# Birds

Birds provide especially interesting test cases for questions concerning the value of habitat edges and corridors. We consider each of these features in turn.

## *Habitat Edges*

Fragmentation often increases, rather than decreases, species diversity by creating habitat for generalists and species dependent on edges. This is an excellent example supporting the axiom that species diversity per se is not, or at least should not be, the goal of conservation biology. With respect to the potential threats of habitat fragmentation, the species of interest are generally those restricted to the original vegetation type (Opdam 1991). The creation of edges, usually as a by-product of habitat fragmentation, almost invariably has a variety of biotic and abiotic effects (Saunders et al. 1991; Murcia 1995) and by definition reduces the amount of the original habitat present. Inevitably this will have adverse effects on many of the original populations. Exactly what effects there may be, beyond a straightforward numerical response, are less clear (for example, Andrén 1994; Simberloff 1995).

Andrén (1994), in a recent review, concluded that the effects of habitat fragmentation go beyond that of simple habitat loss and are complemented by the effects of patch size and isolation in highly fragmented landscapes with less than 30 percent suitable habitat remaining. In other words, modest amounts of fragmentation may be relatively benign. But this is certainly not always the case. Highways offer a clear counterexample. Although roads and roadsides can provide valuable habitat for wildlife and even act as dispersal corridors (Bennett 1991), they may adversely affect population densities and reproductive output, serve as dispersal sinks, and ultimately reduce the reproductive potential of a population below sustainable levels. In willow warblers (*Phylloscopus trochilus*) in the Netherlands, for example, dispersal distances of yearling males breeding along a busy road were greater than those living away from the road, but overall population density and reproductive success were significantly lower in the road zone (Foppen and Reijnen 1994; Reijnen and Foppen 1994).

Two proximate causes of the apparent lower habitat quality of road areas are noise (Reijnen and Foppen 1994) and the direct mortality of roadkills (Bennett 1991). The potentially significant effects of the latter are well documented by Mumme (1994), who found that annual mortality rates of Florida scrub jays (*Aphelocoma coerulescens*) increased almost 100 percent among birds bordering a two-lane county road compared to those living away from the road. Mortality rates were exceptionally high among immigrants with no experience living next to the road, suggesting that birds may eventually learn to avoid vehicles, or alternatively that only roadwise individuals survive more than a short time in proximity to the highway. The high mortality among inexperienced birds means that roads act as dispersal sinks, essentially luring birds away from high-quality habitat to an untimely death. The increased mortality rates among birds living near the road were great enough to significantly depress the overall reproductive output of the population, leading to the conclusion that roads and roadside mortality constitute a significant conservation problem for this threatened Florida endemic (Mumme 1994; Chapter 9 in this volume).

Recent work documenting nesting success of migratory birds related to forest fragmentation suggests that other kinds of threats may also increase with forest fragmentation long before overall habitat loss is severe. Robinson et al. (1995) found significant inverse correlations between both the incidence of nest predation and parasitism by brown-headed cowbirds (*Molothrus ater*) on a series of neotropical migrants and the degree of habitat fragmentation. The latter was measured by the percentage of forest cover within a 10-km radius for nine study areas in the midwestern United States, making the study unique in its large geographic scale. But damaging edge effects of nest predation have been documented on smaller geographic scales as well (for example, Gibbs 1991; Robinson and Wilcove 1994; for reviews see Rothstein and Robinson 1994 and Wiens 1995). These results suggest that edge effects may adversely affect many forest species beyond that due to habitat loss, even when habitat loss per se is minor.

Ultimately, severe fragmentation guarantees at least the local extinction of specialists dependent on extensive tracts of undisturbed habitat. Other less drastic long-term effects remain to be determined, but fragmentation is likely to alter many selective factors influencing surviving populations. Where brood parasitism increases as a consequence of forest fragmentation, for example, there will be increased selection for behaviors reducing the effectiveness of this source of nesting failure, including more effective egg recognition and rejection of parasitic eggs (Rothstein and Robinson 1994). Fragmentation also offers interesting opportunities to test hypotheses concerning the effects of edge permeability and habitat geometry on territoriality and space-use patterns of animals (Stamps et al. 1987a, 1987b). Specifically, insofar as remaining fragments resemble true islands, fragmentation may result in selection for reduced intraspecific aggression, reduced territory size, increased territorial overlap, and increased acceptance of subordinates on territories (Stamps and Buechner 1985). It will be fascinating to see whether such changes occur in ecological time fast enough to be observed and documented by researchers.

### Corridors

Because many birds fly and many migrate long distances between their breeding and wintering ranges, one might predict that individuals of a high proportion of species would quickly and frequently traverse large stretches of unsuitable habitat to investigate and perhaps colonize habitat patches. In this case, corridors would be relatively unimportant.

The extensive colonizing abilities of some species of island birds do fit this pattern (for example, Diamond 1974). Others do not. Despite the prodigious flight capabilities of many species, birds offer excellent examples of species that are apparently so sedentary, habitat-specific, and specialized in their foraging requirements that even short stretches of unsuitable habitat appear to form formidable barriers. Habitat corridors are likely to facilitate dispersal of such species between larger habitat patches, even if long-distance dispersal events occasionally take place.

Critical testing of the value of corridors to avian metapopulations is in a primitive state. Indeed, the difficulties of unambiguously determining whether corridors are necessary, or even desirable, for a particular species are formidable, especially given the potentially deleterious side effects (Simberloff and Cox 1987; Nicholls and Margules 1991; Hobbs 1992). This dearth of critical data is paradoxical given that the natural history of many birds has been extensively studied and avian dispersal has long been a topic of intense interest. Yet the degree to which our extensive knowledge of within-population avian demography can be used to infer the extent to which metapopulations are connected by dispersal events is at best unclear. This can result in a situation worse than not knowing anything, since the extensive data from population studies may yield the misleading impression that gene flow on a larger scale is known when in fact it is not.

Dispersal patterns of the acorn woodpecker (*Melanerpes formicivorus*) exemplify this paradox. Acorn woodpeckers are generally permanently resident, cooperative breeders common throughout oak woodland habitats of California and much of the southwestern United States. Extensive population study of marked individuals over a 25-year period in California indicated that many birds inherit their natal territory and a high proportion of the remainder disperse no farther than one or two territories away to eventually breed (Koenig and Mumme 1987). Data from this and other cooperatively breeding species have been interpreted as evidence that overall dispersal is low, with corresponding implications for the evolution of their highly social behavior and for their conservation biology. Because populations are so habitat-specific and dispersal distances apparently so limited, for example, such a species might be expected to be highly dependent on habitat corridors consisting of oak woodlands to disperse between habitat patches more than a few kilometers apart.

As it turns out, other observations clearly suggest that such a conclusion would be erroneous. Individuals of this species are frequently observed outside their normal range and are known to have colonized various islands and sites up to nearly 200 km away from their primary range within historical times (Koenig et al. 1995). Furthermore, birds persist in small, isolated mountain ranges throughout the Southwest despite evidence indicating that such populations would regularly go extinct in the absence of immigration from other, disjunct subpopulations (Stacey and Taper 1992). These lines of evidence indicate that acorn woodpeckers are good dispersers even in the absence of habitat corridors.

This discrepancy arises in part because of the different geographic scales that are typically investigated by researchers focusing on marked individuals compared to what is necessary to determine the degree to which subpopulations separated by tens or hundreds of kilometers are connected by dispersal. One approach to circumventing the problem is to resort to genetic estimates of gene flow. Edwards (1993), using mitochondrial DNA, was able to obtain evidence for detectable dispersal events between subpopulations of grey-crowned babblers (*Pomatostomus temporalis*) separated by over 1000 km.

Although information on long-distance dispersal unobtainable by observation is gained by such techniques, the different temporal scale makes interpretation difficult with respect to contemporary conservation goals. Edwards' (1993) study demonstrates that long-distance gene flow occurs, for example, and this result is valuable in terms of understanding the genetic structure of the species. But whether the temporal scale is such that small isolated populations are likely to be "rescued" from extinction regularly enough to maintain subpopulations, as occurs in acorn woodpeckers in the Southwest, is unclear.

In the absence of basic data on dispersal patterns, it is virtually impossible to come to an unambiguous conclusion regarding the value of corridors to avian metapopulations. It is safe to say that corridors will be of little value for some species, particularly those that are used to colonizing patchy habitats. The importance of corridors to species that live in habitats that are currently being degraded and fragmented at high rates is of greater interest and, unfortunately, less clear.

Moreover, it is not safe to conclude that corridors cannot hurt. Not only can they facilitate the transmission of diseases, parasites, and other catastrophes between subpopulations (Simberloff and Cox 1987; Hess 1994), but they harbor the potential for acting as population sinks, drawing individuals away from scarce high-quality habitats to areas where, because of strong negative edge effects, survival and reproductive success are poor (Soulé and Gilpin 1991). Minimizing such deleterious effects can be accomplished only by increasing the size of corridors to the point where they are no longer simply a conduit between larger habitat patches but themselves constitute high-quality habitat for species of conservation interest, as discussed earlier. An impressive example where large corridors between even larger reserves are being carefully planned in a temperate New Zealand rainforest is described by O'Donnell (1991).

Resolving these issues will require considerable empirical effort. At least two approaches are needed. The first is to implement studies designed to test the biological importance of corridors (Nicholls and Margules 1991). Second, and in many cases more realistic, is to design studies that examine in detail how dispersal is accomplished by species of interest. This can be done, at least in some cases, by careful study of individual movement patterns using radiotelemetry. Such a study of acorn woodpeckers, for example, has demonstrated not only that young birds prior to obtaining reproductive positions in the population make frequent "forays" over distances of several kilometers or more in search of reproductive vacancies but that they do so on a regular, almost daily, basis (Hooge 1995). Again this pattern contradicts demographic data derived from observations of marked individuals. Ideally, detailed behavioral information acquired in this way could be used to examine not only the home ranges of individuals but how they explore their surroundings, the extent to which nonoptimal habitats and corridors are used, and, if so, whether they appear to facilitate movement between habitat patches.

Birds offer many promising opportunities to understand the role of corridors in conservation. But taking advantage of these opportunities requires intimate knowledge about some of the most difficult and least known life-history features of birds. Devising ways to circumvent the formidable technical difficulties and acquire the data needed to answer basic questions about fragmentation, corridors, and reserve design promises to keep avian conservation biologists busy for years to come.

## Mammals

Our knowledge of how mammals respond to habitat edges and how they locate and utilize corridors is in its infancy, for these topics are not yet widely perceived as important. In a recently published volume (Grüm 1994), 25 papers were devoted to the subject of responses by voles (Arvicolidae) to patchy environments. Only one of these papers presented data on edge- and corridor-related behavior (Liro and Szacki 1994), however, and only one (Lidicker 1994) mentioned these topics as subjects for future investigation. The following comments are therefore more of a call to action than a review of what we know about the subject.

Mammals, as well as many other animals, can be thought of as responding behaviorally to edges in at least seven ways (Figure 5.3). These range from not even approaching edges (response 1) to moving freely across habitat patch boundaries (response 7). How organisms behave toward edges will strongly influence their chances of finding corridors and then using them. A species' response to edges may also vary with the type of edge, that is, the kind of adjacent habitat (Yahner 1988). Most investigators have used live trapping or telemetry to infer behavioral responses. Direct observations have been little used for obvious reasons, but perhaps they can be more profitably exploited in the future.

Beier (1993, 1995, Chapter 13 in this volume) has used radiotelemetry to study exploratory movements of mountain lions (*Felis concolor*) in southern California. He has demonstrated the ability of this species to locate and explore corridors as well as the importance of corridors to the viability of remnant populations. Others have investigated this subject by using experimental landscapes (Andreassen et al. 1990, 1996; La Polla and Barrett 1993) or agricultural mosaics (Henderson et al. 1985; Lovejoy et al. 1986; Liro and

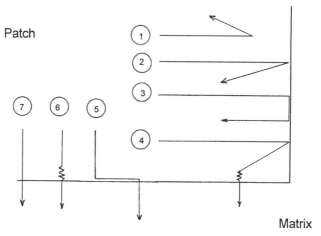

Figure 5.3. Seven behavioral responses to habitat-patch edges. 1: avoidance of edge zones. 2: approach to edge with immediate retreat to patch interior. 3: following of edge before return to interior. 4: rejection of edge followed by second cautious approach and subsequent outward movement. 5: following of edge before departure. 6: cautious approach to edge leading directly to entering matrix. 7: crossing edge without inhibitions.

Szacki 1987, 1994; Merriam and Lanoue 1990; Wegner and Merriam 1990; Laurance 1991*a*, 1995; Kozakiewicz et al. 1993; Bennett et al. 1994; Kozakiewicz and Szacki 1995; Merriam 1995).

A potentially fruitful line of investigation is to explore whether there are life-history correlates of edge behavior. If generalizations are found, it will be possible to predict successfully the behavior of unstudied species and thus make management and conservation programs more effective (Laurance 1991*b;* Laurance and Yensen 1991). Of course, some species are known to be edge or ecotonal species. For them, edges are optimal habitat, and they are thus less likely to be threatened by habitat fragmentation. Such edge species may pose management or conservation problems, however, in that they might be predators, competitors, or introduced species that threaten nonedge species in various ways. Similarly, other species can be classified as habitat generalists. For them, edges and corridors are irrelevant, at least at a scale important to most small mammal species, as such species view their environment in a finer-grain manner (Figure 5.3, response 7). But on the scale of urban development and major highways, even these species may respond to edges and require corridors (Beier 1993). Habitat specialization, while remaining an important concept, clearly is relative and must be defined in terms of specific attributes relevant to focal species.

It is the habitat specialists that, almost by definition, respond negatively to habitat edges (Figure 5.3, responses 1–6) and are of primary concern in an increasingly fragmented world. Are there any life-history generalizations that we can anticipate for this assemblage? Certainly body size will be a relevant variable. For a given spatial scale, large-bodied species will tend to perceive a landscape as fine-grained and hence will appear to be habitat generalists. Related to size is vagility. Larger species are generally more vagile, which means that independently of habitat selection, they will encounter edges more often, find corridors more easily, and cross unfavorable or dangerous areas more quickly.

Like habitat relationships, trophic specializations will also be important. Again, specialists are more likely than generalists to find themselves in a metapopulation structure requiring them to confront edges. Mean density levels and demographic patterns of density change should be explored for their relevance. Populations exhibiting chronically high numbers may be expected to produce mostly saturation dispersers. Such individuals (social subordinants) may respond to edges and corridors differently than presaturation dispersers. Lukyanov (1994) found that in the central Urals the bank vole (*Clethrionomys glareolus*) exhibits a negative relationship between population density and dispersal probability. Laurance (1990) has shown that, in comparisons among species of arboreal marsupials in Queensland tropical rainforest, abundance in unfragmented forest is not a good predictor of use of corridors or of success in fragments. Rather, it is a species' behavioral response to edges and its willingness to use secondary forest that are critical in this regard. He also found that a species' social system can be important. Complex sociality provides a centripetal force that inhibits dispersal across nonhabitat. Such species also tend to be attracted to inhabited patches and to avoid empty ones. Successful dispersal is generally by groups of individuals.

Species exhibiting irruptive density dynamics will exhibit episodes of massive long-distance dispersal interrupted by long periods of local isolation. In such cases, short-term study of movement behavior may underestimate the species' ability to maintain a viable population structure. Successful metapopulation management will then depend on understanding the causes of density irruptions to ensure that they are not inadvertently eliminated. Little is known about this kind of population dynamics, but there are examples of species exhibiting irruptive behavior: the long-haired rat, *Rattus villosissimus* (Newsome and Corbett 1975), house mouse, *Mus musculus* (Newsome 1969; Pearson 1963), gray squirrel, *Sciurus carolinensis* (Hamilton 1953), pericote, *Phyllotis xanthopygeus* (= *darwini*; Pearson 1975), and Scandinavian lemming, *Lemmus lemmus* (Henttonen and Kaikusalo 1993).

A final life-history trait to be considered is that of competitive ability. Poor competitors are more likely to be pushed into a fugitive status than good competitors. Thus they may be good colonists and opportunists and in that sense preadapted to survival as a metapopulation. So long as there are sufficient patches in the landscape, they may succeed well as fugitives or "weeds." Good competitors will be less motivated to leave a good patch, but if they do they should have good survival rates and the ability to immigrate into other patches, including those inhabited by a subordinate competitor. Laurance (1994) reports that two species of forest-dwelling *Rattus* tend not to coexist in the same small patch of habitat.

A second area of investigation that relates to use of edges and crossing of interpatch matrix is the question of how fragmentation affects the home range concept in mammals. The idea that individual mammals characteristically possess a home range—a limited, even if dynamic, chunk of real estate—has been a basic precept of mammalogy at least since Burt's (1943) classic paper on the subject. Modern views incorporate the possibility that a home range may have a core area of high-intensity use, possibly defended, and a border area characterized by a declining probability of use as one proceeds away from the core.

The home range concept has been recently challenged for fragmented landscapes by Kozakiewicz and Szacki (1995). They suggest that under these conditions the only species of small mammals to persist are those that can either greatly expand their home range to include multiple fragments (and thus to compromise the whole notion of home range) or become essentially nomadic. To the extent that this is true, significant demographic and social consequences can be anticipated. Related to this is the question of the extent to which an animal can use habitat patches that are individually too small by structuring a home range composed of multiple fragments (Dunning et al. 1992; Noss and Csuti 1994, figure 9.10). This is one of the questions being addressed by the group working with the root or tundra vole in experimental landscapes at Evenstad, Norway (Andreassen et al. 1990; Ims et al. 1993; Berg 1995; Halle 1995). Liro and Szacki (1994) and Kozakiewicz and Szacki (1995) suggest that the connectivity among habitat fragments involves pathways used by many individuals, not a random or diffuse series of movement routes. Such pathways enhance olfactory communication among individuals and may or may not be discernible to human observers by habitat features. Possibly because of this network of trails, small mammals can move up to 6 km in a few days (Liro and Szacki 1994).

## Lessons

Perhaps the clearest message from this review is that our understanding of how terrestrial vertebrates respond to habitat edges and to corridors connecting patches is in its infancy. Yet it is increasingly being recognized that this information is vital to a host of questions in landscape ecology and conservation biology (for example, Merriam 1991; Lidicker 1995*a;* Chapter 4 in this volume). Our future success in landscape management will depend on our understanding of those community discontinuities we refer to as edges, whether our goal is the production of food and fiber resources, fresh water supplies, recreation, biodiversity protection, or various combined objectives.

Viewing terrestrial vertebrates as a whole, we can surmise that for a given body size birds are generally most vagile, amphibians and reptiles least, and mammals in between. To the extent that vagility correlates with facility to cross habitat edges and use corridors, we might predict that birds would generally be most successful in coping with anthropogenic fragmentation of habitat and amphibians and reptiles the least. Yet amphibians and reptiles can survive for long periods without food and often are cryptic, two traits that would facilitate long-distance movement across nonoptimal terrain. Moreover, many amphibians are adapted for distance travel because of their need to find localized aquatic habitats for breeding. It may be, therefore, that mammals are the most vulnerable as a group to fragmentation.

Despite these generalities, we also know that to apply them to particular species is risky indeed. Potential vagility does not always translate into realized vagility, as illustrated by certain tropical forest birds that are reluctant even to approach edges, much less fly across nonforest gaps such as rivers (Diamond 1972, 1973). Moreover, as demonstrated by the acorn woodpecker, a pattern of restricted local movements can mislead us regarding the potential for long-distance dispersal. The same is true for the aquatic salamander, *Siren intermedia,* which has very small home ranges but is quick to colonize new ponds (Gehlbach and Kennedy 1978). Small mammals with home ranges measuring tens of meters across can occasionally travel several kilometers in a single night. On the other hand, eastern chipmunks (*Tamias striatus*) weighing 70 to 100 g are much less likely to approach edges or use corridors than the much smaller (15–25 g) white-footed mouse (*Peromyscus leucopus;* Henderson et al. 1985; Merriam and Lanoue 1990). Slender salamanders (*Batrachoseps* spp.) may move less than 2 m in their lifetime (Hendrickson 1954; Anderson 1960), whereas the red-bellied newt (*Taricha rivularis*) may move several kilometers every year and has been documented to travel 400 m in one day (Twitty 1959; Twitty et al. 1966). Many more examples could be given, but this should be sufficient to warn that, while generalities are useful, their application to specific cases may be misleading.

Because single life-history features such as body size or vagility are inadequate for predicting edge-related behavior for particular species, we must continue to explore other life-history features for improved predictability. Various combinations of traits might be particularly useful in this regard. Moreover, we must continually add to our knowledge of how individual species respond behaviorally and demographically to habitat discontinuities. Part of this effort should be directed to use of corridors: defining the relevant qualities of corridors; measuring transit times; and relating all of these to sex; age, season, and population densities.

An important task for future investigation is to explore how habitat fragmentation relates to the role of species in their communities. A significant beginning has been made with respect to predation and parasitism. It is already clear that such coactions can be influenced by edge effects, size of patches, and the nature of the interpatch matrix. In some cases the intensity of interspecific competition can also be affected. Other topics that seem worthy of investigation are possible influences of community dominance, trophic level (including the number of trophic levels maintained within a patch, Oksanen and Schneider 1995), rarity, and habitat or food specializations.

Of long-term interest is the possibility that terrestrial vertebrates may exhibit a trade-off between competition and colonization abilities as has been hypothesized for vascular plants by Tilman (1990, 1994). If this is in fact the case, we might anticipate that fragmentation would lead to selection for colonizing ability at the expense of competitive strength. This could have important consequences for community structure and function and, in particular, might enhance the success of introduced species.

We conclude that the subject outlined in this chapter has an important future. There is vast potential for contributions to life-history theory, to population, community, and landscape ecology, and to conservation biology. Terrestrial vertebrates will surely play a major role in this effort.

## Acknowledgments

We thank Dale R. McCullough for inviting us to write this chapter and for his helpful comments on the manuscript. H. W. Greene assisted with the herpetological literature, and L. N. Lidicker provided expert assistance in manuscript preparation.

## REFERENCES

Anderson, P. K. 1960. Ecology and evolution in island populations of salamanders in the San Francisco Bay region. *Ecological Monographs* 30:359–385.

Andreassen, H. P., S. Halle, and R. A. Ims. 1996. Optimal width of movement corridors for root voles: Not too narrow and not too wide. *Journal of Applied Ecology* 33:63–70.

Andreassen, H. P., H. Steen, and R. A. Ims. 1990. Felles avkomspleie registrert hos fjellrotte: En ny type sosial atferd for norsk pattedyrfauna. *Fauna* 43:71–73.

Andrén, H. 1994. Effects of habitat fragmentation on birds and mammals in landscapes with different proportions of suitable habitat: A review. *Oikos* 71:355–366.

Beier, P. 1993. Determining minimum habitat areas and habitat corridors for cougars. *Conservation Biology* 7:94–108.

———. 1995. Dispersal of juvenile cougars in fragmented habitat. *Journal of Wildlife Management* 59:228–237.

Bennett, A. F. 1991. Roads, roadsides and wildlife conservation: A review. Pages 99–118 in D. A. Saunders and R. J. Hobbs, eds., *Nature Conservation 2: The Role of Corridors*. Chipping Norton, NSW: Surrey Beatty & Sons.

Bennett, A. F., K. Henein, and G. Merriam. 1994. Corridor use and the elements of corridor quality: Chipmunks and fencerows in a farmland mosaic. *Biological Conservation* 68:155–165.

Berg, K. W. 1995. Space use responses of root voles (*Microtus oeconomus*) to a habitat fragmentation gradient. Candidate in Science thesis, University of Oslo, Norway.

Bock, B. C., A. S. Rand, and G. M. Burghardt. 1985. Seasonal migration and nesting site fidelity in the green iguana. *Contributions in Marine Science* (University of Texas), Supplement 27:435–443.

Buechner, M. 1987. Conservation in insular parks: Simulation models of factors affecting the movement of animals across park boundaries. *Biological Conservation* 41:57–76.

Burt, W. H. 1943. Territoriality and home range concepts as applied to mammals. *Journal of Mammalogy* 24:346–352.

Clobert, J., M. Massot, J. Lecomte, G. Sorci, M. de Fraipont, and R. Barbault. 1994. Determinants of dispersal behavior: The common lizard as a case study. Pages 183–206 in L. J. Vitt and E. R. Pianka, eds., *Lizard Ecology: Historical and Experimental Perspectives*. Princeton: Princeton University Press.

Diamond, J. M. 1972. Biogeographic kinetics: Estimation of relaxation times for avifaunas of southwest Pacific islands. *Proceedings of the National Academy of Sciences* (U.S.A.) 69:3199–3203.

———. 1973. Distributional ecology of New Guinea birds. *Science* 179:759–769.

————. 1974. Colonization of exploded volcanic islands by birds: The su-
pertramp strategy. *Science* 184:803–806.

Dunning, J. B., B. J. Danielson, and H. R. Pulliam. 1992. Ecological
processes that affect populations in complex landscapes. *Oikos* 65:
169–175.

Edwards, S. V. 1993. Long-distance gene flow in a cooperative breeder detected in genealogies of mitochondrial DNA sequences. *Proceedings of the
Royal Society of London* B 252:177–185.

Foppen, F., and R. Reijnen. 1994. The effects of car traffic on breeding bird
populations in woodland. II: Breeding dispersal of male willow warblers
(*Phylloscopus trochilus*) in relation to the proximity of a highway. *Journal
of Applied Ecology* 31:95–101.

Forman, R. T. T., and M. Godron. 1986. *Landscape Ecology.* New York:
Wiley.

Gehlbach, F. R., and R. S. Baldridge. 1987. Live blind snakes (*Leptotyphlops
dulcis*) in eastern screech owl (*Otus asio*) nests: A novel commensalism.
*Oecologia* 71:560–563.

Gehlbach, F. R., and S. E. Kennedy. 1978. Population ecology of a highly
productive aquatic salamander (*Siren intermedia*). *Southwestern Naturalist* 23:423–430.

Gibbons, J. W. 1970. Terrestrial activity and the population dynamics of
aquatic turtles. *American Midland Naturalist* 83:404–414.

————. 1986. Movement patterns among turtle populations: Applicability
to management of the desert tortoise. *Herpetologica* 42:104–113.

Gibbons, J. W., and D. H. Bennett. 1974. Determination of anuran terrestrial activity patterns by a drift fence method. *Copeia* 1974:236–243.

Gibbons, J. W., and R. D. Semlitsch. 1987. Activity patterns. Pages 396–421
in R. A. Siegel, J. T. Collins, and S. S. Novak, eds., *Snakes: Ecology and
Evolutionary Biology.* New York: Macmillan.

Gibbs, J. P. 1991. Avian nest predation in tropical wet forest: An experimental study. *Oikos* 60:155–161.

Gilpin, M., and I. Hanski, eds. 1991. Metapopulation dynamics: Empirical
and theoretical investigations. *Biological Journal of the Linnean Society*
42:1–336.

Gregory, P. T., J. M. Macartney, and K. W. Larsen. 1987. Spatial patterns and
movements. Pages 366–395 in R. A. Siegel, J. T. Collins, and S. S.
Novak, eds., *Snakes: Ecology and Evolutionary Biology.* New York:
Macmillan.

Grüm, L., ed. 1994. Polish ecological studies. *Polish Ecological Studies*
20:69–596.

Halle, S. 1995. Diel pattern of locomotor activity in populations of root voles, *Microtus oeconomus*. *Journal of Biological Rhythms* 10:211–224.

Hamilton, W. J., Jr. 1953. Migrants and emigrants. *New York State Conservationist* June–July:10–11.

Henderson, M. T., G. Merriam, and J. Wegner. 1985. Patchy environments and species survival: Chipmunks in an agricultural mosaic. *Biological Conservation* 31:95–105.

Hendrickson, J. R. 1954. Ecology and systematics of salamanders of the genus Batrachoseps. *University of California Publications in Zoology* 54(1):1–46.

Henttonen, H., and A. Kaikusalo. 1993. Lemming movements. Pages 157–186 in N. C. Stenseth and R. A. Ims, eds., *The Biology of Lemmings*. London: Academic Press.

Hess, G. R. 1994. Conservation corridors and contagious disease: A cautionary note. *Conservation Biology* 8:256–262.

Hobbs, R. J. 1992. The role of corridors in conservation: Solution or bandwagon? *Trends in Ecology and Evolution* 7:389–392.

Hooge, P. N. 1995. Dispersal dynamics of the cooperatively breeding acorn woodpecker. Ph.D. thesis, Department of Zoology, University of California, Berkeley.

Ims, R. A., J. Rolstad, and P. Wegge. 1993. Predicting space use responses to habitat fragmentation: Can voles *Microtus oeconomus* serve as an experimental model system (EMS) for capercaillie grouse in boreal forest? *Biological Conservation* 63:261–268.

Kiester, A. R. 1985. Sex-specific dynamics of aggregation and dispersal in reptiles and amphibians. *Contributions in Marine Science* (University of Texas), Supplement 27:425–434.

Koenig, W. D., and R. L. Mumme. 1987. *Population Ecology of the Cooperatively Breeding Acorn Woodpecker*. Princeton: Princeton University Press.

Koenig, W. D., P. B. Stacey, M. T. Stanback, and R. L. Mumme. 1995. Acorn woodpecker (*Melanerpes formicivorus*). No. 194 in A. Poole and F. Gill, eds., *The Birds of North America*. Philadelphia: Academy of Natural Sciences.

Kozakiewicz, M., and J. Szacki. 1995. Movements of small mammals in a landscape: Patch restriction or nomadism? Pages 78–94 in W. Z. Lidicker, Jr., ed., *Landscape Approaches in Mammalian Ecology and Conservation*. Minneapolis: University of Minnesota Press.

Kozakiewicz, M., A. Kozakiewicz, A. Ludowski, and T. Gortat. 1993. Use of space by bank voles (*Clethrionomys glareolus*) in a Polish farm landscape. *Landscape Ecology* 8:19–24.

La Polla, V. N., and G. W. Barrett. 1993. Effects of corridor width and presence on the population dynamics of the meadow vole (*Microtus pennsylvanicus*). *Landscape Ecology* 8:25–37.

Laurance, W. F. 1990. Comparative responses of five arboreal marsupials to tropical forest fragmentation. *Journal of Mammalogy* 71:641–653.

———. 1991*a*. Ecological correlates of extinction proneness in Australian tropical rainforest mammals. *Conservation Biology* 5:79–89.

———. 1991*b*. Edge effects in tropical forest fragments: Application of a model for the design of nature reserves. *Biological Conservation* 57:205–219.

———. 1994. Rainforest fragmentation and the structure of small mammal communities in tropical Queensland. *Biological Conservation* 69:23–32.

———. 1995. Extinction and survival of rainforest mammals in a fragmented tropical landscape. Pages 46–63 in W. Z. Lidicker, Jr., ed., *Landscape Approaches in Mammalian Ecology and Conservation*. Minneapolis: University of Minnesota Press.

Laurance, W. F., and E. Yensen. 1991. Predicting the impacts of edge effects in fragmented habitats. *Biological Conservation* 55:77–92.

Lidicker, W. Z., Jr. 1975. The role of dispersal in the demography of small mammals. Pages 103–128 in F. B. Golley, K. Petrusewicz, and L. Ryszkowski, eds., *Small Mammals: Their Productivity and Population Dynamics*. Cambridge: Cambridge University Press.

———. 1994. A spatially explicit approach to vole population processes. *Polish Ecological Studies* 20:215–225.

———, ed. 1995*a*. *Landscape Approaches in Mammalian Ecology and Conservation*. Minneapolis: University of Minnesota Press.

———. 1995*b*. The landscape concept: Something old, something new. Pages 3–19 in W. Z. Lidicker, Jr., ed., *Landscape Approaches in Mammalian Ecology and Conservation*. Minneapolis: University of Minnesota Press.

Liro, A., and J. Szacki. 1987. Movements of field mice *Apodemus agrarius* (Pallas) in a suburban mosaic of habitats. *Oecologia* 74:438–440.

———. 1994. Movements of small mammals along two ecological corridors in suburban Warsaw. *Polish Ecological Studies* 20:227–231.

Lovejoy, T. E., R. O. Bierregaard, A. B. Rylands, J. R. Malcolm, C. E. Quintela, L. H. Harper, K. S. Brown, Jr., A. H. Powell, G. V. N. Powell, H.O.R. Schubart, and M. B. Hays. 1986. Edge and other effects of isolation on Amazon forest fragments. Pages 257–285 in M. E. Soulé, ed., *Conservation Biology: The Science of Scarcity and Diversity*. Sunderland, Mass.: Sinauer Associates.

Lukyanov, O. 1994. Analysis of dispersal in small mammal populations. *Polish Ecological Studies* 20:237–242.

Marlow, R. W., and K. Tollestrup. 1982. Mining and exploitation of natural mineral deposits by the desert tortoise, *Gopherus agassizii. Animal Behaviour* 30:475–478.

Merriam, G. 1991. Corridors and connectivity: Animal populations in heterogeneous environments. Pages 133–142 in D. A. Saunders and R. J. Hobbs, eds., *Nature Conservation 2: The Role of Corridors.* Chipping Norton, NSW: Surrey Beatty & Sons.

————. 1995. Movement in spatially divided populations: Responses to landscape structure. Pages 64–77 in W. Z. Lidicker, Jr., ed., *Landscape Approaches in Mammalian Ecology and Conservation.* Minneapolis: University of Minnesota Press.

Merriam, G., and A. Lanoue. 1990. Corridor use by small mammals: Field measurements for three experimental types of *Peromyscus leucopus. Landscape Ecology* 4:123–131.

Mumme, R. L. 1994. Demographic consequences of roadside mortality in the Florida scrub jay. Abstract. American Ornithologists' Union Annual meeting, Missoula, Montana.

Murcia, C. 1995. Edge effects in fragmented forests: Implications for conservation. *Trends in Ecology and Evolution* 10:58–62.

Newsome, A. E. 1969. A population study of house-mice temporarily inhabiting a South Australian wheatfield. *Journal of Animal Ecology* 38: 341–359.

Newsome, A. E., and L. K. Corbett. 1975. Outbreaks of rodents in semi-arid and arid Australia: Causes, preventions, and evolutionary considerations. Pages 117–153 in I. Prakash and P. K. Ghosh, eds., *Rodents in Desert Environments.* The Hague: Junk Publishers.

Nicholls, A. O., and C. R. Margules. 1991. The design of studies to demonstrate the biological importance of corridors. Pages 49–61 in D. A. Saunders and R. J. Hobbs, eds., *Nature Conservation 2: The Role of Corridors.* Chipping Norton, NSW: Surrey Beatty & Sons.

Noss, R. F., and B. Csuti. 1994. Habitat fragmentation. Pages 237–264 in G. K. Meffe and C. R. Carroll, eds., *Principles of Conservation Biology.* Sunderland, Mass.: Sinauer Associates.

O'Donnell, C. F. J. 1991. Application of the wildlife corridors concept to temperate rainforest sites, North Westland, New Zealand. Pages 85–98 in D. A. Saunders and R. J. Hobbs, eds., *Nature Conservation 2: The Role of Corridors.* Chipping Norton, NSW: Surrey Beatty & Sons.

Oksanen, T., and M. Schneider. 1995. Predator-prey dynamics as influenced by habitat heterogeneity. Pages 122–150 in W. Z. Lidicker, Jr., ed., *Landscape Approaches in Mammalian Ecology and Conservation.* Minneapolis: University of Minnesota Press.

Opdam, P. 1991. Metapopulation theory and habitat fragmentation: A review of holarctic breeding bird studies. *Landscape Ecology* 5:93–106.

Pearson, O. P. 1963. History of two local outbreaks of feral house mice. *Ecology* 44:540–549.

———. 1975. An outbreak of mice in the coastal desert of Peru. *Mammalia* 39:375–386.

Reijnen, R., and R. Foppen. 1994. The effects of car traffic on breeding bird populations in woodland. I: Evidence of reduced habitat quality for willow warblers (*Phylloscopus trochilus*) breeding close to a highway. *Journal of Applied Ecology* 31:85–94.

Risser, P. G. 1995. The status of the science examining ecotones. *BioScience* 45:318–325.

Robinson, S. K., and D. S. Wilcove. 1994. Forest fragmentation in the temperate zone and its effects on migratory songbirds. *Bird Conservation International* 4:233–249.

Robinson, S. K., F. R. Thompson III, T. M. Donovan, D. R. Whitehead, and J. Faaborg. 1995. Regional forest fragmentation and the nesting success of migratory birds. *Science* 267:1987–1990.

Rothstein, S. I., and S. K. Robinson. 1994. Conservation and coevolutionary implications of brood parasitism by cowbirds. *Trends in Ecology and Evolution* 9:162–164.

Saunders, D. A., and R. J. Hobbs, eds. 1991. *Nature Conservation 2: The Role of Corridors*. Chipping Norton, NSW: Surrey Beatty & Sons.

Saunders, D. A., R. J. Hobbs, and C. R. Margules. 1991. Biological consequences of ecosystem fragmentation: A review. *Conservation Biology* 5:18–32.

Schwartz, E. R., C. W. Schwartz, and A. R. Kiester. 1984. The three-toed box turtle in central Missouri. II: A nineteen-year study of home range, movement, and population. Missouri Department of Conservation, Terrestrial Series No. 12, Jefferson City, Missouri.

Simberloff, D. 1995. Habitat fragmentation and population extinction of birds. *Ibis* 137:S105–S111.

Simberloff, D., and J. Cox. 1987. Consequences and costs of conservation corridors. *Conservation Biology* 1:63–71.

Soulé, M. E., and M. E. Gilpin. 1991. The theory of wildlife corridor capability. Pages 3–8 in D. A. Saunders and R. J. Hobbs, eds., *Nature Conservation 2: The Role of Corridors*. Chipping Norton, NSW: Surrey Beatty & Sons.

Stacey, P. B., and M. Taper. 1992. Environmental variation and the persistence of small populations. *Ecological Applications* 2:18–29.

Stamps, J. A., and M. Buechner. 1985. The territorial defense hypothesis and the ecology of insular vertebrates. *Quarterly Review of Biology* 60:155–181.

Stamps, J. A., M. Buechner, and V. V. Krishnan. 1987*a*. The effects of edge permeability and habitat geometry on emigration from patches of habitat. *American Naturalist* 129:533–552.

———. 1987*b*. The effects of habitat geometry on territorial defense costs: Intruder pressure in bounded habitats. *American Zoologist* 27:307–325.

Stenseth, N. C., and W. Z. Lidicker, Jr. 1992*a*. *Animal Dispersal: Small Mammals as a Model*. London: Chapman & Hall.

———. 1992*b*. Presaturation and saturation dispersal 15 years later: Some theoretical considerations. Pages 201–223 in N. C. Stenseth and W. Z. Lidicker, Jr., eds., *Animal Dispersal: Small Mammals as a Model*. London: Chapman & Hall.

Tilman, D. 1990. Constraints and tradeoffs: Toward a predictive theory of competition and succession. *Oikos* 58:3–15.

———. 1994. Competition and biodiversity in spatially structured habitats. *Ecology* 75:2–16.

Twitty, V. C. 1959. Migration and speciation in newts. *Science* 130: 1735–1743.

Twitty, V. C., D. Grant, and O. Anderson. 1966. Course and timing of the homing migration in the newt *Taricha rivularis*. *Proceedings of the National Academy of Sciences* (U.S.A.) 56:864–871.

Wegner, J., and G. Merriam. 1990. Use of spatial elements in a farmland mosaic by a woodland rodent. *Biological Conservation* 54:263–276.

Wenny, D. G., R. L. Clawson, J. Faaborg, and S. L. Sheriff. 1993. Population density, habitat selection and minimum area requirements of three forest-interior warblers in central Missouri. *Condor* 95:968–979.

Wiens, J. A. 1995. Habitat fragmentation: Island vs. landscape perspectives on bird conservation. *Ibis* 137:S97–S104.

Wilcove, D. S. 1985. Nest predation in forest tracts and the decline of migratory songbirds. *Ecology* 66:1211–1214.

Yahner, R. H. 1983. Small mammals in farmstead shelterbelts: Habitat correlates of seasonal abundance and community structure. *Journal of Wildlife Management* 47:74–84.

———. 1988. Changes in wildlife communities near edges. *Conservation Biology* 2:333–339.

# 6

# Using Logistic Regression to Model Metapopulation Dynamics: Large-Scale Forestry Extirpates the Pool Frog

*Per Sjögren-Gulve and Chris Ray*

In his model of metapopulation dynamics, Levins (1969) envisioned a set of identical habitat patches ($T$), of which a certain number ($N$) are inhabited by the species under study. The rate with which this number of occupied patches would change with time ($t$) was modeled as the net result of local extinction and colonization events,

$$\frac{\mathrm{d}N}{\mathrm{d}t} = m \cdot N \cdot \left(1 - \frac{N}{T}\right) - e \cdot N \tag{1}$$

where $m$ is a constant quantifying the colonization rate of patches as a function of $N$ and $e$ is the extinction rate of patch populations.

Provided that $m > e$ in this deterministic model, a stable equilibrium exists ($\mathrm{d}N/\mathrm{d}t = 0$) where a constant number of patches is occupied ($\hat{N}$) despite the local turnover:

$$\hat{N} = T \cdot \left(1 - \frac{e}{m}\right) \tag{2}$$

Accordingly, a major prediction of the Levins model is that even a system of subpopulations characterized by local instability (local extinction and colonization events) may constitute a system that is regionally stable.

The application of this conceptual model to practical conservation seems controversial and has been questioned (Harrison 1994; Thomas 1994; Wilson et al. 1994). Data from empirical studies indicate that local populations do differ in their extinction probabilities due to differences in population size (for example, Pimm et al. 1988; Bengtsson 1989; Schoener

1991; Kindvall and Ahlén 1992), patch size (for example, Fritz 1979; Harrison et al. 1988; Kindvall and Ahlén 1992; Thomas et al. 1992), distance to closest occupied patch (for example, Smith 1980; Sjögren 1991*a;* Thomas et al. 1992), and patch quality (for example, Thomas 1994).

Although some metapopulation models incorporate both environmental heterogeneity and internal patch population dynamics (Hanski and Gyllenberg 1993), patch occupancy models have lagged behind in incorporating environmental heterogeneity. Notable exceptions are the models of Hanski (1991, 1994*a*, 1994*b*). Hanski (1991) used models from island biogeography theory to construct individual patch colonization and extinction probabilities ($m_i$ and $e_i$, respectively) that were functions of distance to colonization source ($D_i$) and patch area ($A_i$):

$$m_i = m_0 \cdot \exp\left(-a \cdot D_i\right) \tag{3}$$

$$e_i = e_0 \cdot \exp\left(-b \cdot A_i\right) \tag{4}$$

This approach is appropriate for systems in which dispersal declines exponentially with distance and population size is positively correlated with patch area, as discussed by Levins (1969). More recently, Hanski (1994*a*, 1994*b*) has proposed a general occupancy model based on empirical species incidence curves. These curves/functions relate species incidence on patches to a measurable patch variable (usually patch area or distance to propagule source). Individual patch colonization and extinction probabilities can be calculated from these curves, based on the values of the patch variables and on the potentially problematic assumption that the metapopulation is at equilibrium. This model, however, allows only one environmental variable to affect each type of patch state transition; for example, extinction is usually a function of patch area alone. Another variant of the Levins approach involves the use of colonization and extinction rates that are functionally dependent on the fraction of occupied patches in the metapopulation ($p$). These rates vary temporally with $p$, but they do not respond to heterogeneity between patches. Gotelli and Kelley (1993) briefly review these models and present a general model to examine the effectiveness of this approach with empirical data.

In all of the above models, difficulties arise when multiple factors govern colonization and extinction (for example, Sjögren Gulve 1994). No general expansion of the Levins model has yet been presented that is appropriate for metapopulations with more complex turnover dynamics. To help solve this problem, we present a straightforward model that combines the use of logistic regression (which identifies determinants of colonization and extinction), logistic models of colonization and extinction probabilities, and a Monte Carlo simulation procedure to model metapopulation dynamics, based on empirical data. A less generalized formulation of this approach was applied to the European nuthatch (*Sitta europaea*) in a Dutch agricultural landscape by Verboom et al. (1991). We apply our more generalized model to a system of pool frogs (*Rana lessonae*) for which empirical data on colonization and extinction patterns exist and for which ponds constitute the patches (Sjögren Gulve 1994 and unpublished data). Our questions were: what habitat variables govern pond-state transitions, and do empirical data (and the metapopulation model derived from these data) indicate that a metapopulation approach is valid for conservation purposes? This example illustrates a framework of statistical analysis and simulation modeling that can be adapted to many systems. The results from the analysis proved useful both for understanding pool frog metapopulation dynamics and for guiding conservation efforts.

## Study System and Methods

In this section we describe the empirical data and analytical techniques used to incorporate real-world complexity within the metapopulation model.

### *Metapopulation and Environment*

The pool frog (Figure 6.1) is known from some 60 Scandinavian localities, all of which are permanent ponds situated along the Baltic coast of east-central Sweden. Special features of this coastal area are its continuing post-Pleistocene land uplift, its lime-rich moraine, and its brackish seawater. (See Sjögren 1991*a* and Sjögren Gulve 1994 for more information.) The pool frogs spend most of the postbreeding season at the water's edge of the pond; less than 1 percent of the adults emigrate to other ponds annually, whereas some 35 percent of the juvenile cohorts become emigrants (Sjögren-Gulve 1996). All frogs hibernate on land close to the breeding pond (distance less than 200 m). Inhabited ponds (Figure 6.2) are characterized by warm local climate and are typically situated in mixed coniferous/deciduous forest less than 2 km from the Baltic Sea (Sjögren Gulve 1994). As local climate (that is, mean water temperature) depends on the pond's degree of exposure to wind and sun (Sjögren Gulve 1994), the surrounding forest probably has several important functions for the frogs. It provides sheltered hibernation sites and breeding ponds, as well as migration/dispersal biotopes in the landscape "matrix."

Figure 6.1. A male pool frog (*Rana lessonae*) basking during a sunny day in mid-May. During the breeding period, which extends from mid-May to late June, the males form night and day choruses in the warmest part of the pond and defend mobile territories (Sjögren et al. 1988). A chorus with five to ten males can be heard 0.5–1 km from the pond at night, and individual males respond to playback or imitated calls.

In the analysis to be presented we found that data about the forest and the occurrence of modern forestry were needed to successfully predict metapopulation dynamics. Forestry has a long historical tradition in Sweden. Along the Baltic coast, the amount of timber harvesting has varied a lot from region to region, depending primarily on land accessibility, transport facilities, and the interest of the landowners. In some areas clear-cutting started in the eighteenth century (to fuel the iron industry), while small-scale management predominated in others. This heterogeneity in local harvest rate was largely maintained until the 1960s and 1970s when modern technology, combined with national subsidies for ditching and construction of forest roads, spread and significantly increased the regional impact of ditching and clear-cutting (Figure 6.3). Besides draining of clear-cuts, a major goal of ditching was to reduce the amount of wet areas in the landscape and thereby increase both present and future productivity. Today much land is still privately owned, usually with more "old-fashioned" forests, but large-scale forestry—with clear-cuts ranging from 5 ha (subjective limit) to more than 1 km$^2$ (mean = 50 ha in the region) and extensive ditching—has greatly affected landscape species composition and hydrology locally.

Figure 6.2. A typical undisturbed pool frog pond with reed (*Phragmites communis*) and sedges (*Scirpus* spp.) and surrounded by mixed coniferous/deciduous forest. The frogs spend most of their time at the water's edge, usually basking or hunting insects. Inhabited ponds are 0.5–1.5 m deep, permanent, and 0.1–1.5 ha in size. They are furthermore characterized by a warm local climate (Sjögren Gulve 1994).

Figure 6.3. Ditching has been widely used in large-scale forestry to drain clear-cut areas and to increase timber productivity and accessibility of wet areas. Ditches range in depth from 1 to 2.5 m (about 2 m in the picture). Even though individual ponds may be largely unaffected by such draining of surrounding areas, dry environs seem to hamper pool frog dispersal, which has severe effects on both local and regional persistence.

## Census Data

The pool frog metapopulation was closely monitored during 1983–1994 by Sjögren Gulve (1994 and unpublished data), in the 1950s by Forselius (1962), and was censused in part by Haglund (1972) in the 1960s. Initially, we used the same habitat variables as Sjögren Gulve (1994), and census data from 1983 to 1990 covering a greater number of ponds, to construct a model of the "core" of the system (102 ponds in the central part of the total census area). These measured variables, and the results from 1983 to 1987, are described in detail by Sjögren Gulve (1994); a variable (*DITCH*) was added in the second part of our analysis.

The presence or absence of the pool frog was censused systematically during optimal (warm) weather conditions according to Sjögren Gulve (1994): in May and June during the breeding period, when males call loudly and their territorial behavior prompts even a single male at a pond to respond to playback/imitation calls, and in August and September when metamorphosing larvae (5–7 cm long) and juveniles/metamorphs (3 cm) are easily seen in sunny and shallow parts of a breeding pond. (See Sjögren Gulve 1994 for details.) If frogs or larvae were not observed in a habitat, the visit was repeated and searching intensified. Based on these observations from 1983, 1987, and 1990, the status of each pond in each census was classified as (0) not occupied by pool frogs, (1) with calling male(s) present during the breeding season but without reproduction, (2) with a reproducing pool frog population, (E) with previously reported occurrence (status 1 or 2: Forselius 1962; Haglund 1972; Sjögren Gulve 1994) but population now extinct, and (C) unoccupied in 1983 or 1987 but colonized (status 1 or 2) in the next census. Ponds of status 1 and 2 were pooled in our analyses.

Habitat variables were measured according to the following methods and schedule. Water alkalinity (*ALK*) ($HCO_3^-$ mol/m³), calcium/lime content (*CAL*) ($[Ca^{2+}]g/m^3$), and presence or absence of pike (*PIKE*) (*Esox lucius*) were measured in 1990 for each pond. Presence or absence of modern forestry (*DITCH*) (ditching and/or clear-cutting of ≥5 ha areas situated <500 m from the pond) at any time between 1970 and 1990 was determined from field observations, aerial photographs, and maps provided by the local forestry authority. Usually clear-cuts were located between the pond and its neighbors rather than surrounding it. Data on distance (in meters) to the closest pool frog pond (*Dloc*) for status C and status E ponds and on distance (in meters) to the closest neighboring extinction site (*Dext*) for status E ponds derive from the census preceding the year of the respective event. The values of *Dloc* and *Dext* for the other ponds pertain to 1987; all pond data on local climate (*TEXP*, estimated mean water temperature [°C] in late May 1987; see Sjögren Gulve 1994) and distance to the Baltic Sea (*Dsea*) are also from 1987. Pond surface *AREA* (in hectares) was measured from maps (see Sjögren Gulve 1994).

Data for logistic regression analysis were available for 54 continually occupied ponds (status 1 or 2 during 1983–1990), 9 ponds that were colonized during 1983–1990 (status C), 25 ponds where extinction occurred sometime during 1962–1990 (status E), and 57 unoccupied ponds (status 0 during 1983–1990). Two ponds where extinction occurred in 1983 and 1987, respectively, were recolonized in 1990 and thus occur among both status E and status C ponds.

## Logistic Regression

In a two-state model of patch occupancy, patch-state transitions (extinction or persistence, for example) can be viewed as binary response variables affected by one or more independent environmental variables. Thus local extinction (versus persistence) might be determined by the (categorical) presence or absence of a predator species and by the (continuous) distance to the nearest extant conspecific population. Logistic regression, which employs binary response variables and both categorical and continuous independent variables, is a widely recognized statistical method for determining the relationship between environmental variables and observed responses (Press and Wilson 1978; Hosmer and Lemeshow 1989)—that is, for generating patch "transition-incidence" functions. Good descriptions of the analysis and its applications are found, for example, in Hosmer and Lemeshow (1989), Engelman (1990), Jongman et al. (1987), and Schoener and Adler (1991). We used the stepwise logistic regression program BMDPLR (Engelman 1990). The following discussion can be generalized to other logistic or logit regression programs—for example, SAS Proc Catmod and Logistic (SAS 1988) and SPSS Logit (Norušis 1990)—by noting that while BMDPLR uses response values of 1 and 0, the other programs use values of 1 and 2.

In logistic regression analysis, the predicted proportion of "success" in the total sample (study system) is assumed to follow the logistic model

$$\frac{\exp[u]}{1 + \exp[u]} \tag{5}$$

which ranges in value from 0 to 1. The logit $u$ is a linear equation comprising any independent variables $(X_i)$ and a constant $(C)$ that contribute significantly to the discrimination between the two response groups in the analysis (that is, minimize the deviance between predicted and observed responses in the total sample):

$$u = C + \beta_1 \cdot X_1 + \beta_2 \cdot X_2 + \cdots + \beta_i \cdot X_i \qquad (6)$$

where $\beta_i$ is the regression coefficient for $X_i$. Qualification of independent variables should be based on improving the model's goodness of fit rather than on high–significance-level criteria, and step selections are preferably based on the maximum-likelihood ratio (Hosmer and Lemeshow 1989; Engelman 1990). The analysis can also include interaction terms (such as $X_i^* X_{i-1}$), but we have ignored this option in our analyses. We used a significance level of 0.10 ($P$ = SLE = SLS = 0.10) for qualification of both the constant $C$ (which started outside the model) and the independent variables, and we allowed a maximum of 100 iterations for estimating the regression parameters. Reporting the goodness of fit of the resulting extinction and colonization models, we use the deviance $\chi^2$, which quantifies the deviance between predicted and observed responses in the total sample. (See Hosmer and Lemeshow 1989 for details on the regression statistics.)

Provided that the response-detection probabilities are 1, Equations (5) and (6) can be used to calculate the conditional probability of "success" (response = 1) for each individual observation ($j$) based on the specific values of the independent variables [$X_i(j)$]. (See Hosmer and Lemeshow 1989.) To determine the conditional probability of patch-state transitions, we must apply two logistic regressions: one to analyze the response of occupied patches, which experience either population extinction (response = 1) or persistence (response = 0), and one to analyze unoccupied patches, which experience either colonization (response = 1) or continued vacancy (response = 0). The logits generated by these analyses ($u_e$ and $u_c$, respectively) can be used to calculate the conditional probabilities of patch extinction ($P_{ext}$) and colonization ($P_{col}$). Using the values of environmental variables at time $t$ for patch $j$, these probabilities are

$$P_{ext}(j,t) = \frac{\exp\left(u_e\left[j,t\right]\right)}{1 + \exp\left(u_e\left[j,t\right]\right)} \qquad (7)$$

$$P_{col}(j,t) = \frac{\exp\left(u_c\left[j,t\right]\right)}{1 + \exp\left(u_c\left[j,t\right]\right)} \qquad (8)$$

Equations (7) and (8) are conditioned on the state of $j$ at $t$.

### The Simulation Model

From Equations (7) and (8), we formulate a stochastic, discrete-time metapopulation model:

$$E\left(\frac{\Delta N}{\Delta t}\right) = \sum_{j=1}^{T-N} P_{col}\left(j,t\right) - \sum_{j=1}^{N} P_{ext}\left(j,t\right) \tag{9}$$

Equation (9) is analogous to Equation (1) when $P_{col}$ is a constant function of $N$ and $P_{ext}$ is constant for all $j$ and $t$. More generally, this model allows both colonization and extinction to depend on several variables, some or all of which may vary temporally and differ between patches. If dispersal is spatially restricted, for example, the probability that a patch becomes colonized is a function of the number of occupied patches within its local neighborhood ($n < N$). This number will vary temporally with metapopulation dynamics. It will also differ between patches, assuming a spatially irregular and/or finite metapopulation.

Simulation of metapopulation dynamics according to Equation (9) is straightforward. We simulated 102 pool frog ponds using METAPOP (Ray et al. 1994). METAPOP accepts patch locations (coordinates), initial states, initial $X_i(j)$ values, and the significant constants and coefficients of regression ($C$ and $\beta_i$ of the colonization and extinction logits) as inputs. If these regression parameters are from analyses made with the SAS and SPSS packages, simply change their sign (negative coefficient becomes positive and vice versa) to use them in Equations (6) to (8).

Simulations began with each pond initialized to the state and $X_i(j)$ values recorded in the 1987 census. At each time step (3.5 years in the pool frog case), occupied ponds were tested for population extinction by comparing a random number, uniformly distributed on $[0,1)$, against $P_{ext}(j,t)$; similarly, unoccupied ponds were tested against $P_{col}(j,t)$. Each pond was tested for only one state transition per time step. Thus populations experiencing extinction were not rescued (in the sense of Forney and Gilpin 1989) by simultaneous recolonization. Rescue is inappropriate because Equations (7) and (8) represent the combined probability of any single- or multiple-state transitions that have occurred during the census interval. For example, $1 - P_{ext}$ is the probability that a patch is occupied at two consecutive censuses, regardless of how many state transitions may have occurred between censuses. After all patches were tested for state transitions, patch states and any independent $X_i(j)$ were updated.

## Results

All observed pond-state transitions are compiled in Table 6.1. A negative trend in metapopulation occupancy was observed from 1983 to 1987, partly due to a reproductive failure in 1985 (Sjögren 1991*b*). This trend was

TABLE 6.1
Changes in pond status from 1983 to 1990

| | Status | | | | | | |
|---|---|---|---|---|---|---|---|
| | Colonization | Extinction | | | | | |
| No. of ponds | 0 → 1 or 2 | 1 → E | 1 → 2 | 2 → 1 | 1 → 1 | 2 → 2 | Net change |
| 1983–1987 | 3[a] | 6 | 2 | 3 | 6 | 54 | −3 |
| 1987–1990 | 12[b] | 8[c] | 5[d] | 14[e] | 5 | 41 | +4 |

*Note:* 0 = absence; E = local extinction (no frogs or larvae present during the latter census and the following year); 1 = pond with calling male(s) but no reproduction; 2 = pond with reproducing pool frog population (Sjögren Gulve 1994). Five more ponds were censused reliably in 1987–1990 than in 1983–1987.

[a] All were status 0 → 1.

[b] 3 were recolonizations (E → 1); 3 were status 0 → 2.

[c] 1 was colonization failure; 1 was status 2 → E.

[d] 3 were successful colonizations (2 initiated prior to 1983).

[e] 5 are now isolated.

reversed by extensive colonization from 1988 to 1990 by young frogs from the cohorts of 1986 and 1988 (Table 6.1). Consequently, the net change in number of occupied ponds from 1983 to 1990 was one.

## A Preliminary Model

Our first logistic regression analyses included the environmental variables used by Sjögren Gulve (1994; that is, all variables except *DITCH*). By these analyses, extinction sites (including both satellite and nonsatellite localities in the sense of Sjögren Gulve 1994) were colder (lower *TEXP:* $\chi^2$ = 19.29, df = 1, $P < 0.0001$) and/or larger (*AREA:* $\chi^2$ = 5.21, df = 1, $P = 0.0225$) than persistence sites, and colonization sites were nearer to occupied ponds (*Dloc:* $\chi^2$ = 57.01, df = 1, $P < 0.0001$). The resulting colonization and extinction logits were

$$u_c = -0.001503 \cdot Dloc \tag{10}$$

and

$$u_e = 21.46 - 1.678 \cdot TEXP + 0.4794 \cdot \ln\left(AREA\right) \tag{11}$$

The logistic regression based on Equation (10) predicted state transitions for unoccupied ponds quite well (goodness of fit: $\chi^2$ = 34.48, df = 65, $P = 0.999$). However, the extinction model (Equation 11 with a significant constant: $\chi^2$ = 18.06, df = 1, $P < 0.0001$) predicted extinctions with much less precision (goodness of fit: $\chi^2$ = 78.71, df = 76, $P = 0.393$). This poor fit was surprising, because the effects of pond temperature and area on pool frog fecundity and survival (Sjögren 1991*a*, 1991*b*) are in agreement with this extinction model.

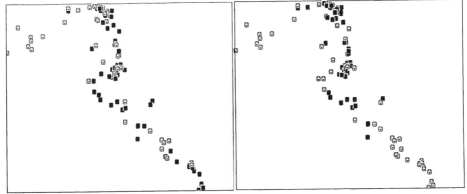

A. Preliminary model                    B. Secondary model

Figure 6.4. Output from METAPOP showing the pond occupancy pattern of pool frogs after five time steps when simulated (**A**) without, and (**B**) with the effect of large-scale forestry (*DITCH*) incorporated into the simulation model. Closed squares denote occupied ponds; open squares denote unoccupied ponds. The preliminary model (**A**) caused rapid and repeated colonization of ponds in regions where no colonizations, but local extinctions, were observed in the field. The secondary model (**B**) both gave higher goodness of fit in the logistic regression analyses and produced a pond occupancy pattern resembling that observed in the field.

Using Equations (10) and (11) in the METAPOP simulation model of 102 ponds resulted in anomalous dynamics. In simulations, vacant ponds in the southern part of the metapopulation were rapidly and repeatedly colonized (Figure 6.4*a*). In reality, only extinctions were observed in this region during 1983–1994. Because this region was particularly subject to modern forestry practices—clear-cutting and extensive draining of environs between ponds—it was decided that a variable describing local forestry should be added to the model.

### A Secondary Model

The analyses were repeated including the variable *DITCH* (see the "Census Data" section), which describes the presence (1) or absence (0) of large-scale modern forestry in the vicinity of each pond. *DITCH* affected both colonization and extinction significantly (Tables 6.2 and 6.3), and goodness of fit was high for both models:

$$u_c = -0.001181 \cdot Dloc - 99.83 \cdot DITCH \qquad (12)$$

(goodness of fit: $\chi^2 = 31.23$, df = 64, $P = 1.000$) and

$$u_e = 26.92 - 2.179 \cdot TEXP + 0.4906 \cdot \ln\left(AREA\right) + 3.553 \cdot DITCH \qquad (13)$$

TABLE 6.2.

Differences in eight environmental variables between ponds colonized by pool frogs during 1983–1990 (status 0 → 1 or 2; response = 1) and noncolonized vacant ponds (response = 0) using stepwise logistic regression

| Variable entered (i) | Colonized $\bar{x} \pm$ SD | Noncolonized $\bar{x} \pm$ SD | $\beta_i$ | SE | $x^2$ | $P$ |
|---|---|---|---|---|---|---|
| 1 *Dloc* | 433 ± 287 | 4095 ± 3336 | −0.0012 | 0.00052 | 21.45 | < 0.0001 |
| 2 *DITCH* | — | — | −99.83 | $8 \times 10^{17\dagger}$ | 3.25 | 0.0712 |
| Not entered: | | | | | | |
| a *ALK*[a] | 1.98 ± 0.77 | 2.31 ± 1.00 | — | — | 1.32 | 0.2510 |
| b *Dsea* | 244 ± 515 | 456 ± 711 | — | — | 1.07 | 0.3016 |
| c *CAL*[a] | 74.47 ± 32.66 | 66.46 ± 32.08 | — | — | 0.97 | 0.3254 |
| d Constant | | | — | — | 0.04 | 0.8427 |
| e *TEXP* | 12.5 ± 0.8 | 12.8 ± 0.8 | — | — | 0.03 | 0.8596 |
| f *PIKE*[b] | — | — | — | — | 0.03 | 0.8733 |
| g ln(*AREA*) | −1.62 ± 1.64 | −0.85 ± 1.56 | — | — | 0.02 | 0.8756 |
| *N* = | 9 | 57 | | | | |

*Note:* Means and SD are shown by group in columns. Regression coefficient ($\beta_i$) and its SE are shown for each entered variable ($P < 0.10$); step selections were based on maximum-likelihood ratio, and a maximum of 100 iterations was used. Goodness of fit: $x^2 = 31.23$, df = 64, $P = 1.000$.

[†] The large SE is due to the absence of *DITCH* = 1 observations among the colonized ponds. (See the "Caveats" section in the text and Hosmer and Lemeshow 1989: chap. 4.5.) We chose the $\beta_{ditch}$ based on 100 iterations and $P_{CONV} = 0.002$ in the BMDPLR analysis.

[a] $N_c = 7$, $N_{nc} = 28$, $x^2$ given with variables 1 and 2 already in the model.
[b] $N_c = 7$, $N_{nc} = 42$, $x^2$ given with variables 1 and 2 already in the model.

(goodness of fit: $\chi^2 = 54.19$, df = 75, $P = 0.967$). Presence of pike was also a significant extinction factor (Table 6.3; Semlitsch 1993). But because pike data were often absent (from 20 of the 102 ponds) and pike presence is significantly associated with large pond area (Sjögren Gulve 1994), we chose a model with better goodness-of-fit statistics including *TEXP*, *DITCH*, and ln(*AREA*) (Table 6.3).

Obviously, presence of large-scale forestry in the nearby landscape increases the risk of local extinction and reduces the probability of recolonization (Figure 6.5). Using Equations (12) and (13) in the simulation model, we found that pool frogs declined and eventually disappeared in regions with such forestry practice (Figure 6.4*b*). This result accords with field observations made through 1994. The following analyses are based on this secondary model.

TABLE 6.3.

Differences in nine environmental variables between permanent ponds with extinct
(status E: response = 1 during 1962–1990) vs. extant (status 1 or 2: response = 0
during 1983–1990) pool frog populations, using stepwise logistic regression

| Variable entered (i) | Extant (0) $\bar{x} \pm SD$ | Extinct (1) $\bar{x} \pm SD$ | $\beta_i$ | SE | $x^2$ | $P$ |
|---|---|---|---|---|---|---|
| 0 Constant | | | 26.92 | 7.34 | 19.36 | <0.0001 |
| 1 *TEXP* | 13.3 ± 0.4 | 12.6 ± 1.0 | –2.179 | 0.566 | 22.21 | <0.0001 |
| 2 *DITCH* | — | — | 3.553 | 0.844 | 24.51 | <0.0001 |
| 3a *PIKE*[a] | — | — | — | — | 4.48 | 0.0343 |
| 3b ln(*AREA*) | –1.04 ± 1.01 | –0.79 ± 1.85 | 0.491 | 0.295 | 3.06 | 0.0802 |
| Not entered: | | | | | | |
| a *Dloc* | 933 ±1117 | 1668 ± 3491 | — | — | 0.62 | 0.4320 |
| b *CAL*[b] | 63.65 ± 26.80 | 61.68 ± 33.29 | — | — | 0.29 | 0.5904 |
| c *Dext* | 1915 ± 2099 | 1588 ± 2890 | — | — | 0.17 | 0.6769 |
| d *Dsea* | 313 ± 400 | 508 ± 601 | — | — | 0.06 | 0.8007 |
| e *ALK*[b] | 2.25 ± 0.80 | 2.45 ± 1.14 | — | — | 0.01 | 0.9425 |
| *N* = | 54 | 25 | | | | |

*Note:* Means and SD are shown by response group in columns. Regression coefficient ($\beta_i$) and its
SE are shown for each variable included in the final model (with $P < 0.10$). ln(*AREA*) was included
instead of its correlate, *PIKE*, in the final model because *PIKE* data were not available for all ponds.
Goodness of fit of the final model (0, 1, 2, 3b): $x^2 = 54.19$, df = 75, $P = 0.967$.

[a] Goodness of fit of model including 0–3a: $x^2 = 44.49$, df = 57, $P = 0.886$. $N_0 = 38$, $N_1 = 23$, $x^2$
to remove given for *PIKE*.

[b] $N_0 = 27$, $N_1 = 23$, $x^2$ given with constant and variables 1–2 already in the model.

## How Well Does the Model Predict Regional Dynamics?

The fundamental measure of metapopulation dynamics is average patch occupancy or fraction of patches occupied (Levins 1970). Among the 102
ponds, observed pond occupancy rose from 41 percent in 1987 to 45 percent
in 1994. Model simulation (10,000 replicates) predicted that an average of
43 percent of the ponds would be occupied in 1994 (45 percent being within
the 25 percent confidence limits of the simulation average). Thus the model
accurately predicted both the trend in pond occupancy and its magnitude
across two census intervals (7 years).

## How Well Does the Model Predict Local Turnover?

The model might predict regional dynamics well without accurate prediction
of individual pond turnover events. We assessed the correspondence between

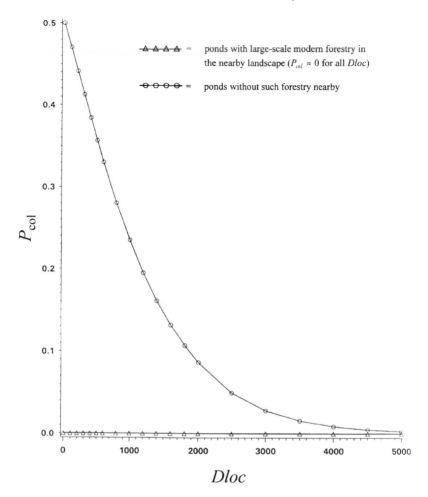

Figure 6.5. Probability of pond colonization ($P_{col}$) in the pool frog system as a function of distance to the closest occupied pond (*Dloc*) and presence or absence of modern forestry (*DITCH*) according to Equations (12) and (8).

predicted and observed local events in two ways. First, we compared state transition probabilities generated by the model with observed state transitions. Figure 6.6 shows the extinction probabilities (generated by Equations 13 and 7) and state transitions (observed) for the 54 pool frog ponds and 25 extinction sites included in the statistical analysis. Few ponds seem to have been assigned inappropriate extinction probabilities. Second, we compared the final (1994) observed and simulated states of each of the 102 ponds. Simulated and observed final states corresponded in more than 50 percent of the simulations for 82 (80.4 percent) of the ponds (Figure 6.7).

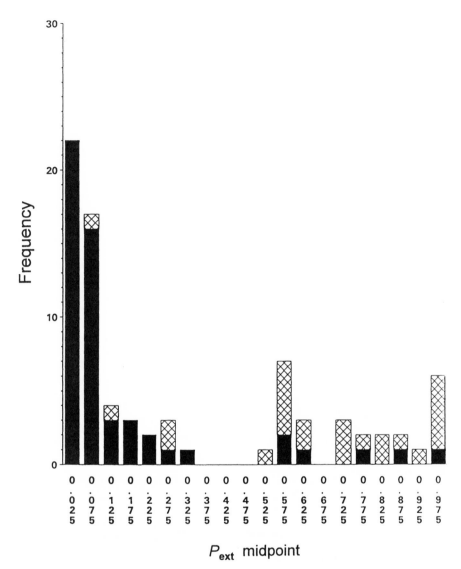

Figure 6.6. The distribution of conditional extinction probabilities $[P_{ext}(j)]$ for the 54 persistence sites (black bars) and 25 extinction sites (checked bars) based on data $[X_i(j)]$ from 1987: $\bar{P}_{ext}(j) \pm$ SD = 0.316 ± 0.336. One locality with $P_{ext}$ = 0.589 and status 1 in 1990 went extinct between 1990 and 1994.

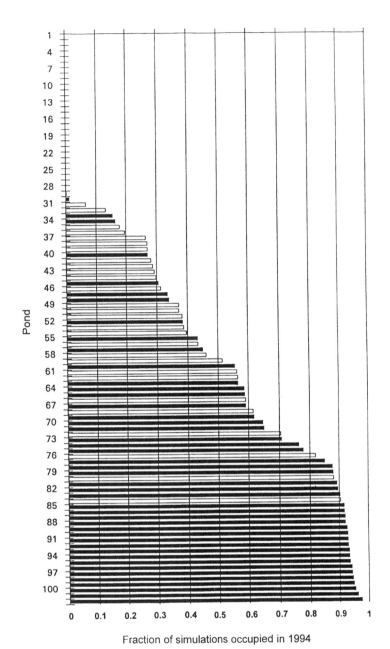

Fraction of simulations occupied in 1994

Figure 6.7. Fraction of simulations that vacant ponds in 1994 (status 0 or E; white bars) and occupied ponds in 1994 (black bars) ended as occupied by pool frogs in simulations from 1987 to 1994 (10,000 replicates), using Equations (12) and (13). Simulated and observed states of 1994 agreed in more than 50 percent of the simulations for 82 of the 102 ponds.

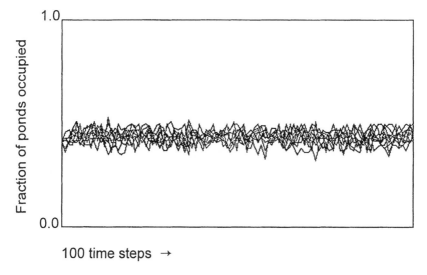

Figure 6.8. Fraction of the 102 ponds occupied by pool frogs during the course of a subset of 10 simulations × 100 census intervals (350 years). No discernible long-term trend in average pond occupancy exists (average = 43.5 percent, range 30–54 percent; data from 100 simulation replicates). Note the absence of a trend also in the data on which the simulations were based (Table 6.1).

### Assessment of a Long-Term Trend

Assuming that Equations (12) and (13) continue to model state transitions accurately into the future, would the present pattern of forestry, if it remained static, make the pool frog metapopulation decline toward extinction? Probably not. Our simulations indicate no long-term trend in pond occupancy; 100 simulations of 100 census intervals (350 years) resulted in an average occupancy of 43.5 percent (range 30–54 percent in Figure 6.8; see Table 6.1). But when we set $DITCH = 1$ for all ponds, indicating that large-scale forestry is practiced throughout the metapopulation, our simulations predict regional extinction with a 99.9 percent probability within 15 census intervals (53 years in Figure 6.9; 1000 replicates).

### Effects of Patch Heterogeneity

The necessity of modeling metapopulation dynamics with full heterogeneity in $P_{ext}(j)$, compared to using a simpler model with one constant extinction probability for all patches, was examined with simulations. We compared 100 simulations in which all 102 ponds had the same $P_{ext}(j)$ ($0.466 = \bar{P}_{ext}(j)$ of the ponds) with the simulations shown in Figure 6.8. The regional pond occupancy became much lower and fluctuated to a greater extent than in the

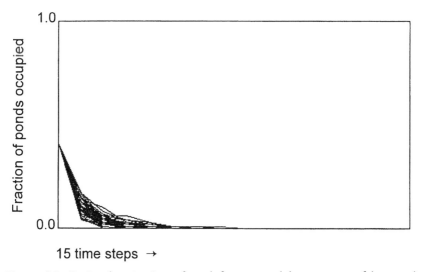

Figure 6.9. Regional extinction of pool frogs caused by presence of large-scale forestry/draining (*DITCH* = 1) between all 102 ponds in the metapopulation, simulated from 1987 (41 percent of the ponds occupied) and onward using Equations (12) and (13). The predicted risk of regional extinction was 0.999 within 15 census intervals (53 years), that is, before the year 2040. Median time to regional extinction was 18 years. (1000 replicates).

simulations shown in Figure 6.8, but it exhibited no discernible long-term trend (average regional occupancy was 26.6 percent; range: 11–45 percent). This indicates unsatisfactory precision in predicting both regional and local dynamics. Part of this imprecision stems from the fact that relatively unsuitable ponds, which nevertheless might be occupied during one census, also contribute to $\bar{P}_{ext}(j)$.

## Discussion

Large-scale forestry obviously has a very negative effect on both local and regional pool frog persistence (Tables 6.2 and 6.3). *DITCH* increases the risk of extinction (Table 6.3 and Equations 13 and 7) and hampers recolonization (Table 6.2 and Figure 6.5). We interpret this pattern as a negative "matrix effect": without altering ponds directly, large-scale draining and clear-cuts create drier and more hostile dispersal habitats over large areas between the ponds. Intervening marshland may disappear altogether. On clear-cuts, great daily variation in both humidity and temperature may slow down the dispersers, and increased mortality is likely from desiccation and predation. Such uncompensated emigration may reduce population growth rate significantly (Fahrig and Merriam 1985); dry areas may as well reduce the propensity of frogs to disperse (Sinsch 1990). Nevertheless, we believe the negative matrix effect is caused more by ditching than by removal of canopy. Large drained areas increase pond isolation even though the geographic distance (*Dloc*) remains the same, explaining why no (re)colonization has been observed in areas with modern forestry. This pattern agrees with the increased risk of extinction following increased patch/population isolation that has been demonstrated for various taxa (for example, Sjögren 1991*a*; Thomas et al. 1992; Fahrig and Merriam 1994).

Previous pool frog analyses (Sjögren 1991*a*; Sjögren Gulve 1994) have stressed the higher risk of extinction at isolated ponds (large *Dloc*). But a different kind of extinction occurs at cold "satellite" ponds (Sjögren Gulve 1994). Satellite ponds are typically situated in the vicinity of a breeding pond and are either recently colonized, recently extinct, or occupied only by calling males. Many of them are repeatedly colonized followed by local extinction. Opportunistic males call from such ponds in the beginning of the breeding season, but later they usually join the adjacent breeding population (P. Sjögren-Gulve, personal observation). The role of such satellite ponds for system dynamics is hard to assess. In most years, they are too cold for reproduction (low *TEXP* in Table 6.3; see Sjögren et al. 1988), but they probably do not constitute population "sinks" unless pike are present. They may, however, serve as dispersal "stepping stones" in the system. The isolated ponds, by contrast, are primarily characterized by larger *AREA* and *PIKE* presence (Table 6.3). At such ponds, predation by pike increases pool frog mortality and the risk of extinction (Semlitsch 1993; see the discussions in Sjögren 1991*a* and Sjögren Gulve 1994). In the present analysis these extinctions have been pooled, which effectively cancels the effect of *Dloc*. The isolation effect of *DITCH*, however, is highly significant. In fact, when modeling satellite and isolation-dependent extinctions separately (Sjögren-Gulve et al. 1996), *DITCH* proves significant in both types of extinctions, again emphasizing the strong matrix effect of large-scale forestry and draining on the pool frog dynamics.

The foregoing results are relevant also for other amphibians. Bradford et al. (1993) found that introduced fishes are associated with the decline and local extirpation of the frog *Rana muscosa* in two California national parks (compare Semlitsch 1993). They argue that, in addition to such local effects, fish predation along the streams and drainage channels that are primary dispersal routes of the frogs reduces population connectivity dramatically, causing a similar matrix effect. Petranka et al. (1993) report that salamanders in the southern Appalachians disappear from clear-cut areas in both wet and dry mature forests. Because of the low dispersal propensity and sensitivity to desiccation in these salamanders, they attribute this decline mainly to increased mortality. Similar patterns have been reported for two salamander and one frog species in the Pacific Northwest by Welsh (1990), who emphasizes the importance of both suitable microclimate and population connectivity for species persistence. Other combinations of deterministic and stochastic (isolation-dependent) causes of amphibian declines are reviewed by Blaustein et al. (1994) with similar conclusions.

### Assessment of Misclassifications

An important part of the biological interpretation of modeling results is to assess the possible causes of individual misclassifications, here defined as a pond for which simulated and observed final states (of 1994) corresponded in less than 50 percent of the simulations. One limitation of the present pool frog model is that local climate (*TEXP*, mean water temperature in late May 1987) is a static extinction factor. Although *TEXP* might well mirror climatic differences between ponds on a relative scale, temperature conditions do vary from year to year (Sjögren 1991*b*), inducing temporal variation in the local extinction probabilities (Equations 13 and 7). The period from 1988 to 1994 was characterized by warmer summer temperatures than 1987 and 1983 to 1986, which probably affected system dynamics and the predictive precision of the model. Seven of the 20 misclassifications (Figure 6.7) were apparently due to the persistence (during 1988–1994) of populations that would have gone extinct during normal years. Additional misclassifications can be explained by the fact that we pooled different types of extinction in a single extinction analysis.

Three types of extinction can be distinguished in the pool frog system (Sjögren-Gulve 1994): deterministic extinctions due to pond loss through draining or successional overgrowth (not included in the present analysis), extinctions at cold satellite ponds, and isolation-dependent extinctions at large, nonsatellite ponds. Using two extinction functions to model the latter two pond types separately, the classifications of five ponds were corrected (Sjögren-Gulve et al. 1996). Four more ponds may have been misclassified because this model does not track internal patch dynamics (population flux within a patch or at a pond). Variation in population size and composition (sex ratio, age structure) within and between patches directly induces variation in local persistence. (See Burgman et al. 1993 for examples.) Among the misclassified ponds, two had strikingly large populations (perhaps due to favorable local climate during 1988–1994) and two had strikingly small populations even though the ponds were neither cold, large, nor particularly isolated. Finally, in six cases, one or more missing factors must have played a role in the observed dynamics.

### Advantages of the Modeling Approach

Predictive models are desirable for conservation because causal models based on the particular ecology and population dynamics of a species are difficult to construct and parameterize. For the Swedish pool frog metapopulation, consisting of many populations in different environs, the task of building and testing a causal model is tremendous. For such complex systems, determining the statistical correlates of population-state transitions allows fast model formulation and prediction of metapopulation dynamics. Identification of these correlates leads to testable hypotheses regarding the causes of extinction and colonization and the relative importance of deterministic and stochastic (or unidentified) processes (see Sjögren Gulve 1994 and Thomas 1994). Occupancy models (reviewed in Hanski 1994b) cannot capture the level of detail of population-dynamic or individual-based models (for example, reviewed in Burgman et al. 1993), but it is arguable whether the details offered by more complex models are useful in guiding the conservative decisions required to prevent population extinction.

Our "transition-incidence" model builds on logistic regression analyses of data from repeated censuses. Another method for determining patch-state transition probabilities for occupancy models relies on species-incidence curves (Hanski 1994a; Hanski and Thomas 1994). Because these curves can be constructed from a single habitat census, this method is appealing. But to generate state transition probabilities from a species-incidence curve, one must assume that the occupancy of the system at the time of the census is equilibrial, maintained by a balance between extinction and colonization rates. This assumption may be difficult to verify, and the effects of violating it may be hard to assess. Models based on transition-incidence curves are free of this assumption: they use logistic regression to determine the correlates of state transitions at the time they occur, rather than assuming that there are static correlates of species incidence that also govern state transition probabilities. To illustrate the difference between these approaches with an example, assume that distance to the nearest occupied pond governs colonization probability [$P_{col}(j,t)$] and that the distance required to maintain a certain $P_{col}(j,t)$ has recently dropped (due to ditching between ponds, for example). In this case, use of a species-incidence curve will not allow discovery of the correct, new $P_{col}(j,t)$ until enough extinctions have occurred that the system is again at equilibrium. Use of a colonization-incidence curve, however, does allow discovery of the new $P_{col}(j,t)$, so long as the time between colonization events is small relative to the rate of decline in $P_{col}(j,t)$ at a given distance. Thus state transition probabilities determined from transition-incidence curves may predict trends in metapopulation occupancy where those determined from species-incidence curves do not.

A central problem in metapopulation analysis is delimiting suitable patches (especially if habitat quality is variable) and defining what objects are part of the functional metapopulation. Compared to the models by Levins (1969, 1970) and Gotelli and Kelley (1993), both transition-incidence and species-incidence metapopulation models benefit from allowing explicit modeling of spatial dynamics. This means that assumptions regarding distance effects can be relaxed. And considering habitat-quality effects, we think the transition-incidence modeling approach is advantageous. First, statistic analysis of the species' occupancy pattern (for example, Sjögren Gulve 1994) should reveal variables that are correlates or determinants of habitat/patch quality. Second, if variation in such a variable is significant in explaining observed metapopulation dynamics ($TEXP$, for example), either or both of the patch-state transition analyses (Tables 6.2 and 6.3) should select this variable, providing a standardized model for how it affects state transitions. Third, all these analyses yield standardized goodness-of-fit statistics. Thus, provided goodness of fit is high, logistic regression can help exclude unsuitable "patches." Alternatively, in cases where habitat quality varies continuously between patches and colonization seems indiscriminate as in the pool frog, careful interpretation of simulation results over a relevant time period may be the best way of delimiting the metapopulation. Regarding the pool frogs, successful experimental introductions have confirmed the existence of isolated but suitable vacant ponds (status 0; Sjögren-Gulve, unpublished data).

### Caveats

Despite its advantages, the model we propose requires careful implementation. First, the requisite logistic regression analyses should be based on sufficient and representative numbers of observations in both response groups. For example, analyzing a low number of extinction sites together with a much greater number of persistence sites reduces the power of the regression to detect significant independent variables affecting extinction. Such a sample may also yield a highly significant constant and high goodness of fit, reflecting the high probability of persistence in the total sample (regardless of the independent variables) rather than anything else. Thus, when there are few observations in one response group, results should be interpreted very carefully.

Second, logistic regression often appears "inert" (relatively unable to score significances) compared to analyses of variance (but see Press and Wilson 1978). To avoid this problem, the significance levels for entering and removing independent variables to and from the regression model in the step-wise procedure (SLE, SLS) may be set to 0.15 or 0.20. (See Hosmer and Lemeshow 1989; Engelman 1990.) In our analyses, SLE = SLS = 0.10 may seem narrow but did not alter the results (Table 6.2 and 6.3).

Third, another potential problem associated with a small sample size is shown in Table 6.2. The fact that none of the nine colonized ponds had $DITCH = 1$ caused numerical problems when BMDPLR fit the logistic regression model. Such zero-frequency problems are less likely to occur when the sample size is large. In this case, we decided to select $\beta_{ditch}$ as specified in Table 6.2 because we believe the overwhelmingly negative impact of $DITCH$ on pond colonization probability is real and not a sampling artifact.

Fourth, regarding sample size and collection of data for the statistical analyses, Markovian assumptions in the modeling procedures must also be considered. Collecting observations over time may violate the assumption of direct effects without time lags. This violation may reduce the effects of factors that change with time, for example, the effect of distance to the closest occupied patch on isolation-dependent extinction in a declining metapopulation. Thus, the time interval between censuses should be short enough to avoid such problems, and to avoid multiple state transitions of a single patch, but long enough to cover at least one generation interval or cohort lifetime of the species. The pool frog generation time ($T_c$), for instance, is 3.7 years (Sjögren-Gulve 1996), but the maximum life span of 8 years still allows single frogs to persist over two censuses ($2 \times 3.5$ years).

Finally, it should be noted that the regression model's goodness-of-fit statistics quantify the deviance between predicted and observed responses in the total sample. The pool frog example shows that simulations may be important in projecting the model further, which can help identify additional colonization and extinction factors. Projecting metapopulation dynamics far into the future is not recommended. Under the assumption of invariant state-transition relationships and environmental conditions, projections can be made to assess metapopulation trends (Figures 6.8 and 6.9), but such results should be interpreted with great caution.

## Lessons

Our modeling approach builds on straightforward analyses that are basic to assessing how observed variables affect metapopulation dynamics. The approach is appropriate for systems where turnover events are not scarce. These could be small or medium-sized systems with high turnover or larger systems with moderate turnover. Because these systems are also the most difficult to parameterize for detailed population models, we feel that transition-incidence modeling is a useful complement to more complex models.

The pool frog example demonstrates the need for a metapopulation approach to assess both local and regional persistence. The matrix effect of large-scale forestry becomes obvious only at the level of the metapopulation, and a metapopulation model is most appropriate for illustrating the categorical effect of such forestry on regional dynamics. The fact that four significant turnover factors may vary independently between individual ponds emphasizes the need for models that address local environmental heterogeneity in a metapopulation context. We suspect that many systems will require similar models.

The results derived from this modeling effort call for limitation and modification of forestry in the pool frog region. This case has been made, and the regional forestry authority and timber company have initiated cooperation with researchers to modify forestry and restore certain areas. Today, extensive ditching is banned and in certain areas clear-cuts are prohibited or restricted in size to less than 2 ha. These measures will probably also benefit other taxa with similar habitat requirements.

## Acknowledgments

We thank Mike Gilpin, Alan Hastings, Henrik Pärn, and Jana Verboom for good discussions at various stages of this project; we thank Rune Persson and Magnus Ohlsson (the local forestry authority in Tierp) and Bo Hjalmarsson (Korsnäs AB) for their cooperation. The project was funded mainly by the Swedish Environmental Protection Agency and also by the Uppland Foundation and Frans von Sydow's fund (grants to Per Sjögren-Gulve).

## REFERENCES

Bengtsson, J. 1989. Interspecific competition increases local extinction rate in a metapopulation system. *Nature* 340:713–715.

Blaustein, A. R., D. B. Wake, and W. P. Sousa. 1994. Amphibian declines: Judging stability, persistence, and susceptibility of populations to local and global extinction. *Conservation Biology* 8:60–71.

Bradford, D. F., F. Tabatabai, and D. M. Graber. 1993. Isolation of remaining populations of the native frog, *Rana muscosa,* by introduced fishes in Sequoia and Kings Canyon National Parks, California. *Conservation Biology* 7:882–888.

Burgman, M. A., S. Ferson, and H. R. Akçakaya. 1993. *Risk Assessment in Conservation Biology.* London: Chapman & Hall.

Engelman, L. 1990. Stepwise logistic regression. Pages 1013–1046 in W. J. Dixon, ed., *BMDP Statistical Software Manual,* vol. 2. Berkeley: University of California Press.

Fahrig, L., and G. Merriam. 1985. Habitat patch connectivity and population survival. *Ecology* 66:1762–1768.

———. 1994. Conservation of fragmented populations. *Conservation Biology* 8:50–59.

Forney, K. A., and M. E. Gilpin. 1989. Spatial structure and population extinction: A study with *Drosophila* flies. *Conservation Biology* 3:45–51.

Forselius, S. 1962. Distribution and reproductive behaviour of *Rana esculenta* L. in the coastal area of N. Uppland, C. Sweden. *Zoologiska Bidrag från Uppsala* 35:517–528.

Fritz, R. S. 1979. Consequences of insular population structure: Distribution and extinction of spruce grouse populations. *Oecologia* 42:57–65.

Gotelli, N. J., and W. G. Kelley. 1993. A general model of metapopulation dynamics. *Oikos* 68:36–44.

Haglund, E. 1972. *Naturvårdsinventering,* vol. VIII: *Tierps kommun, norra delen.* Uppsala: Länsstyrelsen.

Hanski, I. 1991. Single-species metapopulation dynamics: Concepts, models and observations. *Biological Journal of the Linnean Society* 42:17-38.

———. 1994a. A practical model of metapopulation dynamics. *Journal of Animal Ecology* 63:151–162.

———. 1994b. Patch-occupancy dynamics in fragmented landscapes. *Trends in Ecology and Evolution* 9:131–135.

Hanski, I., and M. Gyllenberg. 1993. Two general metapopulation models and the core-satellite species hypothesis. *American Naturalist* 142:17–41.

Hanski, I., and C. D. Thomas. 1994. Metapopulation dynamics and conservation: A spatially explicit model applied to butterflies. *Biological Conservation* 68:167–180.

Harrison, S. 1994. Metapopulations and conservation. Pages 111-128 in P. J. Edwards, R. M. May, and N. R. Webb, eds., *Large-Scale Ecology and Conservation Biology.* Oxford: Blackwell.

Harrison, S., D. D. Murphy, and P. R. Ehrlich. 1988. Distribution of the bay checkerspot butterfly, *Euphydryas editha bayensis:* Evidence for a metapopulation model. *American Naturalist* 132:360–382.

Hosmer, D. W., Jr., and S. Lemeshow. 1989. *Applied Logistic Regression.* New York: Wiley.

Jongman, R. H. G., C. J. F. ter Braak, and O. F. R. van Tongeren. 1987. *Data Analysis in Community and Landscape Ecology.* Wageningen, Netherlands: Pudoc.

Kindvall, O., and I. Ahlén. 1992. Geometrical factors and metapopulation dynamics of the bush cricket, *Metrioptera bicolor* Philippi (Orthoptera: Tettigoniidae). *Conservation Biology* 6:520–529.

Levins, R. 1969. Some genetic and demographic consequences of environmental heterogeneity for biological control. *Bulletin of the Entomological Society of America* 15:237–240.

———. 1970. Extinction. Pages 77–107 in M. Gerstenhaber, ed., *Some Mathematical Questions in Biology.* Providence, R.I.: American Mathematical Society.

Norušis, M. J. 1990. *SPSS Advanced Statistics User's Guide.* Chicago: SPSS.

Petranka, J. W., M. E. Eldridge, and K. E. Haley. 1993. Effects of timber harvesting on Southern Appalachian salamanders. *Conservation Biology* 7:363–370.

Pimm, S. L., H. L. Jones, and J. M. Diamond. 1988. On the risk of extinction. *American Naturalist* 132:757–785.

Press, S. J., and S. Wilson. 1978. Choosing between logistic regression and discriminant analysis. *Journal of the American Statistical Association* 73:699–705.

Ray, C., P. Sjögren, and M. E. Gilpin. 1994. METAPOP (ver. 3.2). Unpublished computer program, Division of Environmental Studies, University of California, Davis.

SAS. 1988. *SAS/STAT User's Guide.* Release 6.03 edition. Cary, N.C.: SAS Institute.

Schoener, T. W. 1991. Extinction and the nature of the metapopulation: A case system. *Acta Oecologica* 12:53–75.

Schoener, T. W., and G. H. Adler. 1991. Greater resolution of distributional complementaries of controlling for habitat affinities: A study with Bahamian lizards and birds. *American Naturalist* 137:669–692.

Semlitsch, R. D. 1993. Effects of different predators on the survival and development of tadpoles from the hybridogenetic *Rana esculenta* complex. *Oikos* 67:40–46.

Sinsch, U. 1990. Migration and orientation in anuran amphibians. *Ethology, Ecology, and Evolution* 2:65–79.

Sjögren, P. 1991*a*. Extinction and isolation gradients in metapopulations: The case of the pool frog (*Rana lessonae*). *Biological Journal of the Linnean Society* 42:135-147.

———. 1991*b*. Genetic variation in relation to demography of peripheral pool frog populations (*Rana lessonae*). *Evolutionary Ecology* 5:248–271.

Sjögren, P., J. Elmberg, and S.-Å. Berglind. 1988. Thermal preference in the pool frog, *Rana lessonae:* Impact on the reproductive behaviour of a northern fringe population. *Holarctic Ecology* 11:178–184.

Sjögren Gulve, P. 1994. Distribution and extinction patterns within a northern metapopulation of the pool frog, *Rana lessonae. Ecology* 75:1357–1367.

Sjögren-Gulve, P. 1996. Metapopulation dynamics and extinction in pristine habitats: A demographic explanation. Unpublished manuscript.

Sjögren-Gulve, P., C. Ray, and H. Pärn. 1996. Complex metapopulation dynamics. Unpublished manuscript.

Smith, A. T. 1980. Temporal changes in insular populations of the pika (*Ochotona princeps*). Ecology 61:8–13.

Thomas, C. D. 1994. Extinction, colonization, and metapopulations: Environmental tracking by rare species. *Conservation Biology* 8:373–378.

Thomas, C. D., J. A. Thomas, and M. S. Warren. 1992. Distribution of occupied and vacant butterfly habitats in fragmented landscapes. *Oecologia* 92:563–567.

Verboom, J., A. Schotman, P. Opdam, and J. A. J. Metz. 1991. European nuthatch metapopulations in a fragmented agricultural landscape. *Oikos* 61:149–156.

Welsh, Jr., H. H. 1990. Relictual amphibians and old-growth forests. *Conservation Biology* 4:309–319.

Wilson, M. H., C. B. Kepler, N. F. R. Snyder, S. R. Derrickson, F. J. Dein, J. W. Wiley, J. M. Wunderle, Jr., A. E. Lugo, D. L. Graham, and W. D. Toone. 1994. Puerto Rican parrots and potential limitations of the metapopulation approach to species conservation. *Conservation Biology* 8:114–123.

# 7

# A Common Framework for Conservation Planning: Linking Individual and Metapopulation Models

*Barry R. Noon and Kevin S. McKelvey*

Many populations exhibit pronounced spatial structure: dispersed areas of high population density embedded in areas of low density, with population centers connected through dispersal. This recognition has led many conservation biologists to embrace the metapopulation concept (Levins 1970) as the appropriate paradigm for reserve design structures (reviewed in Hanski 1991 and Harrison 1994). This concept seems appropriate for those species that have patchy distributions because the critical resources on which they depend are distributed in this fashion. This paradigm may be less applicable, however, to species that historically have had a more or less uniform distribution of individuals across the landscape. If such species are faced with threats to their persistence, is a metapopulation reserve structure appropriate for their conservation? Or is the tailoring of reserve design to a single paradigm similar to attempting to force a square peg into a round hole?

As a species, spotted owls are widely distributed, show extensive geographic variation in their habitat relationships, and, at a landscape scale, have a territory distribution that is spatially variable. Because of threats to their long-term persistence, conservation strategies for all three subspecies of spotted owl have recently been proposed (Thomas et al. 1990; Verner et al. 1992; USDI 1991). Despite striking similarities in the ecologies of these three subspecies and similar threats to their population viability, their conservation strategies appear quite distinct, suggesting that contrasting sets of ecological principles may have been applied in the planning process. Despite appearances, a common hypothesis-testing framework justified the process of conservation planning and a common set of population dynamics principles underlies each conservation strategy (Noon and Murphy 1994).

Conservation plans for the northern spotted owl (Thomas et al. 1990; USDI 1992), the population of California spotted owls in southern California (Verner et al. 1992; LaHaye et al. 1994), and the Mexican spotted owl (USDI 1991) are similar in that all propose a highly structured spatial distribution for the owl population—a population composed of numerous local subpopulations widely distributed across the landscape in the form of a metapopulation (Levins 1970). The metapopulation reserve structure in southern California, Arizona, New Mexico, and Utah is a logical consequence of a historical metapopulation distribution, the result of variation in topographic relief, and the concomitant patchy distribution of suitable habitat across various mountain ranges.

In contrast, the metapopulation reserve structure in northern California, Oregon, and Washington is not a logical consequence of a historical distribution pattern. Rather, the reserve design was imposed primarily by constraints arising from human-induced changes to the landscape and secondarily by confidence in the legitimacy of the metapopulation concept. Since about 1950, harvest of late seral stage forests has created an insular distribution pattern of owl habitat in the Pacific Northwest. Thus, constraints imposed by the current condition of the landscape, the likely pattern and rate of habitat recovery over the next 100 years, and acceptance of economic pressures have prescribed a metapopulation reserve design (Thomas et al. 1990; Murphy and Noon 1992).

The conservation plan for spotted owls in the Sierra Nevada is distinctly different than in other parts of their range. In the Sierra Nevada, Verner et al. (1992) did not propose a discrete reserve design or impose a metapopulation reserve structure. Rather, they recognized that the current distribution of owl territories was more or less evenly distributed throughout this part of its range, and they proposed a dynamic landscape management plan with the goal of maintaining the current distribution. Instead of viewing the dynamics of spotted owls at the scale of local subpopulations, they proposed a strategy operative at the scale of the individual territory.

It is reasonable to ask how one can reconcile two very different conservation strategies for members of the same species, particularly when one strategy proposes millions of acres to be set aside in forest reserves (Thomas et al. 1990), while the other establishes no discrete reserve boundaries and allows active timber management (Verner et al. 1992). Our goal in this chapter is to demonstrate that a common understanding of population dynamics in structured populations was invoked for both the northern and the California spotted owls. We will demonstrate that the original, single-species metapopulation model of Levins (1969, 1970), when generalized to include the added realism of multiple patch searches and variable habitat quality, is generally equivalent to Lande's (1987) spatial model of local population dynamics. We then offer guidelines for deciding which model is most appropriate as a conceptual framework for a specific conservation problem. The choice among model paradigms, as well as the logic and arguments brought to bear, are illustrated by comparing the conservation strategies for the northern and California spotted owls.

## The Metapopulation Model

We have developed a model that incorporates various spatial and temporal scales of biological processes and enables us to compare birth and death rates at the territory scale with colonization and extinction rates at the population scale.

### Original Model

Levins' (1970) metapopulation model describes the dynamics of populations occupying a system of identically sized and evenly spaced habitat patches. The multiple populations are discontinuous at some spatial scale but connected by migration. At a given point in time, each patch is either occupied or unoccupied by a local population—thus the response variable is the proportion of occupied patches equal to $P_t$ at time $t$.

Levins' model represents one extreme of a continuum of metapopulation models that can be arrayed along a gradient of variation in the distribution of increasing habitat patch sizes (Hanski and Gyllenberg 1993). At the opposite extreme from Levins' model are models that assume a large and stable source population that provides colonists to distant but smaller habitat patches. These are a single-species variant of MacArthur and Wilson's (1967) multi-species mainland-island model. In a recent review, Harrison (1994) has argued that available data suggest this paradigm as appropriate for most natural populations. Intermediate to these two models are numerous possibilities that allow for spatial variation in habitat patch size (Hanski and Gyllenberg 1993).

The dynamics of the original Levins model are described by the following differential equation:

$$\frac{dp}{dt} = mp\left(1 - p\right) - ep \qquad (1)$$

where $p$ = fraction of habitat occupied, $e$ = rate of local extinction of occupied patches, and $m$ = colonization rate of empty patches.

Levins' model is similar to the logistic model (Hanski 1991), and Equation (1) can be rewritten as

$$\frac{dp}{dt} = (m-e)p\left[1 - \frac{p}{1-\left(\dfrac{e}{m}\right)}\right]$$

(2)

where $(m - e)$ is equivalent to the rate of increase, and $(1 - e/m)$ is equivalent to the carrying capacity.

The general solution to Equation (2) is

$$p_t = \frac{p_0\left(1 - \dfrac{e}{m}\right)}{p_0 + \left(1 - \dfrac{e}{m} - p_0\right)e^{-(m-e)t}}$$

(3)

Its dynamic behavior (Figure 7.1) is identical to the familiar logistic, showing a stable equilibrium at $\hat{p} = 1 - e/m$.

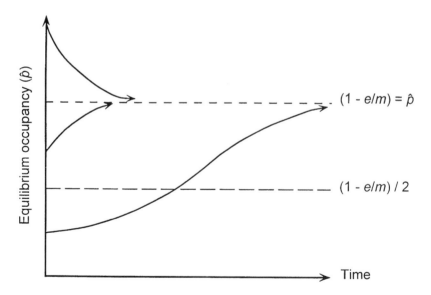

Figure 7.1. General dynamic behavior of the solution equation to Levins' metapopulation model (See Equation 3 in the text.) The graph shows the changes in the occupancy rate ($p$) of a hypothetical metapopulation starting at three different initial conditions.

### *Generalized Model*

Levins' original model (Equation 1) is limited by two simplifying assumptions: all habitat patches are assumed suitable for occupancy; and dispersal success is limited to a single transition, or search, and therefore proportional to $1 - p$. If we relax assumption 1, we can rewrite Equation (1) as

$$\frac{db}{dt} = mp\left[h\left(1 - p\right)\right] - ep \qquad (4)$$

where $h$ = proportion of habitat patches suitable for occupancy. The likelihood of successful dispersal is now proportional to $h(1 - p)$. If we also relax assumption 2, allowing $n$ patches to be searched during dispersal, we can rewrite Equation (4) as

$$\frac{dp}{dt} = mp\left[1 - \left[\left(1 - h\right) + ph\right]^{n}\right] - ep \qquad (5)$$

The likelihood of successful dispersal is now proportional to $1 - [(1 - h) + ph]^{n}$.

These modifications are quite minor. The first simply redefines the constant $m$; the second allows for nonlinear density dependence. Equation (5) collapses to Equation (1) when $h = 1$ and $n = 1$.

To facilitate comparison with Lande's (1987) model (discussed below), we rewrite Equation (5) in discrete form as

$$p_{t+1} - p_t = mp_t\left[1 - \left[\left(1 - h\right) + p_t h\right]^{n}\right] - ep_t \qquad (6)$$

Equation (5) is at equilibrium when

$$p_{t+1} = p_t = \hat{p} \quad \text{or when} \quad 1 - \left[\left(1 - h\right) + \hat{p}h\right]^{n} = \frac{e}{m} \qquad (7)$$

The equilibrium solution is

$$\hat{p} = 1 - \left[1 - \frac{\left(1 - \dfrac{e}{m}\right)^{\frac{1}{n}}}{h}\right] \qquad (8)$$

# The Individual Territory Model

Lande (1987) developed an equilibrium-based model to describe changes in occupancy rate ($p$) as determined by the rates of local extinction and colonization at the scale of individual territories. Thus, in contrast to Levins' model where the unit of suitable habitat is the patch, in Lande's model the unit is the individual territory. Levins' model tracks the colonization and extinction of patches at the scale of local populations; Lande's model tracks birth and death events at the scale of the individual.

Similar to the generalized metapopulation model (Equation 5), Lande's model allows us to determine the proportion of suitable habitat occupied by the population at equilibrium. Assuming a random or uniform distribution of territorial sites, of which a fraction $h$ are suitable, the probability that an obligately dispersing juvenile succeeds in finding a suitable territory in $n$ searches is

$$1 - \left[ \left( 1 - h \right) + ph \right]^n \tag{9}$$

Lande (1987) demonstrated that at demographic equilibrium, the proportion of suitable territories is given by equating the lifetime reproduction of female offspring per female, $R_0$, with unity. Thus $R_0$, given by the Euler-Lotka equation (Lotka 1956), is

$$R_0 = b \sum_0^\infty l_x b_x = 1 \tag{10}$$

where $l_x$ = probability of survival to age $x$ and $b_x$ = fecundity (female offspring/female) at age $x$.

Lande (1987) assumed a two-stage model with juveniles and adults; thus $l_0 = 1$, $l_1 = s_0$, $l_x = s$ ($x \geq 2$), and $b_0 = 0$, $b_x = b$ ($x \geq 1$). Given these assumptions, we can rewrite Equation (8) as

$$R_0 = b \sum_1^\infty s_0 s^{x-1} \tag{11}$$

with solution

$$R_0 = s_0 \left( \frac{b}{1-s} \right) \tag{12}$$

An equilibrium occupancy of suitable territories, conditional on successful female dispersal, occurs when $R_0 = 1$, or when

$$\left[ 1 - \left[ \left( 1 - h \right) + \hat{p}h \right]^n \right] R_0' = 1 \tag{13}$$

$R_0' = b/(1 - s)$ incorporates all the life history except $s_0$, which is now replaced by Equation (9). Thus the first-year survival rate ($s_0$) is equated with the probability of successful juvenile dispersal (Equation 9). The solution to Equation (13), the equilibrium proportion of occupied territories, is

$$\hat{p} = 1 - \left[ \frac{1 - \left(1 - \dfrac{1-s}{b}\right)^{\frac{1}{n}}}{h} \right] \tag{14}$$

The similarity between equations (8) and (14) is obvious—and we will demonstrate that their behavior in response to parameter variation is identical. Specifically, there is a biological proportionality among corresponding model parameters, varying primarily in spatial scale: the population extinction rate of an occupied patch is proportional to the mortality rate of an individual territory holder ($e \propto (1 - s)$); the colonization rate of an unoccupied suitable patch is proportional to the number of potential colonists, or the per-individual birthrate ($m \propto b$).

Lande (1987) referred to the quantity $[1 - (1 - s)/b]^{1/n}$ as a measure of the demographic potential of the population. And in a similar fashion we can think of the quantity $(1 - e/m)^{1/n}$ as a measure of the demographic potential of the metapopulation. To distinguish our discussion from Lande (1987) and to address the effects of varying search ability $n$, we subsequently refer to $b/(1 - s)$ and $m/e$ as measures of a species' colonization potential. The first term reflects colonization at the scale of the territory and is proportional to the ratio of birthrate to deathrate; the second is proportional to the ratio of patch colonization to extinction rates and reflects dynamics at the scale of the local population.

### Population Persistence

We are interested in those parameter values for which $\hat{p} > 0$ and in the stability properties for all equilibrium solutions. As the behavior of the individual territory model has been discussed in detail by Lande (1987) and Lamberson et al. (1992), we refer the reader to those publications. We concentrate here on the dynamics of the metapopulation models (Equations 1, 4, and 5).

From Equation (1) we observe an equilibrium at $1 - e/m$ (Figure 7.1), which is stable only if $m > e$. Equation (4) reaches a steady state at $\hat{p} = 1 - e/mh$, which is stable only if $mh > e$. Thus decreasing the proportion of suitable patches, $h$, lowers the equilibrium patch occupancy rate. The generalized metapopulation model (Equation 5) has an equilibrium given by

Equation (8). Changes in equilibrium patch occupancy $\hat{p}$ (Equation 8) show steep thresholds for persistence as habitat proportion declines (Figure 7.2). As found by Lande (1987) and Lamberson et al. (1992) for the individual territory model, the key inference from Figure 7.2 is that $\hat{p}$ can equal zero when $h > 0$. These steep persistence thresholds are strongly ameliorated by increases in allowable number of searches, but less so by increases in colonization potential ($m/e$; Figure 7.2).

A three-dimensional plot shows the response of $\hat{p}$ to simultaneous variation in habitat proportion ($h$) and search ability ($n$) (Figure 7.3). At extreme levels of habitat limitation ($h < 0.1$), slight increases in $h$ greatly increase $\hat{p}$, but the point at which this occurs is strongly dependent on search ability being in the range of $n = 1 - 5$ (Figure 7.3).

Somewhat surprisingly, the effects of increases in colonization potential $m/e$ affect $\hat{p}$ primarily at low levels of search ability and are most pronounced when $h$ is also low (Figure 7.4). Once search ability $n$ reaches about 10, $\hat{p}$ is only sensitive to extremely low levels of $h$ ($\leq 0.1$; Figure 7.4). If we look just at the effects of variation in $n$ or $p$, for a fixed value of $m/e$, we observe both the critical threshold values and the extreme sensitivity to declines in the density of suitable habitat, $h$ (Figure 7.5). This figure is analogous to Figure 1 in Lande (1987). Examining the $\hat{p}$ response surface with respect to simultaneous variation in habitat proportion ($h$) and demographic potential $m/e$ clearly shows the interaction between these variables (Figure 7.6). Species with

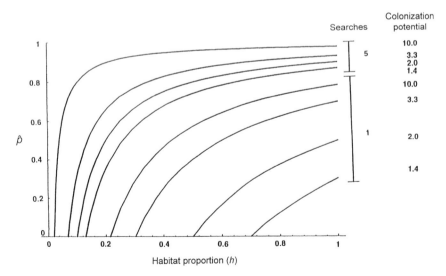

Figure 7.2. Changes in equilibrium occupancy rate $\hat{p}$ of a metapopulation against habitat proportion ($h$) for various levels of search ability ($n$) and colonization potential ($m/e$).

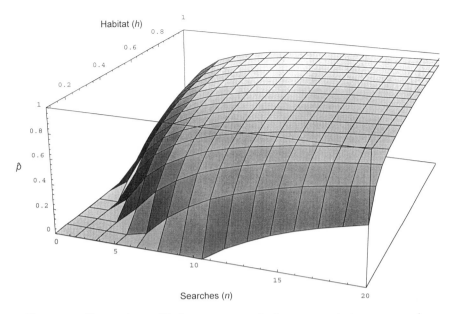

Figure 7.3. Changes in equilibrium occupancy $\hat{p}$ of a metapopulation, portrayed as a response surface, against habitat proportion ($h$) and search ability ($n$).

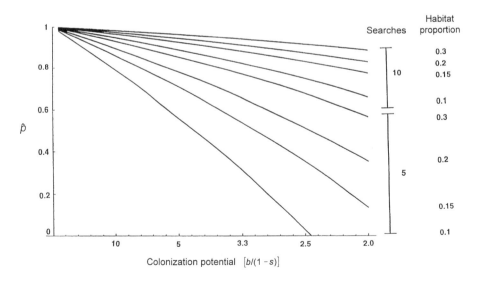

Figure 7.4. Changes in the equilibrium occupancy $\hat{p}$ of a metapopulation against colonization potential $m/e$ for various levels of search ability ($n$) and habitat proportion ($h$).

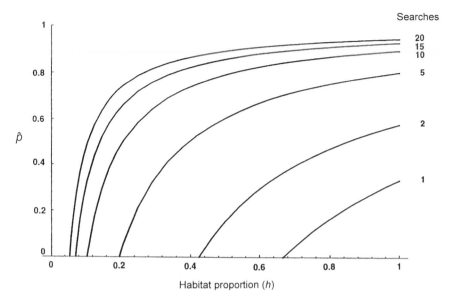

Figure 7.5. Changes in the equilibrium occupancy $\hat{p}$ of a metapopulation against habitat proportion ($h$) for various search capabilities ($n$).

naturally low colonization potential will rapidly be driven toward extinction by increasing habitat fragmentation (left-hand region of the response surface in Figure 7.6).

### Stability of the Equilibrium Points

It is important not only to compute equilibrium occupancy proportions (for both metapopulation and territory models) but also to ask how sensitive these values are to disturbances that may affect habitat amount (changes in $h$), demographic rates (changes in $m/e$ or $b/[1 - s]$), and dispersal behavior (changes in $n$) (Equations 8 and 14). The stability condition for metapopulation equilibrium (Equation 8), $|f'(\hat{p})| < 1$, in terms of habitat proportion is

$$h > 1 - \left(1 - \frac{e}{m}\right)^{\frac{1}{n}} \tag{15}$$

or, in terms of colonization rate,

$$m > \frac{e}{1 - \left(1 - h\right)^{n}} \tag{16}$$

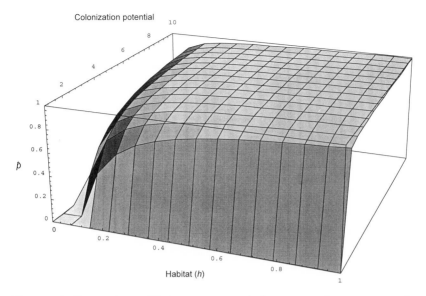

Figure 7.6. Changes in equilibrium occupancy $\hat{p}$ of a metapopulation, portrayed as a response surface, against habitat proportion ($h$) and colonization potential.

According to Equation (15), for fixed demographic potential, increases in search ability ($n$) quickly compensate for declines in habitat proportion ($h$). In the limit of Equation (15) as $n \to \infty$ a stable equilibrium arises so long as $h$ > 0. Equation (16) behaves in a similar fashion relative to increases in search ability. In the limit as $n \to \infty$, for a given habitat proportion, a stable equilibrium arises so long as $m$ > $e$.

The stability response surface (Figure 7.7) shows the combinations of habitat amount ($h$), dispersal ability ($n$), and colonization potential $m/e$ that just meet the stability requirements for a given equilibrium occupancy rate. Points lying above the response surface represent combinations of $h$, $n$, and $m/e$ yielding stable equilibria; points below the surface are unstable. The combination of habitat limitation, limited dispersal ability, and low colonization potential renders a metapopulation particularly extinction-prone (Figure 7.7: upper left).

### Model Comparisons

As previously indicated, the equilibrium equations for the generalized metapopulation model (Equation 8) and for the individual territory model (Equation 14) are functionally equivalent if we assume that $m \propto b$ and $e \propto (1 - s)$. Given this, the dynamic behavior portrayed in Figures 7.2 to 7.7 for Equa-

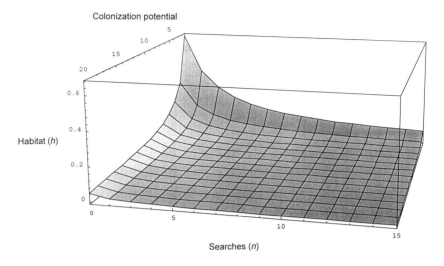

Figure 7.7. Stability response surface for the equilibrium solution to the general metapopulation model. (See Table 7.1.) Axes are habitat proportion ($h$), colonization potential $m/e$, and search ability ($n$). Points in the space lying above the response surface represent combinations of $h$, $m/e$, yielding stable equilibria.

tion (8) are equally applicable to Equation (14). The metapopulation model and its behavior (Table 7.1) can be directly compared with the individual territory model (Table 7.2). Comparing Tables 7.1 and 7.2 emphasizes the contrast in the appropriate scale for biological interpretation—individual birthrates and deathrates at the territory scale; colonization and extinction rates at the reserve scale.

### Model Limitations

There are clear limits to direct extrapolation of our model results to biological populations. The most critical limits arise because we have not considered stochastic fluctuation in the metapopulation rates ($e$ and $m$) and in individual vital rates ($b$ and $s$). Lande (1987) and Lamberson et al. (1992) have explored the effects of stochastic variation in life-history parameters on inferences from the territory model. Importantly, the threshold valuable of habitat ($h$) necessary for population persistence is increased in the presence of serially uncorrelated environmental fluctuation (Lande 1987). Greater environmental variance, however, means a less abrupt threshold into the high-risk zone (Lamberson et al. 1992). The metapopulation model (Equation 5) would be similarly affected; thus our estimates of equilibrium values are optimistic.

TABLE 7.1.
Metapopulation model

| Biological condition | Equilibrium equation | Equilibrium solution ($\hat{p}$) | Stability condition | Model interpretation (persistence requirements) |
|---|---|---|---|---|
| All patches suitable; one colonization event | $e - m(1 - \hat{p}) = 0$ | $1 - \dfrac{e}{m}$ | $m > e$ | Patch colonization rate > patch extinction rate |
| A proportion $h$ of patches suitable; one colonization event | $e - mh(1 - \hat{p}) = 0$ | $1 - \dfrac{e}{hm}$ | $hm > e$ | Colonization rate of suitable patches > patch extinction rate |
| A proportion $h$ of patches suitable; $n$ colonization events (searches) possible | $e - m\left[1 - \left[(1-h) + h\hat{p}\right]^n\right] = 0$ | $1 - \left[\dfrac{1 - \left(1 - \dfrac{e}{m}\right)^{\frac{1}{n}}}{h}\right]$ | $1 - h - \left(1 - \dfrac{e}{m}\right)^{\frac{1}{n}} = 0$  or  $m > \dfrac{e}{1 - (1-h)^n}$ | For a fixed landscape and colonization potential, inital increases in search ability strongly increase the level and stability of $\hat{p}$; for a fixed landscape and search ability, increases in colonization potential only moderately increase stability at $\hat{p}$ |

Source: Levins (1970).

Note: $\hat{p}$ = proportion of suitable patches occupied at equilibrium; $h$ = proportion of patches suitable for occupancy; $n$ = total number of patches that can be searched; $e$ = rate of local extinction of occupied patches; $m$ = colonization rate of empty patches

TABLE 7.2.

Individual territory model

| Biological condition | Equilibrium equation | Equilibrium solution ($\hat{p}$) | Stability condition | Model interpretation (persistence requirements) |
|---|---|---|---|---|
| All territories suitable; one search event | $(1-\hat{p})R_0'=1$ | $1-\dfrac{(1-s)}{b}$ | $\dfrac{b}{1-s}>1$ | Birthrate > deathrate |
| A proportion $h$ of territories suitable; one search event | $h(1-\hat{p})R_0'=1$ | $1-\dfrac{(1-s)}{bh}$ | $\dfrac{bh}{1-s}>1$ | The proportion of suitable territories $h$ must be greater than the ratio of average deathrates to birthrates (inversely related to colonization potential) |
| A proportion $h$ of territories suitable; dispersal can involve up to $n$ searches | $\left[1-\left[(1-h)+\hat{p}h\right]^n\right]R_0'=1$ | $1-\left[\dfrac{1-\left[1-\left(\dfrac{1-s}{b}\right)^{\frac{1}{n}}\right]}{h}\right]$ | $b\,\dfrac{\left(1-(1-h)^n\right)}{1-s}>1$ | For a fixed distribution of territories and $R_0'$, initial increases in search ability strongly increase the level and stability of territory occupancy; for a fixed distribution of territories and search ability, increases in $R_0'$ only moderately stabilize local occupancy. |

*Source:* Lande (1987).

*Note:* $\hat{p}$ = proportion of suitable territories occupied at equilibrium; $h$ = proportion of total "territories" that are suitable; $n$ = total number of territories that can be searched; $R_0'$ = index of colonization potential equal to $b/(1-s)$; $b$ = birthrate; $s$ = survival rate

Levins (1969) has also demonstrated that metapopulation dynamics described by Equation (1) show lower persistence in the presence of environmental fluctuations in extinction rate ($e$). The key point here is that the strong compensation of increased search ability for low $h$ and colonization potential $m/e$ (as in Figure 7.4) have critical limits set by local demographic and regional environmental stochasticity. Thus increased search ability will compensate for declines in the proportion of suitable patches ($h$) only to limits set by demographic stochasticity (which establishes a minimum size for stable, local populations) and environmental variation, which also affects the minimum size of local populations but in addition affects the proportion of occupied patches ($p_t$) necessary to reduce the risk that all local populations will simultaneously experience environmental perturbations or catastrophic events (Den Boer 1981). For an example of how these stochastic factors were incorporated into a metapopulation model for butterflies, see Hanski and Thomas (1994). General insights into minimum population sizes for local populations are provided by Lande (1993); a practical example for the spotted owl is found in Lamberson et al. (1994). Significantly, the critical population size necessary to escape the deleterious effects of demographic and environmental stochasticity is significantly reduced as population growth rate (or demographic potential) increases (Lande 1993).

One other important limitation of our models is that they do not incorporate an Allee effect—the reduction in colonization rate (metapopulation model) or fecundity (territory model) that arises from the difficulties of finding mates (Allee 1938). For the territory model, Lande (1987) explored in detail the effects of uncertainty in finding of mates. He found that difficulty in finding a mate decreased the equilibrium occupancy rate $\hat{p}$ and increased the habitat threshold ($h$) necessary for persistence. If the Allee effect were included, the compensatory growth response at low occupancy rates ($m - e$; Equation 2) would not be demonstrated. Rather, "growth" rate would show a depensatory response, which would raise the habitat threshold ($h$) necessary for persistence (that is, the curves in Figure 7.5 would shift to the right).

Finally, the Levins model assumes equal-sized patches within the metapopulation but does not explicitly consider variation in local population sizes across patches. When this additional complexity is included, there is the possibility of multiple stable equilibria where $\hat{p} > 0$ (Hastings 1991).

## Comparing the Models

We have demonstrated that the generalized metapopulation model (modified from Levins 1970) and the individual territory model (Lande 1987) are functionally equivalent. Both show steep, nonlinear thresholds to persistence set by habitat proportion; extinction in the presence of suitable, unoccupied habitat; and strong compensation for habitat limitation and low colonization potential with increases in search ability. They differ, however, in spatial scale: the traditional metapopulation model is dynamic at a population level, and the basic unit of analysis is the habitat patch; the territory model is dynamic at the individual level, and the basic unit of analysis is the territory or home range. They also differ in temporal scale: the metapopulation model assumes that within-patch dynamics occur much faster than between-patch dynamics; the territory model has no such distinction. The metapopulation model is more phenomenological; the territory model is more process-oriented.

Both models are sensitive to changes in habitat quality or configuration for species that have both low vagility and low colonization potential (Figure 7.7). Species with these life-history attributes will be "sensitive" species—that is, when faced with habitat loss and fragmentation they will be extinction-prone. While generating reliable estimates of demographic parameters is often a costly and lengthy process, both vagility and fecundity often can be crudely estimated from existing natural history information. Estimation of these parameters for a large number of species may allow rapid ranking of species' extinction likelihood (that is, a coarse filter) without the need to collect exhaustive data.

Which model offers the appropriate conceptual framework for conservation depends on several species-specific considerations. First, for a given species, a patchy distribution of habitat and individuals at a local scale (due, for example, to fine-grained fragmentation) must first be distinguished from patchy distributions of populations (Hanski and Thomas 1994). The former distributions show frequent connectivity within the lifetime of an average individual. The latter demonstrate infrequent connectivity within the average lifetime of local populations; successful between-patch colonization may therefore be a very low probability event on a per-individual basis.

Second, it is important to consider the species' historic pattern of distribution and its suitable habitat. Was the species naturally distributed as a metapopulation prior to habitat loss and fragmentation, or was it more or less continuously distributed across the landscape? If the former, we would expect the evolution of pronounced dispersal ability and decreased sensitivity to regional-scale fragmentation. If the latter, the species will probably demonstrate more limited dispersal capabilities and greater sensitivity. Valid insights depend on assessment at the correct scale. For a given species, insights into the appropriate scale at which habitat and population distributions should be estimated are provided by consideration of the historical effects of selection for dispersal ability.

Given the first two considerations, the third concern is the importance of population-level or patch-level dynamics relative to individual or site (territory) dynamics. For a species with both a current and a historical metapopulation distribution, we would expect patch dynamics to be at least as important as site dynamics. In this case, the metapopulation simplification may be justified unless local populations are below the critical minimum set by demographic uncertainty. If $h = 1$ and patch populations are large enough to be insensitive to demographic variation (say, >20 females; Richter-Dyn and Goel 1972; Lande 1993; Lamberson et al. 1994), then for individuals dispersing within the patch we expect an equilibrium $\hat{p} > 0$ even if search is limited to a single site (Doak 1989; Lamberson et al. 1994; McKelvey et al. in review).

For patches that are only partially suitable because they are heavily fragmented or have a great deal of edge, $h$ will be less than 1 and the patch dynamics will be a mixture of the two scales of interaction: individual and local population. The overall population dynamics of the patch may still be stable, but local carrying capacity will be reduced, largely as a consequence of within-patch search inefficiencies (Noon and McKelvey 1992; McKelvey et al. 1993).

For populations whose dynamics depend on the interaction of the metapopulation and on processes operative at the scale both of the patch and of the individual territory, the dynamics will operate in the following manner:

1. Territory-level dynamics will probably dominate the carrying capacity of large patches. This is because $m$ and $e$ rates defined at this scale represent infrequent events—perhaps an order of magnitude less frequent than those operating at the patch level. Because of these numerical relationships, exterior immigration will seldom provide sufficient support to affect the local equilibrium significantly.

2. As habitat declines within a patch ($h \to 0$), the equilibrium population will decline. This will have two effects at the patch level. Patch level $e$ will increase because of an increased probability of stochastic extinction, and patch level $m$ will decrease because there will be fewer dispersers. Parameters $m$ and $e$ are therefore functions of the carrying capacities of the patches and are static only if the conditions within the patches (and hence their carrying capacities) are constant.

For these reasons, once a patch-oriented reserve structure has been chosen, the maintenance of well-distributed, high-quality habitat within each patch is of the utmost importance.

If the habitat is extremely fragmented at both the local and the regional scales, then local territorial dynamics will dominate and determine population processes. The stability of the system will become extremely sensitive to changes in both the birthrates and the deathrates at a local scale and also among-territory dispersal ability. (See Noon and McKelvey 1992 for an example.) In this case, the metapopulation paradigm may lack the resolution and realism necessary for us to understand and manage the dynamics of such a system.

How useful, then, is the metapopulation paradigm for single-species management and the conservation of biodiversity? The answer depends on the species-specific considerations cited above. In many cases, focusing research and management directly on demographic processes may be most relevant, and Lande's (1987) individual territory model may be a more useful paradigm. The models clearly exist along a continuum, however, and to some degree it is possible to scale up directly from the individual to the population. For example, when at least one local population attains the size at which demographic and environmental uncertainties have negligible impacts on persistence, then overall population viability may be more dependent on among-population dynamics (patch extinction and colonization rates) than within-population dynamics (birthrates and deathrates).

## Integrating the Individual and Metapopulation Models

The decision to focus on either within-population or among-population dynamics becomes uncertain in heterogeneous landscapes. In largely continuous landscapes where suitable habitat occurs in large blocks, for example, the management emphasis should be on maintaining habitat quality within blocks. Between-habitat dynamics are less important since the individual subpopulations are locally stable. In heavily fragmented landscapes, by contrast, with only residual small blocks of suitable habitat remaining, the primary emphasis is on maintaining connectivity among local populations. The emphasis shifts because local populations are too small to be demographically stable. Most real landscapes, however, are a mix of these conditions, and the trade-offs between a within-reserve or among-reserve management emphasis are unclear (Harrison 1994).

To clarify the trade-offs among reserve size, spacing, and habitat quality for spotted owl management, we used the simulation model of Lamberson et al. (1994), which includes the dynamics of both Lande's (1987) individual territory model and Levins' (1969) metapopulation model. In this model, a given landscape consists of a regular array of equal-sized, equal-spaced, circular reserves (following Levins 1969). Each reserve, in turn, is composed of a number of pair sites, of which a proportion $h$ are suitable (Lande 1987). By varying the number of reserves of a given size, reserve spacing is affected; by varying the proportion of sites within a reserve that are suitable habitat, habitat quality is affected. The likelihood of successful travel between reserves assumed a constant risk per unit of distance and was modeled as a declining exponential function (Lamberson et al. 1994). The number of searches ($n$) per reserve was limited to that expected from a random walk (0.41 × reserve size; Lamberson et al. 1994). We set the maximum number of sites searched ($n_{max}$) to 20, an estimate based on the study of dispersing spotted owls (Thomas et al. 1990; Murphy and Noon 1992). This number allowed owls to search two or more reserves.

This simulation allowed us to increase the reality of the dynamics by including demographic uncertainty and an Allee effect (uncertainty in mate finding). Average extinction ($e$) and colonization ($m$) rates were estimated directly for a large metapopulation (≥250 reserves), each year for 1000 years, for many combinations of reserve spacing and habitat quality. Thus, the extinction/colonization dynamics of the metapopulation arose from the dynamics of individual territories and were a function of local colonization potential $b/(1 - s)$, habitat quality within a reserve ($h$), and dispersal ability ($n$). Given these changes, the expected equilibrium occupancy rate within reserves was proportional to Equation (14); the expected equilibrium for the total metapopulation was proportional to Equation (8) with $n \infty n_{max}$.

From the simulation results, we estimated the stability curves ($m/e = 1.0$) for four reserve sizes (Figure 7.8). For a given reserve size, combinations of within-reserve habitat quality and reserve spacing falling below the function ($m/e > 1.0$) have $\hat{p} > 0$; values above the function ($m/e < 1.0$) have $\hat{p} = 0$. The results clearly demonstrate the reserve design trade-offs and provide insight into the appropriate paradigm (local or metapopulation dynamics) for management. The more horizontal the curve, the more the within-reserve population dynamics are independent of reserve spacing. This pattern is most pronounced for reserves with 40 pair-sites that are locally stable so long as $h > 0.5$ (Figure 7.8). Thus, the management focus for a few large reserves is the maintenance of local habitat quality. In contrast, small reserves are seldom locally stable, and colonization among reserves is important even when most of the habitat within a reserve is suitable (Figure 7.8). Thus, the management focus for many small reserves is to maintain metapopulation connectivity by facilitating colonization. Importantly, no reserve structure—regardless of size or spacing—is viable if fewer than 30 percent of the sites within the reserve are suitable habitat (Figure 7.8).

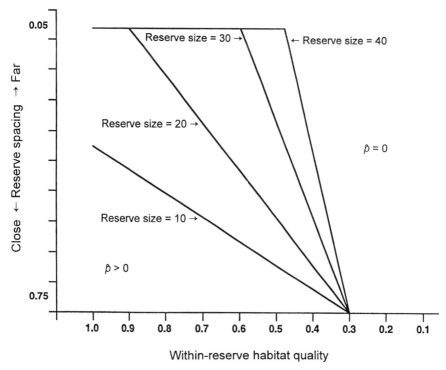

Figure 7.8. Stability response functions for the solutions to the combined individual territory and metapopulation models. Functions illustrate the trade-offs between reserve spacing (proportion of the landscape within reserve boundaries) and within-reserve habitat quality (proportion of sites stocked with suitable habitat) as a function of reserve size.

# The Spotted Owl Example

Conservation plans may differ widely in their design and implementation. Essentially there are two extremes: on the one hand are plans that designate fixed reserve boundaries—a spatially explicit distribution of habitat blocks to support locally stable subpopulations given the condition of facilitated dispersal among subpopulations (a metapopulation structure); on the other hand are plans that restrict allowable management activities within certain habitats to retain their suitability or render them suitable at some future time without specifying the boundaries of a fixed reserve system.

The former strategy is most appropriate for species currently in decline because of habitat loss and fragmentation. For these species, the manager's first responsibility is to arrest the population decline by stabilizing both the amount and the distribution of suitable habitat. (See the discussion in Lande 1987, Lamberson et al. 1992, and Lamberson et al. 1994.) In contrast, special management plans may be adequate for species that do not currently demonstrate significant population declines but are exposed to a subtle, landscapewide degradation of their habitat, such as that induced by fine-grained fragmentation. This can occur, for example, if nesting structures become limiting or of low quality because they are being systematically removed under current management plans. Under this scenario a static reserve system is not drawn on maps; rather, a dynamic reserve system that provides for a shifting mosaic distribution of suitable territories through time must be planned. Each type of plan is spatially explicit because at any point in time the manager must be confident that the landscape is providing the amount and distribution of suitable habitat needed for population persistence.

### Northern Spotted Owl

Currently the range of the northern spotted owl (Figure 7.9) approximates its historic distribution. The current distribution of both habitat and owls, however, is much reduced from historic levels. The area of suitable owl habitat prior to extensive logging is unknown, but by the early 1980s more than 80 percent of the late seral stage forest had been harvested (Booth 1991). Most of the remaining habitat is on national forest lands. Based on then-current management plans, owl habitat was projected to be completely harvested on national forest lands within 60 years (Mulder et al. 1989).

The primary silvicultural method practiced in the Pacific Northwest has been clear-cut harvest, and most harvest on public lands has occurred since 1950. As a consequence, the landscape is very patchy with sharp contrasts between old and very young forest. The Interagency Scientific Committee (ISC) conservation strategy had two key objectives: first stabilize the owl population by instituting a policy that would eventually lead to no net loss of habitat; then provide sufficient habitat in a spatial configuration that would allow for a balance between local extinction and recolonization events (Thomas et al. 1990). The second objective was achieved by concentrating currently suitable habitat as close as possible to existing reserve lands (national parks and wilderness areas) and designating a large number of habitat conservation areas (HCAs).

HCAs were widely distributed across the landscape. They were selected to have a size and spacing that would provide for locally stable populations, contingent upon some level of migration among local populations (Murphy and Noon 1992). Thus the ISC, in its reserve design, invoked the metapopulation paradigm and imposed this spatial structure on the remaining population of northern spotted owls. Significantly, this structure would become progressively more pronounced as suitable habitat outside the reserve systems was lost to timber harvest.

Figure 7.9. Northern spotted owl, Six Rivers National Forest, California. Photo by R. J. Gutiérrez.

*California Spotted Owl*

Forests in the Sierra Nevada have been markedly affected by human activity within the last 150 years (McKelvey and Johnston 1992). A combination of logging and natural attrition of the old forest led to a decline in the number of large, old trees (particularly pines); broke up the patchy mosaic of the natural forest; and encouraged the development of dense, understory conifer regeneration. The result was a rather uniform, landscapewide loss of those old forest elements (large, standing live and dead trees and large, downed logs) strongly associated with the habitat use patterns of spotted owls. Verner et al. (1992) viewed this decline in habitat quality as a fine-grained fragmentation effect expressed at the scale of the individual territory.

Based on the current Forest Service land-management plans, loss of old-growth forest elements was projected to continue, resulting in forests susceptible to fire disturbance and nearly devoid of large, old trees. Given these projections, Verner et al. (1992) proposed interim (5–10 year) guidelines that restricted silvicultural activities in habitats selected by spotted owls. These restrictions, invoked at a landscape scale but implemented at a local scale, would retain the large-tree components in harvested stands to greatly accelerate the rate at which these stands would become suitable habitat in the future. They assumed that the locations of suitable territories would shift dynamically across the landscape and that the guidlines retaining large trees would ensure that an adequate amount and distribution of suitable habitat sites would always be available.

## Why Two Different Strategies?

Both the northern and California spotted owls select habitats, at both landscape and home-range scales, that retain old-growth forest characteristics. Consequently, timber harvest of old-growth forests, or their components, is responsible for our concern over both subspecies' long-term persistence. Given the wide acceptance by the scientific community of the metapopulation reserve paradigm for the northern subspecies, why was a different strategy adopted for the California spotted owl? There are several reasons why the individual territory paradigm was the preferred conceptual model, leading to a focus on birthrates and deathrates operative at a local scale (Verner et al. 1992).

First, during the past 50 years the number and distribution of northern spotted owls may have been reduced by as much as 50 percent from pretwentieth-century levels (Thomas et al. 1990). No evidence of similar declines in number or distribution exists for California spotted owls, despite the fact that forests in the Sierra Nevada have been logged for the past 100 years. Currently, spotted owls in the Sierra Nevada are widely and rather evenly distributed throughout the conifer zone.

Second, the primary silvicultural method in the Pacific Northwest, west of the Cascade crest, has been clear-cutting. As a result, Thomas et al. (1990) opted for a strategy that clearly separated habitat reserves (HCAs) from areas where logging could occur. In contrast, for many decades selective tree harvest has been the predominant method over most of the range of the California subspecies. The widespread distribution of California spotted owls indicates that some level of selective harvest does not render habitat unsuitable for owls if sufficient amount and distribution of old-growth components are retained.

Third, because clear-cutting practices have predominated in the Pacific Northwest, particularly over the past 50 years, habitat within the range of the northern spotted owl is generally either undisturbed and suitable or cut within the past 50 years and unsuitable. This has not been the case over most of the range of the California subspecies because cutting practices did not convert forest to nonforest.

Fourth, the combined effect of the second and third factors on the subspecies has resulted in strikingly different distributions of habitat and owls in the two regions. In the Pacific Northwest, owls and their habitats are more patchy in distribution. In turn, the suitable patches are often widely separated by areas of unsuitable habitat. In contrast, the distribution of spotted owls in the Sierra Nevada is comparatively uniform in both the conifer zone and the adjacent foothill riparian/hardwood forest.

Fifth, fire is not a major threat to forests west of the Cascade crest in Washington or Oregon (Agee and Edmonds 1992). The Sierran mixed conifer forests, however, where most spotted owls occur, are drier and, given a history of fire exclusion, very prone to catastrophic fires. A metapopulation strategy there could deal with the uncertainties of logging but not fire.

Sixth, the northern spotted owl is considerably more numerous than the California subspecies. Thus, even if northern spotted owls were eliminated outside of HCAs, Thomas et al. (1990) believed that such a reduction in total population would not preclude obtaining a stable metapopulation, rangewide, within 100 years. An HCA strategy in the Sierra Nevada, coupled with their relatively uniform distribution there, could lead to a substantial population loss resulting in small, local populations.

Collectively, the foregoing considerations led Verner et al. (1992) to reject the metapopulation paradigm and instead to propose an interim landscape-level conservation strategy that would retain the old-growth forest components needed by owls for roosting and nesting at the scale of the individual territory.

# REFERENCES

Agee, J. K., and R. L. Edmonds. 1992. Forest protection guidelines for the northern spotted owl. College of Forest Resources, University of Washington, Seattle.

Allee, W. C. 1938. *The Social Life of Animals.* New York: Norton.

Booth, D. E. 1991. Estimating prelogging old-growth in the Pacific Northwest. *Journal of Forestry* 89:25–29.

Den Boer, P. J. 1981. On the survival of population in a heterogeneous and variable environment. *Oecologia* 50:39–53.

Doak, D. 1989. Spotted owls and old growth logging in the Pacific Northwest. *Conservation Biology* 3:389–396.

Hanski, I. 1991. Single-species metapopulation dynamics: Concepts, models and observations. *Biological Journal of the Linnean Society* 42:17–38.

Hanski, I., and M. Gyllenberg. 1993. Two general metapopulation models and the core-satellite species hypothesis. *American Naturalist* 142:17–41.

Hanski, I., and C. D. Thomas. 1994. Metapopulation dynamics and conservation: A spatially explicit model applied to butterflies. *Biological Conservation* 68:167–180.

Harrison, S. 1994. Metapopulations and conservation. Pages 111–128 in P. J. Edwards, R. M. May, and N. R. Webb, eds., *Large-Scale Ecology and Conservation Biology.* Oxford: Blackwell.

Hastings, A. 1991. Structured models of metapopulation dynamics. *Biological Journal of the Linnean Society* 42:57–71.

LaHaye, W. S., R. J. Gutiérrez, and H. R. Akçakaya. 1994. Spotted owl metapopulation dynamics in southern California. *Journal of Animal Ecology* 63:775–785.

Lamberson, R. H., R. McKelvey, B. R. Noon, and C. Voss. 1992. A dynamic analysis of northern spotted owl viability in a fragmented forest landscape. *Conservation Biology* 6:505–512.

Lamberson, R. H., B. R. Noon, C. Voss, and K. S. McKelvey. 1994. Reserve design for territorial species: The effects of patch size and spacing on the viability of the northern spotted owl. *Conservation Biology* 8:185–195.

Lande, R. 1987. Extinction thresholds in demographic models of territorial populations. *American Naturalist* 130:624–635.

———. 1993. Risks of population extinction from demographic and environmental stochasticity and random catastrophes. *American Naturalist* 142:911–927.

Levins, R. 1969. The effects of random variation of different types on population growth. *Proceedings of the National Academy of Sciences* (U.S.A.) 62:1061–1065.

———. 1970. Extinction. Pages 77–107 in M. Gerstenhaber, ed., *Some*

*Mathematical Questions in Biology.* Providence, R.I.: American Mathematical Society.

Lotka, A. J. 1956. Elements of Mathematical Biology. New York: Dover.

MacArthur, R. H., and E. O. Wilson. 1967. *The Theory of Island Biogeography.* Princeton: Princeton University Press.

McKelvey, K. S., and J. D. Johnston. 1992. Historical perspectives on forests of the Sierra Nevada and the transverse ranges of southern California: Forest conditions at the turn of the century. Pages 225–246 in J. Verner et al., eds., The California spotted owl: A technical assessment of its current status. General Technical Report PSW-GTR-133. Albany, Calif.: USDA Forest Service, Pacific Southwest Research Station.

McKelvey, K. S., J. Crocker, and B. R. Noon. In review. A spatially explicit life-history simulator for the northern spotted owl. *Ecological Modelling.*

McKelvey, K., B. R. Noon, and R. H. Lamberson. 1993. Conservation planning for species occupying fragmented landscapes: The case of the northern spotted owl. Pages 424–450 in P. Kareiva, J. Kingsolver, and R. B. Huey, eds., *Biotic Interactions and Global Change.* Sunderland, Mass.: Sinauer Associates.

Mulder, B. S., J. A. Bottorff, S. M. Chambers, J. T. Dowhan, A. Farmer, K. Franzreb, J. F. Gore, E. C. Meslow, J. D. Nichols, M. Scott, M. L. Shaffer, and S. R. Wilbur. 1989. The northern spotted owl: A status review supplement. Portland: USDI Fish and Wildlife Service, Region 1.

Murphy, D. D., and B. R. Noon. 1992. Integrating scientific methods with habitat conservation planning: Reserve design for northern spotted owls. *Ecological Applications* 2:3–17.

Noon, B. R., and K. S. McKelvey. 1992. Stability properties of the spotted owl metapopulation in southern California. Pages 187–206 in J. Verner et al., eds., The California spotted owl: A technical assessment of its current status. General Technical Report PSW-GTR-133. Albany, Calif.: USDA Forest Service, Pacific Southwest Research Station.

Noon, B. R., and D. D. Murphy. 1994. Management of the spotted owl: Experiences in science, policy, politics, and litigation. Pages 380–388 in G. K. Meffe and C. R. Carroll, eds., *Principles of Conservation Biology.* Sunderland, Mass.: Sinauer Associates.

Richter-Dyn, N., and N. S. Goel. 1972. On the extinction of a colonizing species. *Theoretical Population Biology* 3:406–433.

Thomas, J. W., E. D. Forsman, J. B. Lint, E. C. Meslow, B. R. Noon, and J. Verner. 1990. A conservation strategy for the northern spotted owl. Report of the Interagency Scientific Committee to address the conservation of the northern spotted owl. Portland: USDA Forest Service; USDI Bureau of Land Management/Fish and Wildlife Service/National Park Service.

USDI. 1991. Mexican spotted owl (*Strix occidentalis lucida*): Status review. Endangered Species Report 20. Albuquerque: USDI Fish and Wildlife Service.

————. 1992. Recovery plan for the northern spotted owl: Draft. Portland: USDI Fish and Wildlife Service.

Verner, J., K. S. McKelvey, B. R. Noon, R. J. Gutiérrez, G. I. Gould, Jr., T. W. Beck, and J. W. Thomas. 1992. The California spotted owl: A technical assessment of its current status. General Technical Report PSW-GTR-133. Albany, Calif.: USDA Forest Service, Pacific Southwest Research Station.

# 8

# Applying Metapopulation Theory to Spotted Owl Management: A History and Critique

*R. J. Gutiérrez and Susan Harrison\**

Metapopulation theory and models, as this volume attests, have attained great popularity in conservation biology. Nonetheless, considerable confusion still exists over the meaning and applicability of this theory. The spotted owl (*Strix occidentalis*) is the species for which the metapopulation approach to management has been developed most thoroughly, and the owl's case history illustrates much of the important variation and complexity inherent in metapopulation dynamics.

The proliferation of interest in spotted owls began when wildlife biologists were asked to estimate the present and future viability of the northern subspecies (*S. o. caurina*) (see Figure 7.9), whose ecology was little known prior to the mid-1970s (Forsman et al. 1984; Gutiérrez et al. 1995). The context of this challenging assignment was this: the owl was widespread, but it was expected to decline over the next 50 to 100 years because of the rapid loss of ancient forests to logging. The resulting research led to the listing of the northern spotted owl as a threatened subspecies (USDI 1990) and revealed further evidence for population declines (USDI 1992; Burnham et al. 1994). The potentially enormous economic and social impacts of owl conservation plans made for a politically charged issue (Simberloff 1987; Thomas and Verner 1992; Gutiérrez et al. 1995). But these plans also led to advances in conservation theory and techniques (Lande 1987; Lamberson et al. 1992; Murphy and Noon 1992) and to growing interest in the California (*S. o. occidentalis*) and Mexican (*S. o. lucida*) spotted owl subspecies; the latter was listed as threatened in 1993 (USDI 1993).

As we shall see, the spatial structure of spotted owl populations made the application of metapopulation theory to their management inevitable. In

---

\* Order of authorship determined by an owl hooting contest.

Chapter 7, Noon and McKelvey describe in detail the metapopulation models that have been developed for spotted owl management on western U.S. public forests. Our chapter has a complementary aim: we describe the natural history of the owl as it relates to metapopulation structure and trace the history of the use of metapopulation ideas in owl management. With this foundation, we discuss the meaning and value of the metapopulation approach in general. The term "metapopulation" has been given a variety of meanings, however, so we will begin by distinguishing among various types of metapopulation structure relevant to this discussion.

## What Are Metapopulations?

In the broadest sense, metapopulations simply are sets of subdivided populations in which rates of mating, competition, and other interactions are much higher within than among populations. Subdivision may be indicated (and created) by the fact that a species' habitat consists of patches that are much farther apart than the species typically disperses. Subdivision also may be indicated by demographic evidence—such as changes in population size that are partly or wholly uncorrelated between different populations—or by evidence for genetic differentiation among populations, usually measured by Wright's $F_{ST}$ or related statistics. (See Barrowclough and Gutiérrez 1990 for the estimation of $F_{ST}$ for the spotted owl.) Molecular genetic evidence is increasingly being used to detect population subdivision. (See Barrowclough et al. 1996 for a molecular analysis of the spotted owl.) In this broad sense, most species with large and discontinuous geographic ranges form metapopulations, but the consequences for demography and population persistence are ambiguous.

More narrowly defined, metapopulations are subdivided populations with demographically significant exchange among them, meaning that migration or dispersal among populations leads to the stabilization of local population fluctuations, the prevention of local extinctions (the "rescue effect"), the colonization of new habitats or habitats made vacant by local extinctions, or all three. When a species shows metapopulation structure in this sense, the important implication is that its viability is highly sensitive to landscape structure (that is, the distribution of habitat in space and time). If humans alter the landscape such that patches of habitat are too few or too far from one another, individual populations or even the whole metapopulation may go extinct.

In theory, this type of metapopulation structure will occur when habitat patches are separated by distances the species is physically capable of traveling (that is, patches are connected demographically) but exceed the distances most individuals move in their lifetime (that is, patches support separate populations). If habitats are so close together that most individuals visit many patches in their lifetime, the system will tend to behave as a single continuous population. If habitats are so far apart that dispersal between them virtually never occurs, the system will behave as a set of completely separate populations, like the populations of mammals on desert mountaintops (Brown 1971) or possibly like the Mexican spotted owl in parts of Mexico (Gutiérrez et al. 1995).

Experimental manipulations of landscape structure provide the best evidence for metapopulation structure in the more narrow sense, but they are seldom feasible with habitats of large animals. Indirect evidence may be found in distributional patterns, when species occur more frequently on less-isolated than more-isolated patches of their habitat (Opdam 1991; Laan and Verboom 1990; Sjögren 1991; Chapter 6 in this volume; Thomas et al. 1992). Genetic evidence can sometimes give us an idea of the relative amounts of dispersal among a number of populations, but it cannot yield absolute estimates of dispersal, and it is beset by the problem that dispersal and unique historical events may produce similar genetic patterns (Slatkin 1985). Although genetic evidence reveals "broad-sense" metapopulation structure, it cannot tell us whether metapopulation dynamics are critical to persistence in ecological time.

In the original and perhaps narrowest sense, metapopulations are subdivided systems in which the turnover of local populations is frequent; these could be called extinction-and-colonization metapopulations. In the "classic" model (Levins 1970) all local populations are equally susceptible to local extinction; therefore, persistence depends on there being enough populations, patches, and dispersal to ensure an adequate rate of recolonization (Hanski and Gilpin 1991). Evidence for this pattern is seen in pool frogs, *Rana lessonae* (Sjögren 1991), and a fritillary butterfly, *Melitaea cinxia* (Hanski et al. 1994). In other cases, however, there may be a mixture of small extinction-prone populations and large persistent ones (Schoener and Spiller 1987; Harrison et al. 1988; Harrison and Taylor 1996); viability of such "mainland-island" metapopulations is less sensitive to landscape structure. For organisms such as owls, which are long-lived, even with intensive monitoring relatively few local extinctions or colonizations will be observed directly. Thus the frequency and pattern of local extinctions are subjects for modeling and guesswork.

Two other related concepts are relevant to spotted owls. The first, which could be called spatially structured population dynamics, arises from the effects of territorial behavior within a continuous habitat and corresponds to the "individual-territory model" for the spotted owl (Lande 1987, 1988;

Chapter 7 in this volume). Animals with large and exclusive home ranges tend to interact only with their neighbors, resulting in dynamics (and genetic structure) considerably different from those of freely mixing populations. An important aspect of these dynamics is that birthrates and/or deathrates are affected by the structure of habitat at a fine scale; juveniles dispersing in search of breeding territories are likely to suffer higher mortality in a fragmented landscape. As Noon and McKelvey describe in Chapter 7, this process can be modeled under a formal framework identical to that of a metapopulation model if the "extinction" and "colonization" of individuals on territories are substituted for those of whole populations.

The other related concept is source/sink dynamics, referring to species that occupy both high-quality habitats (sources) where populations grow and produce emigrants, and low-quality habitats (sinks) where populations cannot sustain themselves in the absence of immigration. In the original model by Pulliam (1988), sinks are used by territorial animals only when source habitats are fully occupied; sinks serve as valuable reservoirs for the occasional repopulation of sources. Alternatively, the term "sink" may be used to denote habitats detrimental to the viability of populations. For example, the model of Lamberson et al. (1992) showed that the logged matrix surrounding forested reserves for spotted owls may reduce the owls' population growth rate if naive juveniles attempt to establish territories in this matrix rather than searching for better-quality habitat.

## What Are the Spotted Owl's Metapopulation Characteristics?

Here we summarize aspects of spotted owl biology relevant to two central questions of owl conservation: what is the owl's natural metapopulation structure, and how can viable metapopulations of the owl be created or maintained while forests are managed for other objectives? Metapopulation structure is the product of habitat distribution, local population dynamics, and dispersal, so we focus on these attributes. The three subspecies, northern, California, and Mexican, are genetically differentiated, use different habitats, and may vary in their dispersal behavior (Gutiérrez et al. 1995). Moreover, there is great variation across the owl's range not only in the natural spatial distribution of owl populations (Figure 8.1) but also in the type and intensity of habitat perturbations, both natural and human caused. Geographic variation in natural and present-day metapopulation structure is therefore an essential part of the story.

Spotted owls are primarily forest dwellers; numerous studies have documented their selection for old-growth conifer forests throughout their range

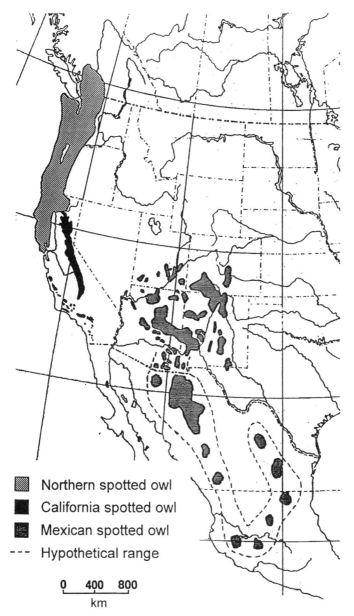

Northern spotted owl
California spotted owl
Mexican spotted owl
--- Hypothetical range

0    400   800
km

Figure 8.1. Distribution of three subspecies of spotted owls showing potential metapopulation structure.

(for example, Forsman et al. 1984; Ganey and Balda 1989; Solis and Gutiérrez 1990; Gutiérrez et al. 1992). They also are known to inhabit partly logged redwood (*Sequoia sempervirens*) forest (Folliard 1993), closed-canopy oak forests in the Sierra Nevada and southern California (Gutiérrez et al. 1992), and deep, narrow canyons with little tree cover in Utah (Rinkevich 1991). All these environments share certain features: open understories that favor low flight and hunting; high structural diversity; tree cavities, old hawk nests, or rock ledges that serve as nest sites; presence of the owl's primary prey, flying squirrels (*Glaucomys sabrinus*) in the north and usually woodrats (*Neotoma* spp.) elsewhere; and cool temperatures relative to surrounding habitats. The owl's narrow thermal neutral zone makes them easily heat stressed, which appears to be a major factor in their selection of microhabitats (Barrows 1981; Forsman et al. 1984; Solis and Gutiérrez 1990; Ganey et al. 1993).

In the Pacific Northwest, the elevational range of the owl is approximately 20 to 1400 m; in the Sierra Nevada and southern California, about 300 to 2600 m; and in the Southwest, roughly 1800 to 2700 m, with some variation owing to topography, prey distribution, and the presence of perennial streams, an apparent requisite in arid habitats. Thus, the owl's habitat is naturally almost continuous from southern British Columbia to the southern Sierra Nevada and becomes insular in southern California and the Southwest, where montane forests are surrounded by desert, grassland, shrub-steppe, or sparse lower-elevation woodland.

Northern spotted owls historically were distributed almost continuously across their range, with the exception of the Willamette Valley in Oregon and small-scale discontinuities owing to fire (Figure 8.1). They are now largely extirpated from the western Washington lowland and northwestern Oregon physiographic provinces, and remaining populations are considerably fragmented due to habitat loss from logging (Thomas et al. 1990). There are currently more than 8000 known owls (Gutiérrez et al. 1995), but demographic analyses indicate a rangewide decline of 7 percent per year (Anderson and Burnham 1992; Burnham et al. 1994).

The historic distribution of the California spotted owl (Figure 8.2) in the Sierra Nevada probably was continuous, and it now remains essentially so, although logging has altered it at a fine scale (Figure 8.1; Bias and Gutiérrez 1992; Verner et al. 1992). Fire suppression in the Sierra has led to an increase in dense conifer forest, which owls may inhabit if large residual old trees are present (Gutiérrez et al. 1992). This has led to speculation that owl habitat in some areas of the Sierra Nevada is more abundant today than prior to European settlement (USDA 1995). This assertion is unsupported, however:

Figure 8.2. California spotted owl, San Bernardino Mountains, California.

not only do the earliest studies of Sierran forests postdate the arrival of Europeans and their livestock (McKelvey and Johnston 1992), but we have little knowledge of pre-European fire frequencies or owl densities. The Sierran population is estimated at more than 3050 individuals, and demographic trends are presently ambiguous (Noon et al. 1992). In contrast to the continuous populations of California spotted owls in the Sierra Nevada, the distribution in the south is discontinuous at a large scale. The California spotted owl in southern and central coastal California occurs as a series of discrete populations of generally less than 100 pairs (Figure 8.1; Noon et al. 1992; LaHaye et al. 1994).

The Mexican spotted owl appears to exist in several large populations (more than 100 pairs) on the Mogollon Rim, Sacramento Mountains, and Sierra Madre Occidental as well as a series of small isolated populations spanning approximately 2400 km from Utah and Colorado in the north to Michoacan, Mexico, in the south (Figure 8.1). At present, 777 individuals are known to exist, but only two populations have been thoroughly censused, and neither the number nor sizes of Mexican spotted owl populations are well known (USDI 1995).

Spotted owls form long-term pair bonds and occupy very large home ranges, usually greater than 800 ha; they are long-lived and have high annual adult survival rates (0.8–0.95) and low reproductive rates (about one fledgling per year, with variable annual nesting frequencies) (Gutiérrez et al. 1995). Juveniles leave their natal territory when they fledge in autumn, and they appear to traverse unsuitable habitat readily in their search for territories and mates (Gutiérrez et al. 1985, 1995). The distance and frequency of these moves are the focus of intensive study.

Juvenile northern spotted owls are known to move distances of up to 150 km, sometimes across nonforest, but have never been known to move between study sites (about 150 km) in northwestern California and the Sierra Nevada (Thomas et al. 1990; Franklin 1992; Zabel et al. 1992). Molecular evidence suggests that in the past, however, gene flow occurred between the subspecies (Barrowclough et al. 1996). For the California owl in the Sierra Nevada, little is known of juvenile movement; only two banded juveniles have been recaptured in a 9-year study (Moen and Gutiérrez 1995). In an 8-year study in southern California, more than 90 percent of California spotted owls ($n = 745$) were banded in two populations less than 10 km apart in the San Bernardino Mountains and Mount San Jacinto, and no moves between these populations have been detected (LaHaye et al. 1994; R. Gutiérrez, personal observation). Of 172 Mexican spotted owls banded in a 4-year study, several long-distance moves were detected. This is surprising given the low sampling intensity. Two juveniles dispersed to different mountain ranges (22 and 56 km distant) and established territories, and one bird banded as a mated female was found dead 187 km from her territory 7 months later, having moved across at least two mountain ranges and 25 km of treeless desert grassland (Gutiérrez et al. 1996). Hence the three subspecies may have different tendencies regarding dispersal distances and behavior, perhaps a reflection of the landscape structure in which they evolved. Further research on dispersal will be needed to validate these apparent patterns.

Genetic studies have provided a very general picture of large-scale population structure. Allozyme variation was low and revealed differentiation between only the Mexican owl and the other two subspecies (Barrowclough and Gutiérrez 1990). However, mitochondrial DNA (mtDNA) sequencing revealed differences among all subspecies and among geographic regions within the northern subspecies (Barrowclough et al. 1996). Paradoxically, the San Bernardino and San Jacinto populations of the California subspecies were identical at all base pairs studied (more than 1000), even though these were the same populations between which an intensive banding study revealed no dispersal (LaHaye et al. 1994). This apparent lack of dispersal may be an artifact of a relatively short study (less than 5 years) or an effect of suburban development across the most likely dispersal paths.

In summary, then, spotted owls within continuous areas of forest can be described as forming "spatially structured populations" as defined earlier. In other words, population densities are extremely low, each individual interacts only with a few neighboring individuals, and the likelihood of recruiting to the breeding population may be limited by the large distances between potential territory sites. Superimposed on this local population structure, the southern California populations and the Mexican spotted owl also form broad-sense metapopulations (that is, subdivided populations) at a larger scale, owing to the discontinuity of their habitat (Figures 8.1 and 8.3). The northern and the Sierran California spotted owl are not naturally subdivided at this larger scale, but the northern owl is becoming increasingly more subdivided because of the coarse-grained fragmentation of Pacific Northwest forests.

Both the direct evidence (leg banding) and the indirect (genetic) evidence on juvenile dispersal are sparse and inconsistent, however, leaving the magnitude and importance of dispersal among populations poorly known throughout the owl's range. This constitutes the major gap in our understanding of the owl's metapopulation structure, as well as in our ability to prescribe management strategies. We turn now to a brief history of the use of metapopulation concepts in spotted owl conservation.

Figure 8.3. Natural metapopulation structure imposed on spotted owls by elevation gradients in southern California. Three mountain ranges that harbor spotted owl populations are shown here. In the foreground are the San Bernardino Mountains; in the middle is Mount San Jacinto; in the distance is Palomar Mountain. Photo by William S. LaHaye.

## How Have Metapopulation Concepts Been Applied in Spotted Owl Management?

Metapopulation theory was first applied to the spotted owl by Shaffer (1985); prior to this, population viability analyses for the owl had focused on viable population sizes. (See Simberloff 1987 for a review.) Shaffer described the metapopulation model of Levins (1970) and explained how it might apply to the owl at the level of individuals "colonizing" and going "extinct" on territories. The U.S. Forest Service aimed to establish a network of spotted owl management areas (SOMAs); Shaffer recommended that SOMAs have the same average size and median spacing as present-day occupied patches. His paper helped to turn thinking away from effective population size and toward a focus on the spatial arrangement of habitat for the first time in population viability analysis. The plan that subsequently emerged, however, called for SOMAs one to three territories in size and 16 to 32 km apart (USDA 1988)—a plan labeled a "prescription for extinction" by a U.S. federal judge in 1991 (*Seattle Audubon Society* v. *Evans,* case C89-160WD).

Quantitative modeling of spotted owl metapopulations began with Lande

(1987, 1988), whose models (described in Chapter 7 of this volume) illustrated that populations of territorial animals will collapse when the proportion of the landscape that is suitable habitat falls below a critical threshold. For the owl, Lande (1988) estimated that this threshold might lie near 20 percent, which happens to be the proportion of the northern spotted owl's range that is currently old-growth forest (see also Johnson 1992). Doak (1989) added the larger-scale dynamics created by habitat fragmentation to the local spatial dynamics modeled by Lande (1987, 1988) and showed the value of clustering suitable habitat into the largest contiguous blocks possible. Lande's and Doak's models were not spatially explicit—in their models, every patch or territory is equally accessible to all others—so these models were not capable of exploring the effects of habitat geometry on population viability.

Spatially explicit simulation models for the northern spotted owl were introduced in the Interagency Scientific Committee's report (Thomas et al. 1990) and are described in Lamberson et al. (1992), McKelvey et al. (1993), and Chapter 7 in this volume. The "territory cluster model" incorporated distances between patches in relation to juvenile dispersal distances and specific patch sizes in terms of number of territories, as well as detailed local dynamics within and between adjacent territories. Results suggested that a viable metapopulation could be maintained on a network of habitat conservation areas greater than 20 territories in size and less than 19.2 km apart (Thomas et al. 1990). Besides producing these recommendations for patch size and spacing, the models demonstrated other key behavior likely in the owl population—most notably a long lag between the time when habitat was fragmented and the time when the resulting decline in the owl population was observable.

The same model was applied to the California spotted owl in southern California, with the important distinction that the model depicted the real landscape rather than a hypothetical one (Noon and McKelvey 1992). This model strongly emphasized the importance of dispersal among populations, and the results suggested that the largest population in the region (San Gabriel/San Bernardino Mountains) plays a key role in metapopulation viability. LaHaye et al. (1994) studied a different spatially explicit model of the owl in southern California, however, and found that persistence time was not very sensitive to the rate of dispersal among populations, especially in contrast to the overwhelming effect of the current rapid decline ($\lambda = 0.85$) in the large San Bernardino population. In other words, the set of populations in southern California may not be functioning as an interconnected metapopulation to any significant extent. Molecular genetic data are ambiguous because all individuals examined from the San Bernardino Mountains and Mount San Jacinto are genetically identical (Barrowclough et al. 1996), seemingly contradicting the leg-banding studies that revealed very little dispersal. Again, recent demographic patterns may not reflect historic patterns.

The use of spatially explicit models to devise management plans for the spotted owl received a setback in legal decisions in 1991 and 1992 (*Seattle Audubon Society* v. *Evans,* case C89-160WD; *Seattle Audubon Society* v. *Moseley,* case 92-479WD) when a federal judge ruled that the U.S. Forest Service had not fully disclosed the environmental risks of the Interagency Scientific Committee's plan (Thomas et al. 1990; USDA 1992*a*). Questioning the reliability of the "habitat conservation area" strategy proposed by Thomas et al. (1990), the litigants pointed out that results of the spatial models were extremely sensitive to assumptions about habitat geometry and, especially, juvenile dispersal behavior (Harrison et al. 1993). Another key concern was the growing evidence for a rangewide decline in the northern spotted owl population (Anderson and Burnham 1992). Following these adverse rulings, the U.S. Forest Service devised an "ecosystem management" plan that was very similar to the Thomas et al. (1990) strategy, with the exception of increased riparian protection, but contained no reference to owl population models (Thomas 1993).

For the California spotted owl in the Sierra Nevada, neither the population viability study by Verner et al. (1992) nor the U.S. Forest Service's recently proposed management plan (USDA 1995) contains any population models. The viability study considers the possibility that certain areas at the margins of the owl's range—for example, the forests on the east side of the Sierra Nevada—may be "sink" habitats in the detrimental sense described earlier (Noon et al. 1992). Although there is no published evidence that these marginal populations are sinks, this conjecture was adopted as a basis for giving lesser protection to east side and Tahoe Basin forests (USDA 1995).

The recent recovery plan for the Mexican spotted owl adopts an entirely different approach to inferring metapopulation structure (USDI 1995). Using statistical associations between the presence of owls and habitat characteristics, a map of presumed suitable habitat was created. The map was then analyzed to determine landscape connectivity—measured as the average of the maximum width of each habitat patch—under a range of assumptions about how far apart two patches must be to be considered separate. The results showed that when owls are assumed to be able to readily disperse 40 km (that is, two patches less than 40 km apart are effectively one patch), there is a dramatic improvement in the inferred connectivity of the landscape relative to shorter assumed dispersal distances. Since 40 km is in the range of known dispersal distances for juvenile Mexican spotted owls, it was concluded that the Mexican spotted owl indeed forms a metapopulation in the sense that its populations are subdivided, yet interact with one another, and that their viability is highly sensitive to the arrangement (and removal) of patches. However, this conclusion was not used as the basis for any specific reserve design.

## Lessons

Spotted owl population dynamics and population viability can be understood properly only with a metapopulation approach. The older approach to viability analysis was to identify a target size for the total population by using nonspatial models that emphasize environmental stochasticity. This simply does not work for long-lived territorial species, distributed over large areas at low densities, and in habitats that may be naturally (or anthropogenically) patchy at multiple spatial scales. Early models of the spotted owl showed that the risks of extinction were primarily dependent on the numbers, sizes, and spacing of habitat patches and the interaction of these variables with demography and dispersal. Thus the case of the spotted owl led to major shifts in the thinking of conservation biologists about population viability. Moreover, the first detailed metapopulation models for the spotted owl illustrated the inadequacy of previous management plans and provided an essential stimulus for newer plans to retain more old-growth forest. If a metapopulation approach had been taken earlier in the process, perhaps much expense, time, and litigation could have been saved.

Nonetheless, attempts to use metapopulation models to devise specific management plans for the spotted owl and other species have encountered great difficulty. The basic problem is that metapopulation processes take place at such large scales of time and space that it is virtually impossible to gather all the data needed by the models at the level of precision required for strong conclusions. In the absence of highly detailed and accurate information on habitat distribution, on local abundance and demography, and (especially) on dispersal behavior, models require numerous assumptions to which their results are highly sensitive (Harrison et al. 1993; Doak and Mills 1994; Harrison 1994). Uncertainties in model results are generally well explained by the modelers themselves (see Lamberson et al. 1992; LaHaye et al. 1994), but they are often overlooked in the agency reports that cite them in support of specific plans (for example, USDA 1992a, 1992b, 1994, 1995).

If great uncertainties remain for the spotted owl, despite the unprecedented research efforts, it is unlikely that metapopulation models of wildlife ordinarily will be precise enough to devise "safe" strategies of forest fragmentation, though they can sometimes expose unsafe plans fairly clearly. Models will be helpful only for reasonably well-studied species; they cannot substitute for good data on species distribution and abundance. Furthermore, a plan devised specifically for the spotted owl cannot be transferred directly to other species in other forests.

Perhaps the greatest usefulness of metapopulation models has been to demonstrate some of the general, qualitative behavior we may expect in wildlife populations as their habitat undergoes fragmentation. Metapopulation models show that populations will collapse when fragmentation exceeds a certain threshold, that this critical level is impossible to determine with precision, and that by the time population declines are evident, it may be too late to reverse them. These conclusions provide strong support for exercising caution and conservatism in forest management if maintaining viable wildlife populations is truly a priority.

## Acknowledgments

We thank J. Dunk, L. George, D. Kristan, and J. Verner for reading various drafts of this chapter. Research on spotted owl population dynamics has been supported by the California Department of Fish and Game, U.S. Forest Service (Contract 53-91S8-4FW20), Pacific Southwest Experiment Station, Rocky Mountain Forest and Range Experiment Station (Contract 53-82FT-4-07), U.S. Fish and Wildlife Service (Unit Cooperative Agreement 14-16-0009-1547, Work Order 28), U.S. Bureau of Land Management (Agreement 1422-B950-A5-0014), U.S. National Park Service (Cooperative Agreement CA 8480-3-9005), and Southern California Edison grants to R. J. Gutiérrez.

### REFERENCES

Anderson, D. R., and K. P. Burnham. 1992. Demographic analysis of northern spotted owl populations. Appendix C in *Recovery Plan for the Northern Spotted Owl*. Portland: U.S. Department of the Interior Fish and Wildlife Service.

Barrowclough, G. F., and R. J. Gutiérrez. 1990. Genetic variation and differentiation in the spotted owl (*Strix occidentalis*). *Auk* 107:737–744.

Barrowclough, G. F., R. J. Gutiérrez, and J. G. Groth. 1996. Genetic structure of spotted owl (*Strix occidentalis*) populations based on mitochondrial DNA sequences. In preparation.

Barrows, C. W. 1981. Roost selection by spotted owls: An adaptation to heat stress. *Condor* 83:302–309.

Bias, M. A., and R. J. Gutiérrez. 1992. Habitat associations of California spotted owls in the central Sierra Nevada. *Journal of Wildlife Management* 56:584–595.

Brown, J. H. 1971. Mountaintop mammals: Nonequilibrium insular biogeography. *American Naturalist* 105:467-478.

Burnham, K. P., D. R. Anderson, and G. C. White. 1994. Estimation of vital rates of the northern spotted owl. Appendix J in *Final Supplemental Environmental Impact Statement on Management for Old-Growth Forest Related Species Within the Range of the Northern Spotted Owl.* Portland: U.S. Department of Agriculture and U.S. Department of the Interior.

Doak, D. F. 1989. Spotted owls and old growth logging in the Pacific Northwest. *Conservation Biology* 3:389–396.

Doak, D. F., and L. S. Mills. 1994. A useful role for theory in conservation. *Ecology* 75:615–626.

Folliard, L. 1993. Nest site characteristics of northern spotted owls in managed forests of northwestern California. M.S. thesis, University of Idaho, Moscow.

Forsman, E. D., C. Meslow, and H. M. Wight. 1984. Distribution and biology of the spotted owl in Oregon. *Wildlife Monographs* 87.

Franklin, A. B. 1992. Population regulation in northern spotted owls: Theoretical implications for management. Pages 815–827 in D. R. McCullough and R. H. Barrett, eds., *Wildlife 2001: Populations.* London: Elsevier Applied Sciences.

Ganey, J. L., and R. P. Balda. 1989. Distribution and habitat use of Mexican spotted owls in Arizona. *Condor* 91:355–361.

Ganey, J. L., R. P. Balda, and R. M. King. 1993. Metabolic rate and evaporative water loss of Mexican spotted owls. *Wilson Bulletin* 105:645–656.

Gutiérrez, R. J., A. B. Franklin, and W. S. LaHaye. 1995. Spotted owl. No. 179 in A. Poole and F. Gill, eds., *The Birds of North America.* Philadelphia: Academy of Natural Sciences/American Ornithologists' Union.

Gutiérrez, R. J., M. S. Seamans, and M. Z. Peery. 1996. Intermountain movement by Mexican spotted owls. *Great Basin Naturalist* 56:87–89.

Gutiérrez, R. J., A. B. Franklin, W. S. LaHaye, V. J. Meretsky, and J. P. Ward. 1985. Juvenile spotted owl dispersal in northwestern California: Preliminary analysis. Pages 60–65 in R. J. Gutiérrez and A. B. Carey, eds., *Ecology and Management of the Spotted Owl in the Pacific Northwest.* General Technical Report PNW-185. Portland: USDA Forest Service.

Gutiérrez, R. J., J. Verner, K. S. McKelvey, B. R. Noon, G. N. Steger, D. R. Call, W. S. LaHaye, B. B. Bingham, and J. S. Senser. 1992. Habitat relations of the California spotted owl. Pages 79–98 in J. Verner et al., *The California Spotted Owl: A Technical Assessment of Its Current Status.*

General Technical Report PSW-GTR-133. Albany, Calif.: USDA Forest Service, Pacific Southwest Research Station.

Hanski, I., and M. Gilpin. 1991. Metapopulation dynamics: Brief history and conceptual domain. *Biological Journal of the Linnean Society* 42:3–16.

Hanski, I., M. Kuussaari, and M. Nieminen. 1994. Metapopulation structure and migration in the butterfly *Melitaea cinxia. Ecology* 75:747–762.

Harrison, S. 1994. Metapopulations and conservation. Pages 111–128 in P. J. Edwards, N. R. Webb, and R. M. May, eds., *Large-Scale Ecology and Conservation Biology.* Oxford: Blackwell.

Harrison, S., and A. D. Taylor. 1996. Empirical evidence for metapopulation dynamics: A critical review. In I. Hanski and M. Gilpin, eds., *Metapopulation Dynamics: Ecology, Genetics, and Evolution.* New York: Academic Press.

Harrison, S., D. D. Murphy, and P. R. Ehrlich. 1988. Distribution of the bay checkerspot butterfly, *Euphydryas editha bayensis:* Evidence for a metapopulation model. *American Naturalist* 132:360–382.

Harrison, S., A. Stahl, and D. Doak. 1993. Spatial models and spotted owls: Exploring some biological issues behind recent events. *Conservation Biology* 7:950–953.

Johnson, D. H. 1992. Spotted owls, great horned owls, and forest fragmentation in the central Oregon Cascades. M.S. thesis, Oregon State University, Corvallis.

Laan, R., and B. Verboom. 1990. Effect of pool size and isolation on amphibian communities. *Biological Conservation* 54:251–262.

LaHaye, W. S., R. J. Gutiérrez, and H. R. Akçakaya. 1994. Spotted owl metapopulation dynamics in southern California. *Journal of Animal Ecology* 63:775–785.

Lamberson, R. H., R. McKelvey, B. R. Noon, and C. Voss. 1992. A dynamic analysis of northern spotted owl viability in a fragmented forest landscape. *Conservation Biology* 6:505–512.

Lande, R. 1987. Extinction thresholds in demographic models of territorial populations. *American Naturalist* 130:624–635.

———. 1988. Demographic models of the northern spotted owl (*Strix occidentalis caurina*). *Oecologia* 75:601–607.

Levins, R. 1970. Extinction. Pages 77–107 in M. Gerstenhaber, ed., *Some Mathematical Questions in Biology.* Providence, R.I.: American Mathematical Society.

McKelvey, K. S., and J. D. Johnston. 1992. Historical perspective on forests of the Sierra Nevada and the transverse ranges of southern California:

Forest conditions at the turn of the century. Pages 225–246 in J. Verner et al., *The California Spotted Owl: A Technical Assessment of Its Current Status*. General Technical Report PSW–GTR–133. Albany, Calif.: USDA Forest Service, Pacific Southwest Research Station.

McKelvey, K., B. R. Noon, and R. H. Lamberson. 1993. Conservation planning for species occupying fragmented landscapes: The case of the northern spotted owl. Pages 424–450 in P. M. Kareiva, J. G. Kingsolver, and R. B. Huey, eds., *Biotic Interactions and Global Change*. Sunderland, Mass.: Sinauer Associates.

Moen, C. A., and R. J. Gutiérrez. 1995. Population ecology of the California spotted owl in the central Sierra Nevada: Annual results, 1994. Annual Progress Report (Contract 53–91S8–4–FW20). San Francisco: USDA Forest Service, Region 5.

Murphy, D. D., and B. R. Noon. 1992. Integrating scientific methods with habitat conservation planning: Reserve design for northern spotted owls. *Ecological Applications* 2:3–17.

Noon, B. R., and K. S. McKelvey. 1992. Stability properties of the southern California metapopulation. Pages 187–206 in J. Verner et al., *The California Spotted Owl: A Technical Assessment of its Current Status*. General Technical Report PSW–GTR–133. Albany, Calif.: USDA Forest Service, Pacific Southwest Research Station.

Noon, B. R., K. S. McKelvey, D. W. Lutz, W. S. LaHaye, R. J. Gutiérrez, and C. A. Moen. 1992. Estimates of demographic parameters and rates of population change. Pages 175–186 in J. Verner et al., *The California Spotted Owl: A Technical Assessment of Its Current Status*. General Technical Report PSW–GTR–133, Albany, Calif.: USDA Forest Service, Pacific Southwest Research Station.

Opdam, P. 1991. Metapopulation theory and habitat fragmentation: a review of holarctic breeding bird studies. *Landscape Ecology* 5:93–106.

Pulliam, H. R. 1988. Sources, sinks and population regulation. *American Naturalist* 132:652–661.

Rinkevich, S. E. 1991. Distribution and habitat characteristics of Mexican spotted owls in Zion National Park, Utah. M.S. thesis, Humboldt State University, Arcata, Calif.

Schoener, T. W., and D. A. Spiller. 1987. High population persistence in a system with high turnover. *Nature* 330:474–477.

Shaffer, M. L. 1985. The metapopulation and species conservation: The special case of the northern spotted owl. Pages 86–99 in R. J. Gutiérrez and A. B. Carey, eds., *Ecology and Management of the Spotted Owl in the Pacific Northwest*. General Technical Report PNW–185. Portland: USDA Forest Service.

Simberloff, D. 1987. The spotted owl fracas: Mixing academic, applied and political ecology. *Ecology* 68:766–772.

Sjögren, P. 1991. Extinction and isolation gradients in metapopulations: The case of the pool frog (*Rana lessonae*). *Biological Journal of the Linnean Society* 42:135–147.

Slatkin, M. 1985. Gene flow in natural populations. *Annual Review of Ecology and Systematics* 16:393–430.

Solis, D. M., Jr., and R. J. Gutiérrez. 1990. Summer habitat ecology of spotted owls in northwestern California. *Condor* 92:739–748.

Thomas, C. D., J. A. Thomas, and M. S. Warren. 1992. Distributions of occupied and vacant butterfly habitats in fragmented landscapes. *Oecologia* 92:563–567.

Thomas, J. W., team leader. 1993. Forest ecosystem management: An ecological, economic and social assessment. Report of the Forest Ecosystem Management Assessment Team. Portland: USDA Forest Service, Pacific Northwest Research Station.

Thomas, J. W., and J. Verner. 1992. Accommodation with socio-economic factors under the Endangered Species Act—more than meets the eye. *Transactions of the North American Wildlife and Natural Resources Conference* 57:627–641.

Thomas, J. W., E. D. Forsman, J. B. Lint, E. C. Meslow, B. R. Noon, and J. Verner. 1990. A conservation strategy for the northern spotted owl. Report of the Interagency Scientific Committee to address the conservation of the northern spotted owl. Portland: USDA Forest Service; USDI Bureau of Land Management/Fish and Wildlife Service/National Park Service.

USDA. 1988. Final supplement to the environmental impact statement for an amendment to the Pacific Northwest regional guide. Portland: USDA Forest Service.

———. 1992*a*. Final environmental impact statement on management for the northern spotted owl in the National Forests. Portland: USDA Forest Service, Pacific Northwest Region.

———. 1992*b*. Flathead National Forest plan, proposed amendment 16. Draft environmental impact statement. Kalispell, Mont.: USDA Forest Service, Intermountain Region.

———. 1994. Interim habitat management guidelines for maintaining well-distributed viable wildlife populations within the Tongass National Forest. Draft environmental assessment. Juneau: USDA Forest Service, Alaska Region.

———. 1995. Managing California spotted owl habitat in the Sierra Nevada

National Forests of California. Draft environmental impact statement. Sacramento: USDA Forest Service, Pacific Southwest Region.

USDI. 1990. Endangered and threatened wildlife and plants: Determination of threatened status for the northern spotted owl. *Federal Register* 55:26114–26194.

————. 1992. Recovery plan for the northern spotted owl: Draft. Portland: USDI Fish and Wildlife Service.

————. 1993. *Threatened Wildlife of the United States.* Resource Publication 114. Washington, D.C.: USDI Bureau of Sport Fisheries and Wildlife.

————. 1995. Draft recovery plan for the Mexican spotted owl. Albuquerque: USDI Fish and Wildlife Service, Southwestern Region.

Verner, J., K. S. McKelvey, B. R. Noon, R. J. Gutiérrez, G. I. Gould, Jr., T. W. Beck, and J. W. Thomas, technical coordinators. 1992. *The California Spotted Owl: A Technical Assessment of Its Current Status.* General Technical Report PSW-GTR-133. Albany, Calif.: USDA Forest Service, Pacific Southwest Research Station.

Zabel, C. J., G. N. Steger, K. S. McKelvey, G. P. Eberlein, B. R. Noon, and J. Verner. 1992. Home-range size and habitat-use patterns of California spotted owls in the Sierra Nevada. Pages 149–164 in J. Verner et al., *The California Spotted Owl: A Technical Assessment of its Current Status.* General Technical Report PSW-GTR-133, Albany, Calif.: USDA Forest Service, Pacific Southwest Research Station.

# 9

# Classification and Conservation of Metapopulations: A Case Study of the Florida Scrub Jay

*Bradley M. Stith, John W. Fitzpatrick, Glen E. Woolfenden, and Bill Pranty*

Metapopulation theory, now a major paradigm in conservation biology (Harrison 1994; Doak and Mills 1994), can be viewed as island biogeography theory applied to single species (Hanski and Gilpin 1991). Whereas application of island biogeography to conservation followed shortly after its creation (MacArthur and Wilson 1967), application of metapopulation theory lagged far behind its formalization by Levins (1969, 1970). Simberloff (1988) attributes its growing emergence to a shift in ecological and conservation focus from analysis of species turnover to analysis of extinction in small populations of individual species. Describing real-world metapopulations, however, remains problematic.

Harrison (1991) has pointed out ambiguities in the term "metapopulation" and described four different configurations of habitat patches that could be called metapopulations. Reviewing field studies of patchy systems, she found few natural examples that matched Levins' original concept of a metapopulation. Recently Harrison (1994) has argued that metapopulation theory often is not applicable, as in cases where populations are highly isolated, highly connected, or so large as to be essentially invulnerable. She warns: "The metapopulation concept is being taken seriously by managers, and taken too literally could lead to the 'principles' that single, isolated populations are always doomed, or that costly strategies involving multiple connected reserves are always necessary." Doak and Mills (1994) have reviewed the different metapopulation classes described by Harrison and hold that "it will often be difficult or impossible to distinguish between these alternatives, and thus to assess the importance of metapopulation dynamics." They also warn that spatially explicit population models (SEPM) simulating metapopulation dynamics typically use parameters that are difficult to measure in the field. Their list of required data includes within-patch demographic rates and variances, temporal and spatial correlation of vital rates among populations, and dispersal distances and success. Harrison (1994) has also stressed the difficulty of identifying all local populations and suitable habitat and the problem of estimating extinction and colonization rates among patches.

Although still controversial (see Harrison et al. 1993), the metapopulation concept does provide a useful framework for describing the spatial structure of real populations. The concept, after all, is grounded on two of the most robust empirical generalizations in ecology and conservation biology: extinction rates decline with increasing population size, and immigration and recolonization rates decline with increasing isolation (MacArthur and Wilson 1967; Hanski 1994). Our goal in this chapter is to illustrate how these two generalizations can be used to characterize quantitatively the metapopulation structure of a species and to develop "landscape rules" for conserving metapopulations of a declining species.

We begin with the results of a rangewide survey of the Florida scrub jay (*Aphelocoma coerulescens*) conducted in 1992 and 1993. This species is patchily distributed and thus presents a challenging case study for describing metapopulation structure. We offer a method for doing so using a detailed spatial database combined with existing biological information and GIS technology. The technique uses computer-generated buffers, at several distances reflecting the dispersal behavior of the species, to delineate subpopulations with differing degrees of connectivity. Extinction vulnerability of each subpopulation is estimated via a population viability analysis (PVA) model—in our case, that of Fitzpatrick et al. (1991). We propose a simple nomenclature for classifying Florida scrub jay metapopulations based on subpopulation size and connectivity. We conclude by deriving a few metapopulation-based landscape rules that may be incorporated into a statewide framework for conservation plans affecting this rapidly declining species.

## The Statewide Survey

The Florida scrub jay (Figure 9.1), Florida's only endemic bird species, is a disjunct, relict taxon separated by more than 1600 km from its closest western relatives (Woolfenden and Fitzpatrick 1984). This habitat specialist is restricted to a patchily distributed scrub community found on sandy, infertile soils—mostly pre-Pleistocene and Pleistocene shoreline deposits. The vegetation is dominated by several species of low-stature scrub oaks (*Quercus* spp.). Jays rely heavily on acorns for food, especially during the winter, when

Figure 9.1. A Florida scrub jay on sandy substrate typical of its habitat.

they retrieve the thousands of acorns cached in open, sandy areas during the fall (DeGange et al. 1989). Florida scrub jays show a strong preference for low, open habitats with numerous bare openings and few or no pine trees (Breininger et al. 1991). These optimal habitat conditions are maintained by frequent fires (Abrahamson et al. 1984). Jays living in fire-suppressed, over-grown habitats have much poorer demographic performance than jays in optimal conditions (Fitzpatrick and Woolfenden 1986), leading rapidly to local extirpation unless the habitat is burned (Fitzpatrick et al. 1994).

Florida scrub jays are monogamous, cooperative breeders that defend permanent territories averaging 10 ha per family (Woolfenden and Fitzpatrick 1984). They have a well-developed sentinel system: a family member watches for predators while others in the group engage in other activities such as foraging (McGowan and Woolfenden 1989). Young nearly always delay dispersal for at least a year, remaining at home as helpers. Dispersal distances from natal to breeding territories are extremely short for both sexes, averaging less than one territory for males and three and a half territories for females within contiguous habitat (Woolfenden and Fitzpatrick 1984). Dispersal behavior is associated with greatly elevated mortality even within optimal habitat (Fitzpatrick and Woolfenden 1986), and many behavioral adaptations (cooperative breeding, sentinel system, delayed dispersal) suggest that predation is extremely important to both resident and dispersing jays (Koenig et al. 1992).

## Methods

The Florida scrub jay was listed by the U.S. Fish and Wildlife Service (USFWS) as a threatened species in 1987. In 1991 the USFWS notified landowners and county governments that clearing scrub could violate the Endangered Species Act (U.S. Fish and Wildlife Service 1991). At the same time, the USFWS began encouraging counties to develop regional habitat conservation plans (HCP) that could solve local permitting problems by means of a single, biologically based regional plan. To aid in this process, the USFWS partially sponsored the authors, and their cooperators, to conduct an intensive survey during 1992 and 1993 to document the range and sizes of subpopulations throughout the state and inventory potential habitat, whether occupied or not.

Our methods were similar to those used for the northern spotted owl (*Strix occidentalis caurina;* Murphy and Noon 1992). We had extensive information to guide us. Cox (1987) had documented numerous jay localities throughout the state in the early 1980s and had compiled historic records from diverse sources such as museums and Christmas bird counts. The Florida Breeding Bird Atlas (Kale et al. 1992) provided valuable data on jay sightings made by hundreds of volunteers from 1986 to 1991. Virtually all previously known jay localities were revisited for this survey. Information from the public was solicited through notices in magazines, newsletters, and newspapers. Other potential habitat patches were identified from U.S. Soil Conservation Service maps and aerial photographs, on which the white, sandy soil associated with jay habitat forms a distinctive signature.

We used standard surveying techniques based on tape playback of jay territorial scolds (Fitzpatrick et al. 1991) to locate jays in habitat patches. Location and number of individuals in each group were plotted on field maps. The following qualitative habitat data were collected at most patches: occupancy by jays, estimated degree of vegetative overgrowth (1–4 scale), extent of human disturbance (1–4 scale), and ownership status with respect to permanent protection from development. Time constraints prevented us from making quantitative habitat measurements at the thousands of habitat patches we visited. Although our survey goals included attempting to find all known jay families outside of federal lands, we know that a few jays were missed because of limited access to certain private lands. The total number missed, however, is not likely to exceed a few percent of the statewide population.

Federally owned lands were not surveyed for this project. Those with large populations of jays include Cape Canaveral Air Force Station, Merritt Island National Wildlife Refuge, Canaveral National Seashore, and Ocala National Forest. Florida scrub jays at each of these areas are currently under study, so for the statewide summary we used estimates of numbers and locations of jays provided by the biologists conducting those studies.

To archive, map, and analyze the statewide data, we developed a series of map layers by means of a GIS at Archbold Biological Station. PCs and Sun computers running ARC/INFO were used to input all GIS data (ESRI 1990). Habitat patches (both occupied and unoccupied) and jay locations, originally hand-drawn on soil maps or topographic maps (usually at 1:24000 scale), were digitized. Patch characteristics and jay family sizes were entered into accompanying data files. Map layers included current and historic range of jays, current distribution of suitable and potential habitat, and locations and numbers of jay families encountered.

### Results

We estimate that as of 1993 the total population of Florida scrub jays consisted of about 4000 pairs (Figure 9.2; Fitzpatrick et al. 1994). Both total numbers and overall geographic range, however, have decreased dramatically during this century (Cox 1987). In recent decades the species has been extirpated from 10 of 39 formerly occupied counties, and it is now reduced to fewer than ten pairs in five additional counties. (Detailed tabulations are given in Fitzpatrick et al. in press.) Detailed site-by-site comparison of our survey with Cox's (1987) study suggests that the species may have declined as much as 25 to 50 percent during the last decade alone.

The degraded quality of many currently occupied habitat patches suggests that further substantial declines in the jay population are inevitable. Specifically, those jays occupying suburban areas (approximately 30 percent of all territories) are unlikely to persist as these suburbs continue to build out, given the rapid rate at which Florida's human population continues to expand. Furthermore, jays living in fire-suppressed, overgrown habitat (at least 2100 families, or 64 percent of all occupied scrub patches by area) already are likely to be experiencing poor demographic performance (Fitzpatrick and Woolfenden 1986). These jays can be expected to decline further unless widespread restoration of habitat is begun soon.

## A Method for Classifying Metapopulations

The patchy distribution and variable clustering of territories throughout the range of the Florida scrub jay (Figure 9.2) challenge us to expand on traditional metapopulation concepts to describe the spatial structure of this species. In this section we describe our conceptual approach; in the next we apply it to the Florida scrub jay data.

50    0    50
km

Figure 9.2. Distribution of Florida scrub jay groups (small black circles) in 1993. Note the discontinuous distribution and variability in patterns of aggregation.

Harrison's (1991) four classes of metapopulations can be presented graphically (Figure 9.3) as different regions on a plot of degree of isolation against patch size distribution. Thus Harrison's "nonequilibrium" metapopulation is that set of small patches in which each has a high probability of extinction and among which little or no migration occurs. Local extinctions are not offset by recolonization, resulting in overall decline toward regional extinction. The "classical" model developed by Levins (1969, 1970) is a set of small patches that are individually prone to extinction but large enough and close enough to other patches that recolonization balances extinction. "Patchy" metapopulations consist of patches so close together that migration among them is frequent; hence the patches function over the long run as a continuous demographic unit. Finally, the "mainland–island" model has a mixture of large and small patches close enough to allow frequent dispersal from an extinction-resistant mainland to the extinction-prone islands. The lower right-hand side of Figure 9.3 portrays two classes not presented by Harrison (1991). These large patches are either poorly connected ("disjunct") or moderately connected ("mainland–mainland"). Such large populations tend to be of less concern from a conservation standpoint because they are essentially invulnerable to extinction.

Classifying metapopulations, therefore, requires species-specific information on both connectivity (that is, dispersal behavior and barriers) and extinction probabilities (that is, population sizes in patches) across space. A system of small habitat patches might appear to support stable populations of certain species as classical or patchy systems, for example, while other species might be nonequilibrial in the same system because of low density (hence small population sizes) or limited dispersal ability.

Harrison's (1991) diagram of metapopulations represents connectivity among patches by means of a dashed line around those among which dispersal is frequent enough to "unite the patches into a single demographic entity." This boundary can be viewed as a dispersal buffer—an isoline of equal dispersal probability. Any number of patches may be included within a given dispersal buffer of a single subpopulation, provided that fragmentation is sufficiently fine-grained (Rolstad 1991). But dispersal probability normally diminishes continuously (even if steeply) away from a patch, and for most terrestrial species it asymptotically approaches zero at some point farther away than Harrison's single, discrete dispersal buffer. Therefore, we extend Harrison's diagrammatic approach by adding a second buffer to delineate the distance beyond which dispersal is effectively reduced to zero. We maintain that this second buffer functionally identifies separate metapopulations. Although we acknowledge that connectivity should be represented graphically as a continuous surface of dispersal probabilities, discrete boundaries, placed

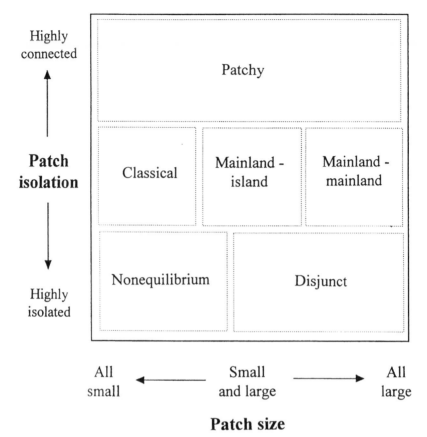

**Patch size**

Figure 9.3. Classification scheme showing different types of metapopulations based on patch size distribution (patches all small in size, mixture of small and large, and all large in size) along the horizontal axis and degree of patch isolation (highly connected to highly isolated) on the vertical axis. Nonequilibrium, classical, mainland–island, and patchy classes are named according to Harrison (1991).

at biologically meaningful (and empirically determined) distances, greatly simplify the description of metapopulations. They also provide explicit, repeatable methodology for comparison or modeling.

Harrison's metapopulation types may be characterized using these two buffers (Figure 9.4). In patchy systems (Figure 9.4*a*), every patch belongs to the same subpopulation, so they are all enclosed within a single, inner dispersal buffer. Classical systems (Figure 9.4*b*) have small subpopulations separately encircled, representing the fact that each may go extinct temporarily or may be rescued before going extinct (Brown and Kodric-Brown 1977) by way

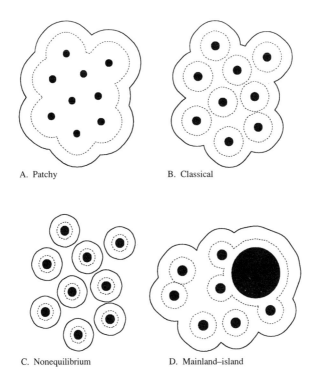

A. Patchy                                 B. Classical

C. Nonequilibrium              D. Mainland–island

Figure 9.4. Schematic depiction of different kinds of metapopulations illustrating use of dispersal-distance buffers to predict recolonization rates among subpopulations. Thin dotted lines separate functional subpopulations based on frequency of dispersal beyond them. Solid lines separate metapopulations based on poor likelihood of dispersal among them. **A.** Patchy metapopulation. All patches are enclosed by a single thin dotted line, indicating that they are sufficiently close to function as a single subpopulation. The thick solid line shows the outer boundary of this patchy metapopulation. **B.** Classical metapopulation. Thin dotted lines enclose small subpopulations (islands) that are extinction-prone but are sufficiently close to neighboring islands for recolonization to occur. All islands are within a single metapopulation, represented by the thick solid line enclosing all patches. **C.** Nonequilibrium metapopulations. Each patch is enclosed by a separate thick solid line, indicating that each functions as a separate metapopulation due to the large interpatch distances. Each metapopulation is highly extinction prone due to its small size. **D.** Mainland–island metapopulation. A large mainland-size patch and several small patches are enclosed by a thin dotted line, indicating that they function as a single subpopulation. Four other small patches are enclosed by separate, thin dotted lines, indicating that they function as separate subpopulations. All patches are within a single metapopulation, represented by the thick solid line.

of colonization from another subpopulation enclosed within the outer buffer. The simplest nonequilibrium systems (Figure 9.4*c*) are represented as bull's-eyes around small, isolated subpopulations. A mainland–island metapopulation (Figure 9.4*d*) has a large subpopulation and several small ones within a single outer buffer.

The important point is this: even more complex patterns may be common in nature, arising from combinations or intermediate cases, and many of these are not easily fit into Harrison's (1991) four metapopulation classes. To deal with such complications, we suggest characterizing metapopulations by describing the sizes of their constituent subpopulations. We propose a simple nomenclature based on three key words—"island," "mainland," and "midland"—to characterize the relative sizes of the subpopulations within a metapopulation. Subpopulations small enough to be highly extinction-prone in the absence of significant immigration are called islands. Those large enough to be essentially invulnerable to extinction are called mainlands. Intermediate-sized subpopulations are neither extinction-prone nor invulnerable to extinction. For lack of a better term, we refer to an intermediate-sized subpopulation as a midland.

Distinctions among these categories need not be completely arbitrary. Species-specific PVA provides an explicit, quantifiable approach for describing subpopulations as extinction-prone, extinction-vulnerable, or extinction-resistant. Our introduction of the midland category helps clarify the importance of turnover, which has been called the "hallmark of genuine metapopulation dynamics" (Hanski and Gilpin 1991). Specifically, turnover is expected in systems with island-sized subpopulations because they have high frequencies of extinction. But systems with midlands rather than islands are perhaps more often characterized by rescue than by recolonization because local extinctions will be rare. Thus, a system of midlands may exhibit little or no turnover even though no real mainlands are present, whereas a system of islands with the same degree of isolation may show high turnover. We agree with Sjögren (1991) in emphasizing the importance of rescue in metapopulation dynamics. Traditional emphasis on turnover probably resulted from the fact that rescue is much more difficult to measure empirically, as turnover requires only presence/absence data.

Harrison's metapopulation classes can be described in terms of this island–midland–mainland nomenclature as follows: a nonequilibrium metapopulation is a system of one or more islands (extinction-prone subpopulations) with a total population size too small to persist. A classical metapopulation is a system of island-sized subpopulations large enough and close enough together and of sufficient total size to allow persistence. Any system containing a midland or mainland (by definition) cannot be a nonequilibrium or classical metapopulation because all subpopulations in the latter systems are extinction-prone. A patchy metapopulation is a set of patches close enough together to form a single subpopulation of sufficient size to persist (that is, a midland or mainland). Mainland–island metapopulations are selfexplanatory.

Explicit reference to midlands—extinction-vulnerable patches of intermediate population size—produces metapopulation types not described in Harrison (1991). Systems with several midlands, for example, or a mainland with several midlands, are possible. We illustrate some of these configurations by applying our nomenclature, quantitatively, to the Florida scrub jay.

## Metapopulation Structure of the Florida Scrub Jay

Application of the foregoing scheme to a species requires choosing two dispersal buffer distances and two threshold values for extinction vulnerability among single populations. Here, for the Florida scrub jay, we based each of these values on empirically gathered biological data. Buffer distances were derived from long-term field studies of marked individuals and from information garnered on the statewide survey regarding occupancy of habitat patches at various distances from source populations. Extinction vulnerability was estimated using a single-population viability model (Fitzpatrick et al. 1991). We then chose thresholds to delineate islands, midlands, and mainlands, much as Mace and Lande (1991) used extinction probabilities to propose IUCN threatened species categories.

### Dispersal Distances

Between 1970 and 1993 we documented 233 successful natal dispersals from the marked population under long-term study at Archbold Biological Station (Figure 9.5; see also Woolfenden and Fitzpatrick 1984, 1986). Unlike the situation for most field studies of birds (such as Barrowclough 1978), many characteristics of our study and the behavior of jays themselves enhance our ability to locate dispersers that leave the main study area. Once established as breeders, for example, Florida scrub jays are long-lived and completely sedentary. Furthermore, we have mapped in detail all scrub habitat within the local range of the species, and we census these tracts periodically in search of dispersed jays. (Such censuses reveal remarkably few banded dispersers among the many hundreds of jays encountered.) Because banded Florida scrub jays from our study usually are tame to humans, both our own searches and casual encounters by local homeowners have high likelihood of exposing any off-site dispersers to us once they become paired on a territory. Indeed, if we assume

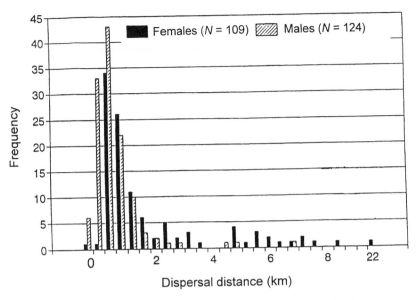

Figure 9.5. Frequency distribution of dispersal distances from natal to breeding territories for color-banded Florida scrub jays at Archbold Biological Station, 1970–1993. About 85 percent of documented dispersals were within 3.5 km, and 99 percent were within 8.3 km. The longest documented dispersal was 35 km.

that immigration and emigration rates are about equal in our study area, evidence suggests that we have succeeded in locating all but a few of the jays that have departed over the 25-year period of our study. Therefore, although some dispersers do escape our detection, our observed dispersal curve (Figure 9.5) can be only marginally biased toward the shorter distances.

About 80 percent of the documented dispersals were within 1.7 km of the natal territory, 85 percent within 3.5 km, 97 percent within 6.7 km, and 99 percent within 8.3 km (Figure 9.5). Data from field studies elsewhere in Florida reveal the same, remarkably sedentary dispersal behavior. The longest dispersal so far documented was a female we discovered pairing 35 km from her natal territory at Archbold in 1994. All the dispersals we have documented around Archbold, including the longest one, involved jays that had moved either through continuous habitat or across gaps no greater than 5 km. To test the generality of this observation, we pooled dispersal information from the seven other biologists currently color-banding Florida scrub jays around the state (D. Breininger, R. Bowman, G. Iverson, R. Mumme, P. Small, J. Thaxton, and B. Toland; unpublished data). Their studies, along with ours, cumulatively have produced about a thousand banded non-breeders that achieved dispersal age (Fitzpatrick et al. in prep.). Collectively these studies have documented only about ten dispersals of 20 km or more, and only a few of these had crossed habitat gaps as large as 5 km. More important in the present context, despite ample opportunity to observe longer-distance movements, not a single example yet exists of a banded Florida scrub jay having crossed more than 8 km of habitat that does not contain scrub oaks. We suspect that this distance is close to the biological maximum for the species.

## Patch Occupancy

The observations just outlined suggest that for habitat specialists such as the Florida scrub jay, dispersal curves measured in relatively contiguous habitat may overestimate the dispersal capabilities of individuals across fragmented systems. Direct behavioral observations strongly indicate that Florida scrub jays resist crossing large habitat gaps. Still, few opportunities exist to observe jays in the act of dispersing; hence the theoretical maximum dispersal distance (that is, the outer dispersal buffer) is extremely difficult to establish directly.

Seeking an indirect measure of dispersal frequencies across habitat gaps, we examined patch occupancy statewide as documented by our 1992 to 1993 survey. We used FRAGSTATS software (McGarigal and Marks 1994) to measure distances between each occupied patch of scrub habitat to its nearest occupied patch. We then measured by hand (FRAGSTATS cannot measure distances between patches of different attributes) the distances between each *unoccupied* suitable patch and the nearest *occupied* patch. For each distance class, the ratio of the count of the occupied-to-occupied distances to the total number of nearest neighbor distances (the sum of occupied-to-occupied and unoccupied-to-occupied distances) yields the proportion of patches occupied at that distance from occupied habitat.

Presumably, declines in patch occupancy with increasing distance to the nearest occupied habitat (Figure 9.6) reflect diminishing recolonization rates following local extinctions. Occupancy remains above zero even at great distances, probably because larger isolated patches rarely experience extinction. This curve provides an empirical approach for delineating subpopulations and metapopulations: a subpopulation buffer is the maximum interpatch distance where occupancy rates remain high; the metapopulation buffer is the smallest interpatch distance where occupancy rates reach their minimum.

For Florida scrub jays (Figure 9.6) patch occupancy is about 90 percent to at least 2 km from a source; it then declines monotonically to around 15 percent at 12 km. (Sample size of isolated patches decreased rapidly beyond 16 km, necessitating the lumping of classes at the larger distances.) We infer from this occupancy curve that successful recolonization is a rare event beyond about 12 km from an occupied patch of habitat. We use this distance to identify metapopulations that have become essentially demographically

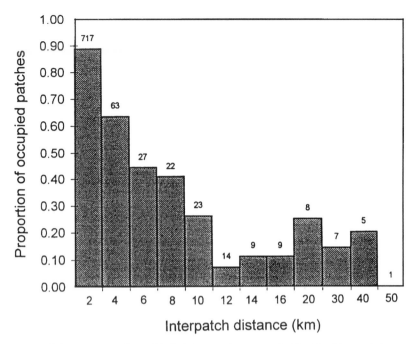

Figure 9.6. Proportion of suitable habitat patches occupied by Florida scrub jays as a function of their distance to the nearest separate patch of occupied habitat. Occupancy rates are high (nearly 90 percent) for patches up to 2 km apart and decline monotonically to 12 km. Note the scale change beyond 16 km.

independent from one another (that is, the outer dispersal buffer).

We selected the distance of 3.5 km (about 2 miles) as an inner dispersal buffer to delineate subpopulations. We chose this figure for several reasons. First, behavioral information from a variety of sources, including radio-tracking data (B. Stith, unpublished data), shows that jays begin to show reluctance to crossing habitat gaps at about this size (and at much smaller gaps where open water or closed-canopy forest are involved). Second, known dispersals of many banded jays included habitat gaps up to 3.5 km, but their frequency declines dramatically thereafter. Third, the observed dispersal curve from Archbold (Figure 9.5) shows that in good habitat more than 85 percent of dispersals by females, and fully 97 percent of those by males, are shorter than 3.5 km. And fourth, patch occupancy data (Figure 9.6) show significant decline in colonization rates at distances above 3.5 km.

## Population Viability Analysis

PVA based on a simulation model incorporating demographic (but not genetic) stochasticity and periodic, catastrophic epidemics (Fitzpatrick et al. 1991; Woolfenden and Fitzpatrick 1991) provided a quantitative method for defining boundaries along the island (extinction-prone), midland (vulnerable), and mainland (extinction-resistant) continuum. (But see Taylor 1995.) Among the several methods for expressing extinction vulnerability (for example, Burgman et al. 1993; Boyce 1992; Caughley 1994) we elect the simple approach of specifying time-specific probability of persistence of populations of a given size.

Our model results indicate that a population of jays with fewer than 10 breeding pairs has about a 50 percent probability of extinction within 100 years, whereas a population with 100 pairs has a 2 to 3 percent probability of extinction in the same time period. These two population sizes—10 and 100 pairs—provide convenient and biologically meaningful values by which to classify subpopulations as islands (<10 pairs), midlands (10–99 pairs), and mainlands (>99 pairs). Although subjectively chosen, these values effectively separate population sizes having fundamentally different levels of protection. These values also receive empirical support from several long-term bird studies (reviewed by Thomas 1990; Thomas et al. 1990; Boyce 1992).

### Metapopulation Structure

We used a GIS buffering procedure (ESRI 1990) to generate dispersal buffers around groups of jays occurring within 3.5 km (for subpopulations) and 12 km (for metapopulations) of each other (Figure 9.7). We buffered jay territories rather than habitat patches because we strongly suspect that dispersing scrub jays cue on the presence of other, resident jays even more strongly than on habitat, so the functional boundaries of occupied patches may be determined by where jay families exist. We modified the resulting buffers in the following areas to reflect the presence of hard barriers to dispersal in the form of open water with forested margins: Myakka River, Peace River, St. Johns River, St. Lucie River, and Indian River Lagoon.

Using a 3.5-km dispersal buffer, we delineated 191 separate Florida scrub jay subpopulations (Figure 9.7). Over 80 percent of these ($N = 152$ islands) are smaller than ten pairs (Figure 9.8), and 70 of them consist of only a single pair or family of jays. Only six subpopulations contain at least 100 pairs (mainlands), leaving 33 midlands (10–99 pairs).

Using a 12-km dispersal buffer, we delineated 42 separate Florida scrub jay metapopulations (Figure 9.7). Again, most of them are small (Figure 9.9). We tabulated the number and type of subpopulations within each metapopulation (Table 9.1) and noted how each metapopulation fits into Harrison's (1991) scheme. Exactly half (21) contain fewer than ten pairs, thereby constituting nonequilibrium systems. Along the north-central Gulf Coast (Figure 9.10), for example, a group of nonequilibrium systems coincides with a heavily developed area containing a burgeoning human population. Only three Florida scrub jay systems qualify as classical metapopulations

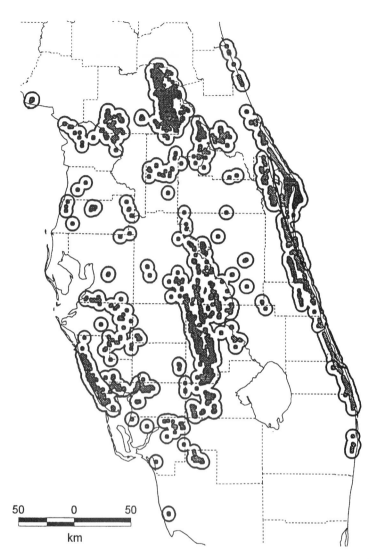

50　　0　　50
km

Figure 9.7. Statewide Florida scrub jay distribution map with dispersal buffers. Shaded areas depict subpopulations of jays within easy dispersal distance (3.5 km) of one another. Thick outer lines delineate demographically independent metapopulations separated from each other by at least 12 km.

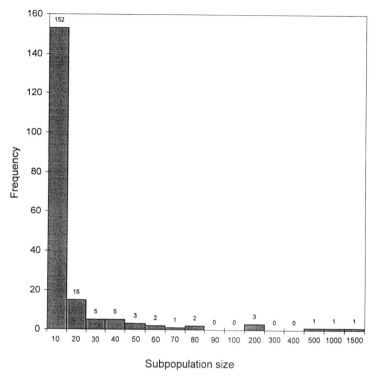

Figure 9.8. Frequency distribution of Florida scrub jay subpopulation sizes. Numbers above the bars indicate the number of jay pairs. Only three subpopulations have more than 400 jay pairs (core populations). Only six subpopulations have more than 100 pairs.

(containing only islands but large enough to support one another following extinctions; Figure 9.11). Another three systems represent patchy metapopulations (containing a single, fragmented subpopulation large enough for long-term persistence).

Five systems approximate mainland–island metapopulations, but each of these examples also contains at least one midland population (for example, the large Lake Wales Ridge system, with 1 mainland, 10 midland, and 39 island populations; Figure 9.12). These five mainland–midland–island systems, plus nine midland–island and one mainland–midland system, do not fit neatly any of Harrison's (1991) metapopulation classes. A total of 33 midland populations exist (mean size = 30.7), and these occur in 18 of the 42 separate metapopulations. Excluding the nonequilibrium systems, true islands are present in 17 systems. Therefore, assuming that dispersal is not inhibited by habitat loss expanding the distances among patches, rescue (on midlands)

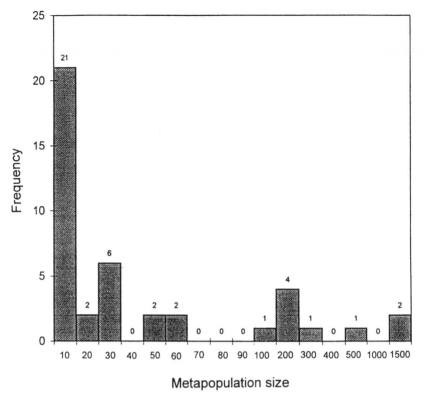

Figure 9.9. Frequency distribution of Florida scrub jay metapopulation sizes. Numbers above the bars indicate the number of jay pairs. Note that 21 metapopulations have ten pairs or less of jays. These represent nonequilibrium metapopulations.

may be at least as important as turnover (on islands) in Florida scrub jay metapopulation dynamics.

Use of empirically derived dispersal buffers and extinction probabilities provides an explicit method for quantitatively describing metapopulation structure. Application of this technique to the Florida scrub jay demonstrates that a species can exhibit a variety of metapopulation patterns across its range. Patterns of aggregation and isolation do not conform to a single metapopulation class in the Florida scrub jay. Such complex spatial structure is probably common in nature, particularly among species with large and widely dispersed populations restricted to a patchy habitat. Such patterns may be further complicated by perturbations of the natural system caused by humans.

### Caveats

We offer several caveats regarding the generality of dispersal-buffer methodology in conservation:

TABLE 9.1.

Summary of 42 Florida scrub jay metapopulations

| Metapopulation type (after Harrison 1991) | Metapopulation type (mainland, midland, island)[a] | | | Size (pairs) | Number of subpopulations |
|---|---|---|---|---|---|
| Mainland–island (?) | Mn | 10Md | 39I | 1247 | 50 |
| | Mn | Md | 5I | 1036 | 7 |
| | Mn | Md | 5I | 466 | 7 |
| | Mn | 2Md | 6I | 237 | 9 |
| | Mn | Md | 5I | 179 | 7 |
| Unknown | Mn | Md | | 126 | 2 |
| | | 4Md | 11I | 120 | 15 |
| | | 2Md | I | 103 | 3 |
| | | Md | 5I | 94 | 6 |
| | | Md | I | 58 | 2 |
| | | 2Md | 3I | 55 | 5 |
| | | Md | I | 50 | 2 |
| | | Md | 2I | 29 | 3 |
| | | Md | 3I | 22 | 4 |
| | | Md | 2I | 18 | 3 |
| Patchy | | Md | | 26 | 1 |
| | | Md | | 22 | 1 |
| | | Md | | 15 | 1 |
| Classical | | | 16I | 49 | 16 |
| | | | 10I | 24 | 10 |
| | | | 6I | 21 | 6 |
| Nonequilibrium | | | 3I | 5 | 3 |
| | | | 3I | 3 | 3 |
| | | | 2I | 7 | 2 |
| | | | 2I | 2 | 2 |
| | | | 2I | 2 | 2 |
| | | | 2I | 3 | 2 |
| | | | 2I | 3 | 2 |
| | | | 2I | 2 | 2 |
| | | | I | 6 | 1 |
| | | | I | 2 | 1[b] |
| | | | I | 1 | 1[c] |

[a] Numerical prefix indicates number of mainlands (Mn), midlands (Md), and islands (I). See the text for nomenclature.

[b] There was a total of four island subpopulations composed of two pairs in one subpopulation.

[c] There was a total of eight island populations composed of a subpopulation of one pair.

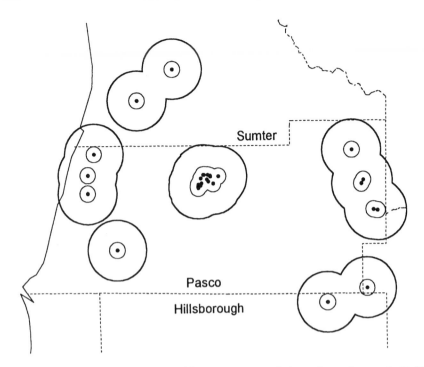

Figure 9.10. Examples of nonequilibrium metapopulations from the north Gulf Coast of Florida. Each of the six metapopulations contains fewer than ten pairs of jays, except for the centrally located system that contains a single, midland-sized sub-population.

1. The technique is best suited for species occupying discrete territories, home ranges, or habitat patches amenable to mapping.
2. The technique is predicated on having a comprehensive survey. Missing data can lead to misleading results, especially regarding connections among metapopulations or subpopulations.
3. The technique presents a static, snapshot view of metapopulations. It does not easily reveal important dynamics among subpopulations, such as those obtainable from a spatially explicit population model. The viability of different configurations is best determined from such models rather than single-population PVAs.
4. Populations in decline or in "sinks" can present an overly optimistic picture (Thomas 1994). Indeed, we suspect that many of the island and midland subpopulations of Florida scrub jays currently are failing to replace themselves demographically, as a result of habitat degradation due to fire suppression. Similarly, there may be abnormally high densities due to the "crowding effect" (Lamberson et al. 1992) following recent habitat losses.

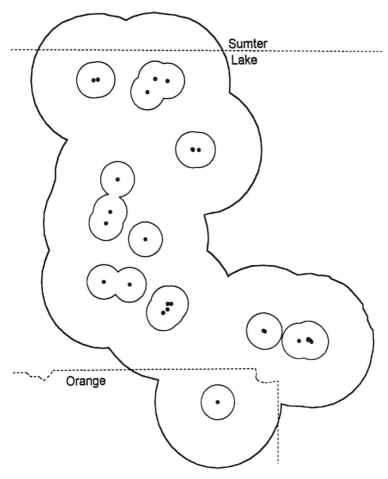

Figure 9.11. Example of a "classical" metapopulation from three counties in central Florida. Note the occurrence of jays in small islands of intermediate distance from one another.

5. The technique relies on numerous simplifying assumptions about dispersal behavior in defining connectivity among patches. Most important, it assumes random movement between patches, equal traversability of interpatch habitats, absence of dispersal biases owing to habitat quality differences at the origin or the destination, and absence of density dependence in behavior. More elaborate applications, of course, could incorporate alternative assumptions about these and other factors.

Another important consideration is the kinds of data to buffer. To create biologically meaningful—but very different—descriptions for the Florida

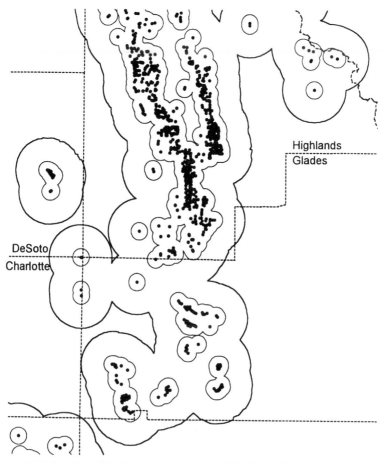

Figure 9.12. Portion of the largest mainland–midland–island metapopulation in interior Florida, consisting of the Lake Wales Ridge and associated smaller sand deposits. The large central subpopulation (enclosed by the thin black line) contains nearly 800 pairs of jays. Small subpopulations to the south and east are within known dispersal distance of the large, central mainland. A small metapopulation to the west (in DeSoto County) contains a single subpopulation of 21 territories. This small system qualifies as a patchy metapopulation, since jays occur in two or more patches, but the patches are so close together that they function as a single demographic unit.

scrub jay, we could have buffered around jay territories (our choice), occupied patches, suitable habitat patches both occupied and unoccupied, or all scrub habitat patches regardless of current suitability. Animals such as Florida scrub jays that are reluctant to become established in unoccupied, suitable habitat (for example, Ebenhard 1991), or that have high conspecific attraction or an "Allee effect" (Smith and Peacock 1990), are best buffered around actual territories or occupied patches. This is because unoccupied sites have a low probability of becoming occupied regardless of their degree of isolation and hence contribute little to the current metapopulation dynamics of the species. Excellent colonizers of empty habitat, however, or species adept at long-distance dispersals via unoccupied stepping stones, probably should be buffered around all habitat patches.

## Conserving Florida Scrub Jay Metapopulations

Human-induced fragmentation and habitat loss already have split the Florida scrub jay into numerous metapopulations that are now effectively isolated from one another. Further habitat loss will have the inevitable effect of driving each metapopulation down an ever-steepening gradient of endangerment: mainland–midland configurations will become midland–island ones; classical configurations will become nonequilibrium ones; and these, in turn, are headed for extinction. At different stages of this process, conservation strategies should vary. For systems still containing mainlands, preserving the mainlands usually overrides other concerns because these large subpopulations have the greatest role in the system's persistence. As mainlands are lost and subpopulations shift toward a classical configuration, conservation emphasis should shift from maintaining area to maintaining connectivity. In this phase, priority should be placed on preserving centrally located patches and minimizing distances among patches to maintain opportunities for recolonization or rescue (Hanski 1994). As both size and connectivity become problematic (approaching nonequilibrium configuration), drastic measures are appropriate—intensive habitat restoration, for example, perhaps coupled with translocation and reintroduction as a substitute for natural dispersal. To stave off the Florida scrub jay's current slide down the endangerment gradient, we propose a set of landscape rules that can be applied to each metapopulation independently.

*Rule 1: Preserve the cores.*    Three large, geographically separate subpopulations still have sufficient size as of 1993 (Figure 9.8) to be highly invulnerable to extinction except in the face of a major catastrophe. Each of these "core" subpopulations contains more than 400 pairs, corresponding to our lowest category of extinction probability (Fitzpatrick et al. in prep.). Each should be prevented from declining below 400 pairs. Moreover, habitat protection should be undertaken to ensure that these large subpopulations are not split into two or more smaller ones (by the creation of new dispersal gaps greater than 3.5 km). Two of these core populations occur on federal land (Ocala National Forest; Merritt Island and Cape Canaveral), and the third is largely on private land in the southern part of the Lake Wales Ridge. We emphasize that even these core populations are not invulnerable to extinction. Epidemics among Florida scrub jays are known to occur, and they can be severe (Woolfenden and Fitzpatrick 1991). Furthermore, the entire Merritt Island–Cape Canaveral population exists only a few meters above sea level, and the effects of a large hurricane or sea level changes could be devastating (for example, Hooper et al. 1990).

*Rule 2: Preserve all potentially viable metapopulations.* The most effective long-term insurance against extinction is to make every effort to spread the risk of catastrophe as widely as possible. To accomplish this, all remaining metapopulations for which habitat exists to support at least ten jay pairs should be preserved. Certain nonequilibrium metapopulations—those with few remaining jays and lacking restorable habitat—probably are not viable in the long run. These may not warrant expensive conservation efforts unless they have special genetic uniqueness or geographic or educational importance (for example, Lesica and Allendorf 1995). Permitting the ultimate loss of these metapopulations should, of course, be accompanied by commensurate mitigation measures carried out in more viable metapopulations. Although our focus here is on Florida scrub jays, it should be pointed out that the jay co-occurs with numerous narrowly adapted, range-restricted scrub endemics (for example, Christman and Judd 1990). Many of these species will simultaneously be preserved if the full jay distribution is maintained; others, however, may require habitat preservation in areas deemed nonviable for the jay.

*Rule 3: Preserve or enhance existing persistence probabilities.* In certain metapopulations, subpopulations cannot tolerate further reductions without destabilizing the system. In others, however, habitat loss can be tolerated without drastically raising the overall probability of extinction. For example, each of the three largest subpopulations that are not "cores" contain over 200 breeding pairs. Subpopulations of this size, if reduced by as much as 50 percent, would incur only a minor increase in long-term extinction probability if the remaining habitat became secure (Woolfenden and Fitzpatrick 1991). As a rule of thumb, we suggest that any further reductions of jay numbers within viable metapopulations be limited to no more than 33 percent of their 1993 numbers and that all habitat loss be accompanied by measures that would protect twice the acreage of extant scrub. Furthermore, we suggest that no metapopulation be allowed to fall below ten pairs of jays. Taken to their ends, these measures alone would ensure that more than two-thirds of the extant jay population would end up under protection.

*Rule 4: Prohibit the splitting of metapopulations.*    Habitat gaps larger than 12 km represent barriers to natural dispersal and recolonization. To maintain all existing metapopulations, therefore, all habitat gaps must be kept well below this 12-km threshold. Failure to do so would split the system and create two smaller, hence less viable, systems. Because coastal populations of Florida scrub jays are distributed in narrow strips parallel to the coastline (dune and shoreline deposits), they are especially vulnerable to being split as a result of elimination of small habitat patches.

*Rule 5: Maintain connectivity within metapopulations.*    Because of their small size, islands (<10 pairs) and midlands (10–99 pairs) require periodic colonization from other occupied tracts to persist. However, occupancy of scrub patches falls below 40 percent if they are isolated from other occupied habitat by as little as 8 km (Figure 9.6). Therefore, habitat patches should be maintained, properly managed, and restored as necessary to avoid isolating any existing island or midland from other subpopulations by distances exceeding 8 km.

In regions with substantial numbers of jays, these few landscape rules would enhance long-term persistence of jays while also providing considerable flexibility for continued human development. If planned according to these rules and accompanied by proper regionally coordinated mitigation, destruction of certain habitat tracts could take place without significant further jeopardy to the Florida scrub jay. In areas with only a few jays remaining, these same landscape rules often will prohibit any further loss of habitat unless accompanied by extensive restoration.

These landscape rules have been applied recently in Brevard County, Florida, where habitat-conservation planning has been undertaken using a countywide approach (O'Connell 1994). Implementation of that plan would provide significant protection for the Florida scrub jay (and other scrub species) while eliminating the economic risks and delays otherwise faced by developers individually seeking USFWS permits to clear habitat occupied by Florida scrub jays.

## Acknowledgments

For stimulating discussions about jays and landscape rules, we are indebted to many colleagues engaged in the study of Florida scrub jays and their conservation, especially Reed Bowman, David Breininger, Grace Iverson, Michael O'Connell, Parks Small, Hilary Swain, Jon Thaxton, and Brian Toland. The following people contributed data for the statewide survey: Reed Bowman, Dave Breininger, Jack Dozier, Florida Game and Freshwater Fish Commission personnel, Grace Iverson, David McDonald, Ron Mumme, Ocala National Forest personnel, Hilary Swain, Jon Thaxton, and Brian Toland. We thank David Wesley and Dawn Zattau, of the U.S. Fish and Wildlife Service, who provided lead funding for the statewide survey and helped stimulate our discussions of habitat-conservation planning. Finally, we thank Jeff Walters, Dale McCullough, and an anonymous reviewer for helpful comments. This work was partially supported by grants from the National Science Foundation (BSR-8705443, BSR-8996276, BSR-9021902).

# REFERENCES

Abrahamson, W. G., A. F. Johnson, J. N. Layne, and P. A. Peroni. 1984. Vegetation of the Archbold Biological Station, Florida: An example of the southern Lake Wales Ridge. *Florida Scientist* 47:209–250.

Barrowclough, G. F. 1978. Sampling bias in dispersal studies based on finite area. *Bird-Banding* 49:333–341.

Boyce, M. E. 1992. Population viability analysis. *Annual Review of Ecology and Systematics* 26:481–506.

Breininger, D. R., M. J. Provancha, and R. B. Smith. 1991. Mapping Florida scrub jay habitat for purposes of land-use management. *Photogrammetric Engineering and Remote Sensing* 57:1467–1474.

Brown, J. H., and A. Kodric-Brown. 1977. Turnover rates in insular biogeography: Effect of immigration on extinction. *Ecology* 58:445–449.

Burgman, M., S. Ferson, and H. R. Akçakaya. 1993. *Risk Assessment in Conservation Biology.* London: Chapman & Hall.

Caughley, G. 1994. Directions in conservation biology. *Journal of Animal Ecology* 63:215–244.

Christman, S. P., and W. S. Judd. 1990. Notes on plants endemic to Florida scrub. *Florida Scientist* 53:52–73.

Cox, J. A. 1987. *Status and Distribution of the Florida Scrub Jay.* Special Publication No. 3. Gainesville: Florida Ornithological Society.

DeGange, A. R., J. W. Fitzpatrick, J. N. Layne, and G. E. Woolfenden. 1989. Acorn harvesting by Florida scrub jays. *Ecology* 70:348–356.

Doak, D. F., and L. S. Mills. 1994. A useful role for theory in conservation. *Ecology* 75:615–626.

Ebenhard, T. 1991. Colonization in metapopulations: A review of theory and observations. *Biological Journal of the Linnean Society* 42:105–121.

ESRI. 1990. *Understanding GIS.* Redlands, Calif.: Environmental Systems Research Institute.

Fitzpatrick, J. W., and G. E. Woolfenden. 1986. Demographic routes to co-operative breeding in some New World jays. Pages 137–160 in M. H. Nitecki and J. A. Kitchell, eds., *Evolution of Behavior.* New York: Oxford University Press.

Fitzpatrick, J. W., B. Pranty, and B. Stith. 1994. Florida scrub jay statewide map, 1992–1993. Final report to the U.S. Fish and Wildlife Service, Jacksonville, Florida.

Fitzpatrick, J. W., G. E. Woolfenden, and M. T. Kopeny. 1991. Ecology and development-related habitat guidelines of the Florida scrub jay (*Aphelocoma coerulescens coerulescens*). Florida Nongame Wildlife Program Technical Report, No. 8. Tallahassee: Office of Environmental Services, Florida Game and Freshwater Fish Commission.

Fitzpatrick, J. W., R. Bowman, D. R. Breininger, M. A. O'Connell, B. Stith, J. Thaxton, B. Toland, and G. E. Woolfenden. Habitat conservation plans for the Florida scrub jay: A biological framework. In preparation.

Hanski, I. 1994. A practical model of metapopulation dynamics. *Journal of Animal Ecology* 63:151–162.

Hanski, I., and M. Gilpin. 1991. Metapopulation dynamics: Brief history and conceptual domain. *Biological Journal of the Linnean Society* 42: 3–16.

Harrison, S. 1991. Local extinction in a metapopulation context: An empirical evaluation. *Biological Journal of the Linnean Society* 42:73–88.

———. 1994. Metapopulations and conservation. Pages 111–128 in P. J. Edwards, R. M. May, and N. R. Webb, eds., *Large-Scale Ecology and Conservation Biology.* Oxford: Blackwell.

Harrison, S., A. Stahl, and D. Doak. 1993. Spatial models and spotted owls: Exploring some biological issues behind recent events. *Conservation Biology* 7:950–953.

Hooper, R., J. C. Watson, and R. E. F. Escano. 1990. Hurricane Hugo's initial effects on red-cockaded woodpeckers in the Francis Marion National Forest. *Transactions of the North American Wildlife and Natural Resources Conference* 55:220–224.

Kale, H. W., II, B. Pranty, B. Stith, and C. W. Biggs. 1992. The atlas of the breeding birds of Florida. Final report to Florida Game and Freshwater Fish Commission, Tallahassee.

Koenig, W. D., F. A. Pitelka, W. J. Carmen, R. L. Mumme, and M. T. Stanback. 1992. The evolution of delayed dispersal in cooperative breeders. *Quarterly Review of Biology* 67:111–150.

Lamberson, R. H., R. McKelvey, B. R. Noon, and C. Voss. 1992. A dynamic analysis of northern spotted owl viability in a fragmented forest landscape. *Conservation Biology* 6:505–512.

Lesica, P., and F. W. Allendorf. 1995. When are peripheral populations valuable for conservation? *Conservation Biology* 9:753–760.

Levins, R. 1969. Some demographic and genetic consequences of environmental heterogeneity for biological control. *Bulletin of the Entomological Society of America* 15:237–240.

———. 1970. Extinction. Pages 77–107 in M. Gerstenhaber, ed., *Some Mathematical Questions in Biology.* Providence, R.I.: American Mathematical Society.

MacArthur, R. H., and E. O. Wilson. 1967. *The Theory of Island Biogeography.* Princeton: Princeton University Press.

Mace, G. M., and R. Lande. 1991. Assessing extinction threats: Toward a reevaluation of IUCN threatened species categories. *Conservation Biology* 2:148–157.

McGarigal, K., and B. Marks. 1994. FRAGSTATS: Spatial pattern analysis program for quantifying landscape structure. Forest Science Department, Oregon State University, Corvallis.

McGowan, K. J., and G. E. Woolfenden. 1989. A sentinel system in the Florida scrub jay. *Animal Behaviour* 37:1000–1006.

Murphy, D. D., and B. R. Noon. 1992. Integrating scientific methods with habitat conservation planning: Reserve design for northern spotted owls. *Ecological Applications* 2:3–17.

O'Connell, M. 1994. Preserving scrub habitat. *Florida Naturalist* 67:9–12.

Rolstad, J. 1991. Consequences of forest fragmentation for the dynamics of bird populations: Conceptual issues and the evidence. *Biological Journal of the Linnean Society* 42:149–163.

Simberloff, D. 1988. The contribution of population and community biology to conservation science. *Annual Review of Ecology and Systematics* 19:473–511.

Sjögren, P. 1991. Extinction and isolation gradients in metapopulations: The case of the pool frog (*Rana lessonae*). *Biological Journal of the Linnean Society* 42:135–147.

Smith, A. T., and M. M. Peacock. 1990. Conspecific attraction and the determination of metapopulation colonization rates. *Conservation Biology* 4:320–323.

Taylor, B. 1995. The reliability of using population viability analysis for risk classification of species. *Conservation Biology* 9:551–558.

Thomas, C. D. 1990. What do real population dynamics tell us about minimum viable population sizes? *Conservation Biology* 4:324–327.

———. 1994. Difficulties in deducing dynamics from static distributions. *Trends in Ecology and Evolution* 9:300.

Thomas, J. W., E. D. Forsman, J. B. Lint, E. C. Meslow, B. R. Noon, and J.

Verner. 1990. A conservation strategy for the northern spotted owl. Report of the Interagency Scientific Committee to address the conservation of the northern spotted owl. Portland: USDA Forest Service; USDI Bureau of Land Management/Fish and Wildlife Service/National Park Service.

U.S. Fish and Wildlife Service. 1991. Floridians advised of scrub jay conservation measures. 18 June 1991 news release, Jacksonville, Florida.

Woolfenden, G. E., and J. W. Fitzpatrick. 1984. *The Florida Scrub Jay: Demography of a Cooperative-Breeding Bird.* Monographs in Population Biology No. 20. Princeton: Princeton University Press.

———. 1986. Sexual asymmetries in the life history of the Florida scrub jay. Pages 87–107 in D. Rubenstein and R. W. Wrangham, eds., *Ecological Aspects of Social Evolution: Birds and Mammals.* Cambridge: Cambridge University Press.

———. 1991. Florida scrub jay ecology and conservation. Pages 542–565 in C. M. Perrins, J. D. Lebreton, and G.J.M. Hirons, eds., *Bird Population Studies: Relevance to Conservation and Management.* Oxford: Oxford University Press.

# 10

# Modelers, Mammalogists, and Metapopulations: Designing Stephens' Kangaroo Rat Reserves

*Mary V. Price and Michael Gilpin*

Both ecology and evolutionary genetics have witnessed a slow but steady movement away from models that consider populations as homogeneous, panmictic units toward models that consider populations as having internal spatial structure and a landscape context. This shift has resulted from conceptual advances in our understanding of what processes are important in nature as well as from technological advances.

Ecologists and evolutionary biologists have long suspected that the spatial context of ecological processes is critical to their outcome and dynamics. As early as the 1930s, Nicholson (1933, 1954; Nicholson and Bailey 1935) recognized that inherently unstable insect host/parasitoid interactions could be stabilized in a spatial mosaic if hosts colonize empty patches before their enemies find and exterminate them. Also in the 1930s and 1940s, Wright (1943; see also Malécot 1969 and Waser 1993) argued that environmental patchiness is critical to genetic differentiation and evolution. Today, the spatial context of interspecific interactions is considered a critical factor that molds patterns of coexistence in natural communities (for example, Kareiva 1994; Tilman 1994), and spatial structure is a central element of most thinking about evolutionary processes (for example, Real 1994).

Space has always been an implicit element of the applied fields of wildlife management and conservation for two simple reasons: harvest and hunting quotas must be established on a district-by-district basis, and the location of reserves must be based on site-specific information. Nevertheless, decisions usually have been based on simple models without spatial structure—either because those were the only models available or because available data provided inadequate estimates for parameters of more complex models. This situation is changing rapidly. The need to assess viability of populations of economic or conservation concern, coupled with advances in computer

technology and in our ability to collect data on animal movements, have stim-
ulated development of spatially explicit population models (Gilpin 1987).

Until recently, a major impediment to spatially explicit modeling was the
need to use analytic approaches involving complex equations and hidden simplifying assumptions (Felsenstein 1975) that were difficult to translate into a
language biologists could understand. The spread of powerful, low-cost computers onto the desks of biologists is rapidly altering the picture: today there is
widespread use of spatially structured models for a variety of ecological and
evolutionary problems (see Chapter 2 in this volume). There are two reasons
for the success of computer-based spatial models. First, they are conceptually
accessible, being built of easy-to-visualize rules for birth, death, and movement of individuals. Second, they permit a system's behavior to be modeled
explicitly and thus are of enormous potential value in management.

Despite their increasing popularity, these models have limitations, as all
models do, and must be used with great caution to make real-life decisions
that matter. Because computers can handle enormous complexity, computer
models normally include far more parameters than do analytic models and
hence demand a more extensive evaluation of how the model's behavior depends on parameter values and assumptions about basic biological processes.
Computer models also make great demands on the amount and quality of the
field data that underpin the choice of biological algorithms and parameter estimates. Furthermore, because the computer can hide the computational complexity behind a user-friendly interface and graphic output, there often is an
uncritical acceptance of the model's predictions, especially by decision makers
who may not be knowledgeable about the limitations inherent in ecological
modeling. These limitations make it essential that construction and implementation of spatially explicit models be done with ample and critical peer review and that there be an iterative feedback process in which model development informs empirical field studies, which in turn inform model revision.

We illustrate these points here with a spatially explicit model that was developed to facilitate the design of reserves for the endangered Stephens' kangaroo rat (*Dipodomys stephensi*). After outlining the conservation context of
the modeling effort, we describe the model and some of its properties. We end
with a critical evaluation of the relationship between modeling and empirical
components of the Stevens' kangaroo rat project and recommendations for
improvement.

## The Stephens' Kangaroo Rat Conservation Effort

Stephens' kangaroo rat (Figure 10.1) is one of 20 species of kangaroo rats, so
named because they hop like tiny kangaroos when moving rapidly. They be-

Figure 10.1. Stephens' kangaroo rat, an endangered species in southern California. Photo by Mark A. Chappell.

long to a rodent family, the Heteromyidae, that is not closely related to the murid rodents that most people think of as "rats" (Williams et al. 1993). Kangaroo rats are found only in arid regions of the United States and Mexico; the greatest diversity of species occurs in California. Species are similar in overall ecology. They are noncolonial, burrowing animals that emerge to forage only at night. Although seeds comprise the bulk of the yearly diet for all but one species, which is a leaf-eater, most kangaroo rats do eat green vegetation and insects on a seasonal basis. Seeds generally are not eaten upon harvest, but instead are transported in cheek pouches to a cache to be consumed later. Caches are either placed in a storage chamber in the burrow or buried about 2 cm deep in a shallow pit dug into the soil surface (Daly et al. 1992; Jenkins and Peters 1992). See Genoways and Brown (1993) and the references there for a complete summary of the biology of kangaroo rats and other heteromyid rodents.

The ecology of kangaroo rats is relatively well studied—partly because they are abundant and easy to catch and keep in captivity but also because they are important components of arid ecosystems of North America. Not only are kangaroo rats important prey for vertebrate predators such as owls, snakes, and mammalian carnivores, but their burrowing activities improve soil fertility and water infiltration (Chew and Whitford 1992) and provide shelter for many other organisms (Vorhies and Taylor 1922). They also can affect plant communities by selectively eating seeds of certain species and by planting them in shallow caches (Price and Jenkins 1986).

Kangaroo rats in general, and individual species in particular, appear to fill unique positions in desert communities. When kangaroo rats are removed experimentally, the other sympatric rodents change in abundance but do not compensate in terms of total consumptive biomass (Brown and Munger 1985; Brown and Heske 1990*a*), and the species composition of the plant community is affected (Brown and Heske 1990*b;* Heske et al. 1993). Kangaroo rat species may also vary in such attributes as microhabitat associations (Brown and Lieberman 1973; Reichman and Price 1993) and caching strategies (Price and Jenkins 1986; Reichman and Price 1993), making it unlikely that even sympatric species completely substitute for one another in their ecological roles. Hence extinction of a species will have unknown, but probably extensive, consequences for local communities.

Stephens' kangaroo rat is a medium-sized kangaroo rat (65 g adult body mass) with a very small geographic distribution centered in western Riverside County, southern California (Bleich 1977; Schmidly et al. 1993). Within its geographic range the species is associated with flat annual grasslands that have been greatly reduced and fragmented by agricultural and urban development in the twentieth century (Price and Endo 1989). This dramatic loss and fragmentation of habitat led in 1988 to the species' protection under the U.S. Endangered Species Act (Kramer 1987, 1988; Price and Endo 1989). In response to the listing, Riverside County started to develop a habitat conservation plan for Stephens' kangaroo rat that would achieve an acceptably low risk of extinction while permitting land development to continue outside a system of reserves (Bean et al. 1991). To decide how much land had to be preserved and what spatial configuration of reserves would be best, the county solicited help from biologists. One of us (Gilpin) was asked to write a spatially explicit computer model to be used for viability assessment and reserve design. The other (Price) was asked to collect field data on demography, dispersal, and habitat requirements. The original intent was to have the fieldwork inform the model, but it soon became clear that the time schedule for making decisions was too short to permit extensive interplay between modeling and fieldwork (Price 1993). We return to this issue later.

## Structure of the Original Stephens' Kangaroo Rat Model

Because Riverside County needed a model that would allow it to determine which land purchases would have greatest value for species persistence, Gilpin chose to base the model on an actual map of potential reserve areas (Gilpin 1993). This approach was made possible by a rangewide survey that had been commissioned by the California Department of Fish and Game. The survey, completed in 1989 (O'Farrell and Uptain 1989), consisted of detailed topographic maps showing the extent of areas containing active Stephens' kangaroo rat burrows. Occupied areas were coded by burrow density, which is correlated with population density.

These maps were entered as a data layer into a GIS database, they were gridded into 10-ha cells (316 m on a side), and each cell was assigned a single habitat value ("saturation density") based on average burrow densities for the area within the cell (Figure 10.2; Gilpin 1993; RECON 1993). The cell size reflected a compromise between computational speed and the scale of Stephens' kangaroo rat movements. With 10-ha cells, the smallest map containing all potential reserves included 9324 cells (84 rows and 111 columns). This map size required over 5 million calculations for a single 200-year simulation run (Gilpin 1993). Moreover, a 10-ha cell size was considered sufficient to represent the dynamics of a population of individuals that is somewhat isolated from the populations in neighboring cells. Approximately 100 core home ranges (the area within which an individual spends 95 percent of its aboveground time) could fit into an area of 10 ha if they did not overlap— Kelly and Price (1993) reported core home ranges of 0.1 ha for Stephens' kangaroo rat—and maximum short-term excursions of individual animals away from their home range center averaged only 60 m (Price et al. 1994), which would not take individuals into the neighboring cell.

Dynamics of the Stephens' kangaroo rat population were then modeled as a several-stage process. In a given year, reproduction occurred first within each 10-ha cell. Then dispersal occurred among neighboring cells. Finally, density dependence was modeled by truncating each cell's population back to the saturation density whenever postdispersal density exceeded this value. The resulting model is a stochastic extension of cellular automata models (Molofsky 1994), which assume that the current state of each cell depends on its past state and that of close neighboring cells. General properties of cellular automata models have been well characterized after two decades of intensive study (Wolfram 1984). Because the Stephens' Kangaroo rat model describes the state of each cell in terms of population size, it differs from individual-based automata models in which the state of a cell reflects the occupancy of a single territory (as in Lamberson et al. 1992).

The seasonal nature of Stephens' kangaroo rat breeding (McClenaghan and Taylor 1993; Price and Kelly 1994) permitted use of a simple discrete-growth algorithm for population dynamics within a cell:

$$N_{t+1} = (\lambda)N_t$$

Lambda ($\lambda$), the annual growth factor for a cell, was assumed to be density-independent and to vary around a mean value $\bar{\lambda}$ in response to two sources of

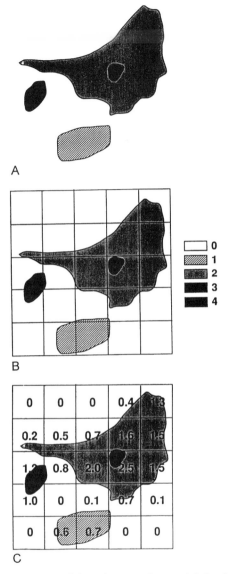

A

B

| | | | | |
|---|---|---|---|---|
| | | 0 | | |
| | | 1 | | |
| | | 2 | | |
| | | 3 | | |
| | | 4 | | |

C

| 0 | 0 | 0 | 0.4 | 1.8 |
|---|---|---|---|---|
| 0.2 | 0.5 | 0.7 | 1.6 | 1.5 |
| 1.2 | 0.8 | 2.0 | 2.5 | 1.5 |
| 1.0 | 0 | 0.1 | 0.7 | 0.1 |
| 0 | 0.6 | 0.7 | 0 | 0 |

Figure 10.2. Steps in preparing cell-based maps of potential Stephens' kangaroo rat reserves. **A**. A typical distribution map entered as a GIS layer from O'Farrell and Uptain (1989). Here 0 = no active burrows; 1 = one active burrow per acre; 2 = two to four active burrows; 3 = four to eight active burrows; 4 = more than eight active burrows. **B.** The same map with 10-ha grids superimposed. **C.** Average density index calculated from the relative proportion of each cell occupied by each density category. The "saturation density," or maximum population size per cell, is derived from the density index.

variation: environmental variation (EV) and demographic stochasticity (DS) (Shaffer 1981, 1987):

$$\lambda = \overline{\lambda} \pm EV \pm DS$$

Environmental variation reflects year-to-year variation in environmental conditions that affect individual reproductive success. An example for the Stephens' kangaroo rat is winter precipitation, which affects seed production and fecundity (Price and Endo 1989; Price and Kelly 1994). Demographic stochasticity reflects variation in $\lambda$ due to purely chance variation among individuals in fecundity or survival. Its effect on population growth is negligible in all but extremely small populations (Shaffer 1987), so Gilpin (1993) assumed that DS was zero unless the population within a cell dropped below a very small value.

Dispersal among cells was assumed to occur after reproduction because juvenile animals often are the primary dispersers in small mammal species (Gaines and McClenaghan 1980), including kangaroo rats (Jones 1987; Jones et al. 1988). Exactly what dispersal algorithm to use was unclear because so little is known about dispersal in small mammal species (Price et al. 1994; Chapter 5 in this volume). This ignorance about dispersal is unfortunate because assumptions—whether dispersal is density-dependent, which and how many individuals disperse, how far they go, what habitats they select, their probability of survival—can critically affect the achieved population growth rate and connectedness of cells, which are important determinants of the dynamics of spatially structured populations (Pulliam 1988; Pulliam and Danielson 1991; Green 1994). As a first guess, Gilpin (1993) assumed that a fixed proportion of the postreproduction population of each cell emigrated to the four adjacent cells and to one randomly chosen additional cell within a 25-cell block around each focal cell (Figure 10.3).

After individuals had dispersed from all cells, the population of each cell was reduced by the number of emigrants lost and augmented by the number of immigrants gained:

$$N_{t+1} = \lambda N_t - E + I$$

Density dependence subsequently reduced the population of the cell to the saturation density of the cell if it exceeded that value:

$$\text{if } N_{t+1} > K \quad \text{then } N_{t+1} = K$$

Density dependence has been observed in one kangaroo rat species (*Dipodomys spectabilis;* Jones 1988) and appears to be a reasonable assumption for kangaroo rats in general. The burrows of these animals are expensive to construct, and survival depends on successful acquisition of a burrow and

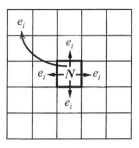

Number of emigrants

Density-independent: $E = (DISP)N$

Density-dependent:  $E = N - K$

Emigrant distribution

Without habitat selection: $e_i = E / 5$

With habitat selection:    $e_i = K_i / \Sigma k_i$

Figure 10.3. Alternative dispersal algorithms. In all cases, the number of emigrants ($E$) from a cell is distributed among the four adjacent cells and to a randomly chosen cell within a 5×5 cell array centered on the focal cell. Density dependence determines the number of emigrants. With no density dependence, the number of emigrants is a constant fraction ($DISP$) of the population of the cell after reproduction. With density dependence, the number of emigrants is the excess in postreproduction population ($N$) over saturation density ($K$). If $N < K$, there are no emigrants. Habitat selection determines the distribution of emigrants among neighboring cells. With no habitat selection, emigrants are distributed equally to the five cells regardless of their intrinsic quality ($K$). With habitat selection, emigrants are distributed among the five cells in proportion to their relative $K$. This prevent emigrants from moving to cells with $K = 0$.

establishment of control over seed stores that will last until the next period of seed production. Stephens' kangaroo rats are particularly sedentary (Price et al. 1994) and appear to occupy the same burrow system (which can be more than 3 m in diameter) for much of their adult lifetime (R. Ascanio, University of California, Riverside, personal communication). In the absence of information about the form of density dependence, Gilpin (1993) chose the simplest form: geometric growth that is truncated once a maximum density is achieved. This maximum, or "saturation density," differs from the various concepts of "carrying capacity" that are commonly associated with density dependence in that it does not vary temporally in concert with food supply and is a ceiling rather than an equilibrium (Pulliam and Haddad 1994).

## Parameterization of the Original Model

The model requires input values for several parameters: the average $\lambda$ ($\bar{\lambda}$), the magnitude of year-to-year variation in $\bar{\lambda}$ due to EV, the magnitude of DS, the dispersal fraction (*DISP*), and the saturation density (*K*) for each cell. In the absence of detailed demographic data for Stephens' kangaroo rat, values had to be based on a mixture of information from other species of kangaroo rat, educated guesses, and preliminary data from field studies of Stephens' kangaroo rat (Gilpin 1993).

Values for the average and temporal variance in $\lambda$ were calculated from distributions of the yearly $\lambda$ achieved by kangaroo rat populations compiled from long-term field studies such as that of Zeng and Brown (1987). Because no information was available on the magnitude of DS, Gilpin (1993) simply assumed that it was no larger than EV and had an effect on $\lambda$ only when the population density in a cell dropped below 2.5 individuals per hectare.

The *DISP* was estimated by asking what fraction of individuals initially placed at random within a 10-ha cell would move into neighboring cells if each individual dispersed a given straight-line distance in a random direction from the initial home-range center (Gilpin 1993). With an average dispersal distance of 50 m, a value obtained from preliminary observations of Stephens' kangaroo rat's movements (Price et al. 1994), about 4 percent of a population would emigrate to each of four adjacent cells (Gilpin 1993). This was used as the dispersal percentage in most simulations.

The final model input is a map showing the saturation densities of cells in potential Stephens' kangaroo rat reserves. Because these densities were estimated from observations over a short time period and during a drought, they surely underestimated maximum densities, and some areas of suitable habitat that were unoccupied temporarily at the time of the survey were erroneously assigned a zero *K*. Hence the maps contained some error.

Given the uncertainties associated with all of these parameter estimates and population growth algorithms, and the fact that all of them affect the time to extinction for model populations (Gilpin 1993), little credence could be placed on exact persistence figures from model runs. Nonetheless, the model was considered a valuable tool for designing reserves under the assumption that it would indicate accurately the *relative* value of different land acquisition options for Stephens' kangaroo rat's expected persistence (RECON 1993). This assumption may or may not be valid. Cellular automata models have notoriously complex dynamics that can change in a nonlinear, threshold manner with changes in model parameters or algorithms, especially those that affect the connectedness of cells (Green 1994). For example, wildfires spread only above a threshold probability that one burning plant ignites its neighbors, diseases die out below a threshold transmission probability, and populations persist in a patchy environment only when the probability that empty patches are successfully colonized is high enough relative to patch extinction rate (Green 1994). The dynamics can become even more complex when stochasticity is included.

### Some Counterintuitive Properties of the Original Model

Some behavior of the original model is intuitive. Persistence was enhanced with large $\lambda$, by large reserve size, by small environmental variation, and by demographic stochasticity that caused some cells during bad years to do relatively well, hence serving as sources for repopulating others when better conditions returned (Gilpin 1993). Some behavior, however, was unexpected. For example, contrary to the usual result that increased connectedness among habitat patches enhances persistence of metapopulations (Hanski 1991), in the original Stephens' kangaroo rat model persistence *decreased* as the dispersal fraction increased (Figure 10.4) and connecting corridors between reserve patches had little effect on persistence (Gilpin 1993).

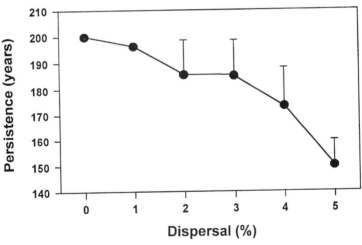

Figure 10.4. Effects of the dispersal fraction on mean persistence in the original Gilpin (1993) analysis. Dispersal = *DISP*/5. Error bars represent one standard error of the mean. Simulations were performed with $\bar{\lambda} = 1.1$, no density-dependent dispersal, no habitat selection during dispersal, environmental stochasticity = 0.8, and demographic stochasticity = 0.8.

These unexpected results suggest that even qualitative results of the original model might be sensitive to the basic biological algorithms it incorporated. For example, the effect of dispersal fraction on persistence could occur because the model assumed that dispersal was "blind"—that is, animals disperse into cells regardless of their suitability as habitat. With blind dispersal, the larger the dispersal fraction, the greater the number of dispersers that are lost from the population each year when they move across a reserve boundary. This assumption contrasts with early models, which assumed that the dispersal that connects subpopulations within metapopulations incurs no cost in terms of population growth rate (Hanski 1991). The "edge effect" created by an assumption of blind dispersal would be more severe for long, thin habitat patches with high ratios of perimeter to center and could artificially inflate the apparent value of reserves with large, contiguous habitat patches, leading to suboptimal land acquisition decisions. And if corridors increase the edge-to-center ratio of a reserve system in addition to connecting otherwise isolated habitat patches, it could also deflate the value of corridors.

## A Modified Model

To evaluate the consequences of our assumptions about dispersal, we modified the original model to include density dependence and habitat selection during dispersal (Figure 10.3). Density dependence was modeled by allowing only excess individuals over saturation density to emigrate; the original model had assumed that a constant proportion of the population dispersed. Habitat selection during dispersal was modeled by allocating emigrants to target cells in proportion to the relative saturation density of those cells; the original model had allocated emigrants equally among target cells. Alternative algorithms are possible. For example, the dispersal fraction could change linearly with increasing density, dispersal distances could themselves be functions of population density (Jones et al. 1988), or dispersal into cells could occur in proportion to the difference between saturation density and present density. Nevertheless, our simple algorithms achieved the desired effects of having density dependence reduce the probability of dispersal when population densities are low and having habitat selection prevent the loss of dispersers into cells lacking suitable habitat.

### Effects of Reserve Shape

We first explored the persistence of hypothetical populations of Stephens' kangaroo rat in four reserves that varied in shape but contained the same number of cells, all with a saturation density of 20. For the reserve with a 1 × 40 array of cells, 68 percent of cell boundaries were on the reserve perimeter.

This "edge percent" was 43 percent for the 2 × 20 array, 30 percent for the 4 × 10 array, and 28 percent for the 5 × 8 array. Average lambda ($\bar{\lambda}$), EV, and DS were held constant, and persistence was recorded for 50 replicate runs of 200 years for each of 16 combinations of four reserve shapes (Edge), two density dependence treatments (DD: Yes or No), and two habitat selection treatments (HS: Yes or No).

Results of a factorial ANOVA (Table 10.1) indicate that DD, HS, and the interaction between DD and HS had significant effects on persistence. Density-dependent dispersal increased persistence (there was lower cost of dispersal). Habitat selection also increased persistence, but more strongly when dispersal was not density-dependent than when it was (Figure 10.5). These effects are intuitive. All else being equal, density-dependent dispersal resulted in fewer dispersers and hence reduced costs of dispersal in terms of population growth rate. By reducing the loss of individuals to unsuitable habitat, habitat selection increased persistence, especially when dispersal was not density-dependent.

TABLE 10.1.

Factorial ANOVA: effect of reserve shape (Edge), density dependence (DD), and habitat selection (HS) on persistence of hypothetical populations of Stephens' kangaroo rat

| Source of variation | df | MS | F | P |
|---|---|---|---|---|
| **A. DS = 0.5** | | | | |
| Edge | 3 | 2,285.22 | 0.76 | 0.5175 |
| DD | 1 | 851,577.75 | 282.68 | 0.0001 |
| Edge × DD | 3 | 1,985.22 | 0.66 | 0.5775 |
| HS | 1 | 15,338.76 | 5.09 | 0.0243 |
| Edge × HS | 3 | 10,827.49 | 3.59 | 0.0134 |
| DD × HS | 1 | 58,910.28 | 19.56 | 0.0001 |
| Edge × DD × HS | 3 | 4,115.52 | 1.37 | 0.2519 |
| **B. DS = 0** | | | | |
| Edge | 3 | 3,584.10 | 1.20 | 0.3088 |
| DD | 1 | 157,472.72 | 52.72 | 0.0001 |
| Edge × DD | 3 | 2,677.92 | 0.90 | 0.4425 |
| HS | 1 | 70,876.12 | 23.73 | 0.0001 |
| Edge × HS | 3 | 4,734.03 | 1.58 | 0.1916 |
| DD × HS | 1 | 22,514.42 | 7.54 | 0.0062 |
| Edge × DD × HS | 3 | 3,269.94 | 1.09 | 0.3505 |

*Note:* Four reserve shapes were compared (1 × 40, 2 × 20, 4 × 10, and 5 × 8 cells), keeping total reserve size constant at 40 cells. Panel A: Fifty replicate simulations were run with $\bar{\lambda}$ = 1.15, EV = 1.0, DS = 0.5, and *DISP* (for runs with DD = No) = 0.02. Panel B: As in panel A but with DS = 0.

The more puzzling result was a significant interaction between reserve shape and habitat selection (Table 10.1) that was complex and nonlinear (Figure 10.6). Without habitat selection during dispersal, persistence was lowest for the reserve with the greatest edge percent (1 × 40), a result that fits with the conventional wisdom that edges between suitable and unsuitable

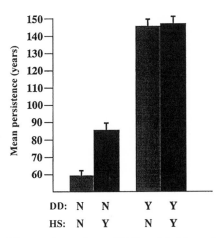

Figure 10.5. Effect of density dependence (DD) and habitat selection (HS) on mean persistence of hypothetical Stephens' kangaroo rat populations. Error bars represent one standard error of the mean; sample size = 100 replicate simulations per treatment.

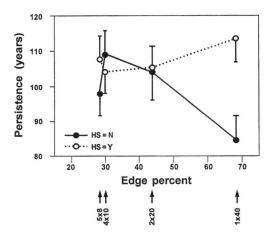

Figure 10.6. Effect of reserve shape and habitat selection on mean persistence of hypothetical Stephens' kangaroo rat populations with $\bar{\lambda}$ = 1.15, EV = 1, DS = 0.5, DISP = 0.02. "Edge percent" refers to the percentage of cell boundaries that are on the perimeter of the reserve. Error bars represent one standard error of the mean; sample size = 100 replicate simulations per shape treatment.

habitat are bad for persistence of populations. With habitat selection during dispersal, however, persistence was highest for the reserve with the greatest edge percent (1 × 40). This significant interaction confirmed our suspicion that the dispersal algorithm could dramatically affect even qualitative conclusions about the best reserve shape.

What might have caused long, thin reserves to do best with habitat selection? One possibility is that the cells of such reserves become uncorrelated by long distances, which enhances the ability of demographic stochasticity to create "refuge" cells during poor years of low $\lambda$. Habitat selection could modify this effect of demographic stochasticity in some way. Another possibility is that habitat selection introduces a "reflecting boundary" at the edge of reserves that funnels dispersers back into the reserve, much as the edge of a ripple tank reflects waves. Because dispersal is spatially restricted, some cells—notably those next to the outer perimeter—might actually accumulate more immigrants than they lose to emigration (Figure 10.7). This would create persistent density gradients among cells even when there is no spatial variation in saturation density and in the absence of demographic stochasticity. The density gradients might be more pronounced in long, thin reserves (Figure 10.7). In a temporally variable environment, "accumulator" cells could rescue adjacent cells whose populations go extinct and hence would increase persistence.

To distinguish these possibilities, we explored the effects of reserve shape as before, but this time we set demographic stochasticity at zero. If the interaction between edge percent and habitat selection occurred because habitat selection modifies the strength of demographic stochasticity, then the interaction should disappear with DS = 0. If, however, habitat selection produces persistent density gradients that are more pronounced in long, thin reserves, then the interaction should still exist in the absence of DS.

As is apparent from Figure 10.8 and Table 10.1B, the qualitative Edge × HS interaction persisted, although it was not statistically significant with sample sizes as small as 50 replicate runs per treatment. Relatively long, thin reserves (2 × 20) still performed better than more nearly square reserves with habitat selection (compare with Figure 10.6) and worse without habitat selection. This suggests that the second hypothesis is correct.

## Effects of Corridors

A surprising result of Gilpin's analysis of the original model is that corridors between reserves had little impact on persistence other than providing additional habitat cells (Gilpin 1993). One possible reason for this result is that, with no density dependence or habitat selection, corridors have two opposing effects. Because they are long and thin, corridors increase the cost of dispersal.

**A. 1 x 40: habitat selection = No**
**weighted average gain = 0.410**

| 0.21 | 0.42 | 0.42 | 0.42 | 0.42 | . . . . | 0.42 | 0.42 | 0.42 | 0.42 | 0.21 |
|------|------|------|------|------|---------|------|------|------|------|------|

**B. 1 x 40: habitat selection = Yes**
**weighted average gain = 0.967**

| 0.50 | 1.44 | 0.98 | 0.97 | 0.97 | . . . . | 0.97 | 0.97 | 0.98 | 1.44 | 0.50 |
|------|------|------|------|------|---------|------|------|------|------|------|

**C. 5 x 8: habitat selection = No**
**weighted average gain = 0.735**

| 0.44 | 0.65 | 0.66 | 0.66 | 0.66 | 0.66 | 0.65 | 0.44 |
|------|------|------|------|------|------|------|------|
| 0.65 | 0.86 | 0.88 | 0.88 | 0.88 | 0.88 | 0.86 | 0.65 |
| 0.66 | 0.88 | 0.90 | 0.90 | 0.90 | 0.90 | 0.88 | 0.66 |
| 0.65 | 0.86 | 0.88 | 0.88 | 0.88 | 0.88 | 0.86 | 0.65 |
| 0.44 | 0.65 | 0.66 | 0.66 | 0.66 | 0.66 | 0.65 | 0.44 |

**D. 5 x 8: habitat selection = Yes**
**weighted average gain = 0.928**

| 0.63 | 0.98 | 0.91 | 0.84 | 0.84 | 0.91 | 0.98 | 0.63 |
|------|------|------|------|------|------|------|------|
| 0.98 | 1.08 | 1.00 | 0.99 | 0.99 | 1.00 | 1.08 | 0.98 |
| 0.91 | 1.00 | 0.93 | 0.93 | 0.93 | 0.93 | 1.00 | 0.91 |
| 0.98 | 1.08 | 1.00 | 0.99 | 0.99 | 1.00 | 1.08 | 0.98 |
| 0.63 | 0.98 | 0.91 | 0.84 | 0.84 | 0.91 | 0.98 | 0.63 |

Figure 10.7. Effect of reserve shape and habitat selection on the expected fraction of emigrants from a cell that are recovered by immigration from neighboring cells, assuming that all cells disperse equal numbers of emigrants. Cells with values greater than 1 are net accumulators of dispersers. The weighted average gain is the average of the fraction of emigrants recovered across all cells in a hypothetical reserve. Note that with no habitat selection during dispersal, interior cells have a greater net gain than peripheral cells. This is because cells outside the reserve contribute no immigrants to peripheral cells. With habitat selection during dispersal, there is greater variation among cells in net dispersal fraction. The weighted average gain is lower for the 1 × 40 reserve than the 5 × 8 reserve in the absence of habitat selection, but higher in the presence of habitat selection.

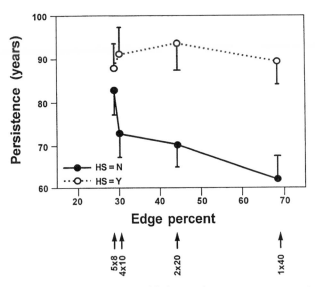

Figure 10.8. Effect of reserve shape and habitat selection on mean persistence of hypothetical Stephens' kangaroo rat populations with $\bar{\lambda}$ = 1.15, EV = 1, DS = 0, and *DISP* = 0.02. "Edge percent" refers to the percentage of cell boundaries that are on the perimeter of the reserve. Error bars represent one standard error of the mean; sample size = 50 replicate runs per treatment. Compare with Figure 10.6.


On the other hand, corridors should increase the time to extinction of an entire reserve system because extinction in one part of the system can be reversed by recolonization from other parts of the system. Perhaps an increased cost of dispersal in Gilpin's (1993) original model obscured the underlying benefit of corridors.

To explore this possibility, we used the modified model to investigate whether moving two isolated habitat patches together to connect them increased persistence of the entire reserve; we used two corridor widths, "fat" and "thin" (Figure 10.9). This analytic method allowed us to explore the effects of connection independent of any effects that corridors might have on reserve shape or on the total amount of habitat in the reserve.

A factorial ANOVA (Table 10.2) indicates that persistence was higher when the two habitat patches were connected than when they were not (Figure 10.10) and that the effect of connection did not depend on corridor width, habitat selection, or density dependence. We are not certain which features of Gilpin's (1993) original analysis obscured the persistence effect that we detected with this analysis.

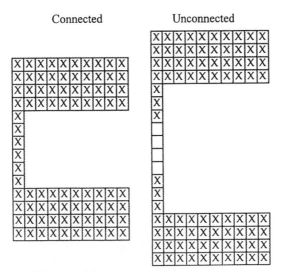

A. Fat corridor

B. Thin corridor

Figure 10.9. Configuration of reserves used to study effects of corridors on persistence of hypothetical Stephens' kangaroo rat populations. Cells containing an "X" have a saturation density of 20; all others have a saturation density of zero.

TABLE 10.2.

Factorial ANOVA: effect on persistence of corridor width
and connection of two reserve patches

| Source of variation | df | MS | F | P |
|---|---|---|---|---|
| Connect | 1 | 13,944.91 | 4.16 | 0.415 |
| Width | 1 | 363.43 | 0.11 | 0.7419 |
| Connect × width | 1 | 4,706.80 | 1.41 | 0.2360 |
| DD | 1 | 803,357.16 | 239.91 | 0.0001 |
| Connect × DD | 1 | 197.23 | 0.06 | 0.8083 |
| Width × DD | 1 | 11,648.70 | 3.48 | 0.0624 |
| Connect × width × DD | 1 | 11,252.23 | 3.36 | 0.0670 |
| HS | 1 | 32,057.20 | 9.57 | 0.0020 |
| Connect × HS | 1 | 4,020.43 | 1.20 | 0.2734 |
| Width × HS | 1 | 212.63 | 0.06 | 0.8011 |
| Connect × width × HS | 1 | 1,431.03 | 0.43 | 0.5134 |
| DD × HS | 1 | 21,385.03 | 6.39 | 0.0116 |
| Connect × DD × HS | 1 | 300.36 | 0.09 | 0.7646 |
| Width × DD × HS | 1 | 2,835.29 | 0.85 | 0.3577 |
| Connect × width × DD × HS | 1 | 0.36 | 0.00 | 0.9918 |

*Note:* Seventy replicate runs were conducted with all combinations of two corridor widths; two connection treatments (Yes or No); two habitat selection treatments; two density-dependent dispersal treatments; and with $\bar{\lambda} = 1.15$, EV = 1, DS = 0.04, and *DISP* = 0.02. Figure 10.9 illustrates the corridor and connection treatments.

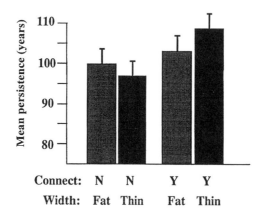

Figure 10.10. Effect of reserve connection ("connect") and corridor width ("width") on mean persistence of hypothetical Stephens' kangaroo rat populations. Error bars represent one standard error of the mean; sample size = 70 replicate simulations per treatment.

## The Interplay with Fieldwork

The analyses we have conducted with the modified Stephens' kangaroo rat model indicate that its predictions about the best reserve shape and the value of corridors are sensitive to assumptions about how animals disperse. Which dispersal algorithm is correct? Despite a several-year field study of Stephens' kangaroo rat dispersal (Price et al. 1994), we do not know. The field study was not designed to determine whether dispersal is density-dependent or directed with respect to habitat quality; instead, our focus was on obtaining the best estimates of the *distance* of dispersal to know what distance between habitat patches results in isolation. This focus arose from our expectation that the primary consequence of dispersal for persistence involves its role in increasing the connectivity of patches, rather than any direct effect on within-cell population dynamics. The possibility of the latter effect became apparent only after we began the field study (Pulliam and Danielson 1991).

This empirical insufficiency points out the importance of having a close interplay between fieldwork and model development. The field study was designed before the model was constructed and before a sensitivity analysis had been performed; hence the importance of discriminating between very different types of dispersal was unforeseen. Had a more thorough analysis of the model been completed prior to starting field studies, we would have used arrays of smaller sampling grids placed in diverse habitat patches, rather than single large grids in homogeneous habitat. The smaller grids would have given us less complete information on dispersal distances but better information on density dependence and habitat selection during dispersal (Price et al. 1994).

## Lessons

Ready availability of computers opens the door to use of biologically realistic, spatially explicit models for making conservation decisions. But the very properties that make these models valuable also give them complex dynamic properties that are challenging to understand. This makes it especially important that empirical research be intimately coupled with model development in an iterative process. Ideally, the model's basic structure will be guided by good natural history information about a species of interest. Subsequent sensitivity analysis should indicate which biological algorithms and parameters critically affect the model's behavior. Results of sensitivity analyses should then guide the design of field studies so that they estimate critical parameter values and distinguish among alternative possible algorithms for biological processes being modeled. In turn, field data can guide algorithm selection as the model is fine-tuned.

In the case of the habitat conservation plan for the Stephens' kangaroo rat, there simply was not enough time to implement such an iterative process (Price 1993). Contracts for fieldwork and modeling were only of a 2-year duration, and there was intense political and economic pressure to complete a habitat conservation plan as soon as possible.

Despite these problems, progress has been made in several regards. We now have a cell-based model that can be easily adapted to species, such as Stephens' kangaroo rat, for which models based on occupancy of individual territories are impractical. We now are more familiar with properties of spatially explicit population models and hence have a better understanding of which biological processes and parameters have large effects on population persistence. In the case of Stephens' kangaroo rat, this knowledge should be of value in guiding development of management plans for reserves, even if the model could not be deployed effectively during the original design of the reserve system. In the case of other endangered species, the knowledge derived from the Stephens' kangaroo rat model should assist in the identification of critical aspects of demography and population ecology that must be understood and incorporated into conservation plans if they are to succeed.

## Acknowledgments

Fieldwork and development of the original computer model were supported by the Riverside County Habitat Conservation Agency, and fieldwork was conducted with permits from the U.S. Fish and Wildlife Service and California Department of Fish and Game. The Academic Senate of the University of California, Riverside, and the A. Starker Leopold Chair (held by D. R. McCullough) made attendance at the symposium possible. We thank many people for assistance and stimulating discussions, in particular R. Ascanio, B. Bestelmeyer, H. Callahan, C. Camarena, R. Goldingay, P. Kelly, and H. Renkin. Nick Waser and two anonymous reviewers made helpful comments on earlier drafts of the chapter.

## REFERENCES

Bean, M. J., S. G. Fitzgerald, and M. A. O'Connell. 1991. *Reconciling Conflicts Under the Endangered Species Act: The Habitat Conservation Planning Experience.* Baltimore: World Wildlife Fund Publications.

Bleich, V. C. 1977. *Dipodomys stephensi. Mammalian Species* 73:1–3.

Brown, J. H., and E. J. Heske. 1990*a*. Temporal changes in a Chihuahuan desert rodent community. *Oikos* 59:290–302.

———. 1990*b*. Control of a desert-grassland transition by a keystone rodent guild. *Science* 250:1705–1707.

Brown, J. H., and G. L. Lieberman. 1973. Resource utilization and coexistence of seed-eating desert rodents in sand dune habitats. *Ecology* 54:788–797.

Brown, J. H., and J. C. Munger. 1985. Experimental manipulation of a desert rodent community: Food addition and species removal. *Ecology* 66:1545–1563.

Chew, R. M., and W. G. Whitford. 1992. A long term positive effect of kangaroo rats (*Dipodomys spectabilis*) on creosotebushes (*Larrea tridentata*). *Journal of Arid Environments* 22:375–386.

Daly, M., L. F. Jacobs, M. I. Wilson, and P. R. Behrends. 1992. Scatter-hoarding by kangaroo rats (*Dipodomys merriami*) and pilferage from their caches. *Behavioral Ecology* 3:102–111.

Felsenstein, J. 1975. A pain in the torus: Some difficulties with models of isolation by distance. *American Naturalist* 109:359–368.

Gaines, M. S., and L. R. McClenaghan, Jr. 1980. Dispersal in small mammals. *Annual Review of Ecology and Systematics* 11:263–196.

Genoways, H. H., and J. H. Brown, eds. 1993. *Biology of the Heteromyidae*. Special Publication Number 10. Shippensburg, Pa.: American Society of Mammalogists.

Gilpin, M. E. 1987. Spatial structure and population vulnerability. Pages 125–140 in M. E. Soulé, ed., *Viable Populations for Conservation*. Cambridge: Cambridge University Press.

———. 1993. A viability model for Stephens' kangaroo rat in western Riverside County. Report no. 12, vol. II: Technical reports. Draft habitat conservation plan for the Stephens' kangaroo rat in western Riverside County, California. Riverside: Riverside County Habitat Conservation Agency.

Green, D. G. 1994. Connectivity and complexity in landscapes and ecosystems. *Pacific Conservation Biology* 3:1–2.

Hanski, I. 1991. Single-species metapopulation dynamics: Concepts, models and observations. *Biological Journal of the Linnean Society* 42:17–38.

Heske, E. J., J. H. Brown, and Q. Guo. 1993. Effects of kangaroo rat exclusion on vegetation structure and plant species diversity in the Chihuahuan desert. *Oecologia* 95:520–524.

Jenkins, S. H., and R. A. Peters. 1992. Spatial patterns of food storage by Merriam's kangaroo rats. *Behavioral Ecology* 3:60–65.

Jones, W. T. 1987. Dispersal patterns in kangaroo rats (*Dipodomys spectabilis*). Pages 119–127 in B. D. Chepko-Sade and Z. T. Halpin, eds., *Mammalian Dispersal Patterns.* Chicago: University of Chicago Press.

————. 1988. Density-related changes in survival of philopatric and dispersing kangaroo rats. *Ecology* 69:1474–1478.

Jones, W. T., P. M. Waser, L. F. Elliott, N. E. Link, and B. B. Bush. 1988. Philopatry, dispersal, and habitat saturation in the banner-tailed kangaroo rat, *Dipodomys spectabilis. Ecology* 69:1466–1473.

Kareiva, P. 1994. Space: The final frontier for ecological theory. *Ecology* 75:1–47.

Kelly, P. A., and M. V. Price. 1993. Home range use of Stephens' kangaroo rats: Implications for density estimation. Report no. 4, vol. II: Technical reports. Draft habitat conservation plan for the Stephens' kangaroo rat in western Riverside County, California. Riverside: Riverside County Habitat Conservation Agency.

Kramer, K. 1987. Endangered and threatened wildlife and plants: Determination of endangered status for Stephens' kangaroo rat. *Federal Register* 52:44453–44454.

————. 1988. Endangered and threatened wildlife and plants: Determination of endangered status for Stephens' kangaroo rat. Final rule. *Federal Register* 53:38465–38469.

Lamberson, R. H., R. McKelvey, B. R. Noon, and C. Voss. 1992. A dynamic analysis of northern spotted owl viability in a fragmented forest landscape. *Conservation Biology* 6:505–512.

Malécot, G. 1969. *The Mathematics of Heredity.* San Francisco: Freeman.

McClenaghan, L. R., and E. Taylor. 1993. Temporal and spatial demographic patterns in *Dipodomys stephensi* from Riverside County, California. *Journal of Mammalogy* 74:636–645.

Molofsky, J. 1994. Population dynamics and pattern formation in theoretical populations. *Ecology* 75:30–39.

Nicholson, A. J. 1933. The balance of animal populations. *Journal of Animal Ecology* 2 (suppl.):132–178.

————. 1954. An outline of the dynamics of animal populations. *Australian Journal of Zoology* 2:9–65.

Nicholson, A. J., and V. A. Bailey. 1935. The balance of animal populations. *Proceedings of the Zoological Society of London* 3:551–598.

O'Farrell, M. J., and C. E. Uptain. 1989. *Assessment of Population and Habitat Status of the Stephens' Kangaroo Rat* (Dipodomys stephensi). Sacramento: California Department of Fish and Game.

Price, M. V. 1993. Involving academics in endangered-species conservation: Lessons from the Stephens' kangaroo rat habitat conservation plan. Pages 21–25 in J. E. Keeley, ed., *Interface Between Ecology and Land De-*

*velopment in California.* Los Angeles: Southern California Academy of Sciences.

Price, M. V., and P. R. Endo. 1989. Estimating the distribution and abundance of a cryptic species, *Dipodomys stephensi* (Rodentia: Heteromyidae), and implications for management. *Conservation Biology* 3: 293–301.

Price, M. V., and S. H. Jenkins. 1986. Rodents as seed consumers and dispersers. Pages 191–246 in D. R. Murray, ed., *Seed Dispersal.* North Ryde, NSW: Academic Press of Australia.

Price, M. V., and P. A. Kelly. 1994. An age-structured demographic model for the endangered Stephens' kangaroo rat. *Conservation Biology* 8:810–821.

Price, M. V., P. A. Kelly, and R. L. Goldingay. 1994. Distances moved by Stephens' kangaroo rat (*Dipodomys stephensi* Merriam) and implications for conservation. *Journal of Mammalogy* 75:929–939.

Pulliam, H. R. 1988. Sources, sinks, and population regulation. *American Naturalist* 132:652–661.

Pulliam, H. R., and B. J. Danielson. 1991. Sources, sinks, and habitat selection: A landscape perspective on population dynamics. *American Naturalist* 137(suppl.):50–66.

Pulliam, H. R., and N. M. Haddad. 1994. Human population growth and the carrying capacity concept. *Bulletin of the Ecological Society of America* 75:141–157.

Real, L., ed. 1994. *Ecological Genetics.* Princeton: Princeton University Press.

RECON. 1993. A population viability analysis of the Stephens' kangaroo rat core reserves in Riverside County, California. Report no. 13, vol. II: Technical reports, draft habitat conservation plan for the Stephens' kangaroo rat in western Riverside County, California. Riverside: Riverside County Habitat Conservation Agency.

Reichman, O. J., and M. V. Price. 1993. Ecological aspects of heteromyid foraging. Pages 539–574 in H. H. Genoways and J. H. Brown, eds., *Biology of the Heteromyidae.* Special Publication 10. Shippensburg, Pa.: American Society of Mammalogists.

Schmidly, D. J., K. T. Wilkins, and J. N. Derr. 1993. Biogeography. Pages 319–356 in H. H. Genoways and J. H. Brown, eds., *Biology of the Heteromyidae.* Special Publication 10. Shippensburg, Pa.: American Society of Mammalogists.

Shaffer, M. L. 1981. Minimum population sizes for species conservation. *BioScience* 31:131–134.

———. 1987. Minimum viable populations: coping with uncertainty. Pages 69–86 in M. E. Soulé, ed., *Viable Populations for Conservation.* Cambridge: Cambridge University Press.

Tilman, D. 1994. Competition and biodiversity in spatially structured habitats. *Ecology* 75:2–16.

Vorhies, C. T., and W. P. Taylor. 1922. *Life History of the Kangaroo Rat* Dipodomys spectabilis spectabilis *Merriam.* Bulletin 1091. Washington, D.C.: USDA.

Waser, N. M. 1993. Population structure, optimal outbreeding, and assortative mating in angiosperms. Pages 173–199 in N. W. Thornhill, ed., *The Natural History of Inbreeding and Outbreeding: Theoretical and Empirical Perspectives.* Chicago: University of Chicago Press.

Williams, D. F., H. H. Genoways, and J. K. Braun. 1993. Taxonomy. Pages 38–196 in H. H. Genoways and J. H. Brown, eds., *Biology of the Heteromyidae.* Special Publication 10. Shippensburg, Pa.: American Society of Mammalogists.

Wolfram, S. 1984. Universality and complexity in cellular automata. *Physica* 10D:1–35.

Wright, S. 1943. Isolation by distance. *Genetics* 28:114–138.

Zeng, Z., and J. H. Brown. 1987. Population ecology of a desert rodent: *Dipodomys merriami* in the Chihuahuan desert. *Ecology* 68:1328–1340.

# 11

# Metapopulation Dynamics of the Mediterranean Monk Seal

*John Harwood, Helen Stanley,*
*Marie-Odile Beudels, and Charles Vanderlinden*

The subfamily Monachinae contains nine seal species, including the most abundant seal in the world: the crabeater seal, *Lobodon carcinophagus.* The three members of the genus *Monachus,* however, are among the most endangered marine mammals. One, *Monachus tropicalis,* the Caribbean monk seal, is probably extinct: the last confirmed sighting was in 1949 (King 1956). *Monachus schauinslandi,* the Hawaiian monk seal, has a population of around 1500; numbers appeared to increase until the early 1980s, but they have declined by 25 percent since then (Anonymous 1994). There is no reliable estimate of the total size of the remaining population of *Monachus monachus,* the Mediterranean monk seal, but a range of 400 to 550 is suggested by figures in Reijnders et al. (1993). Although this species still occurs throughout its historical geographic range, the surviving populations are increasingly fragmented and separated by large areas of apparently unsuitable habitat.

In this chapter we review changes in the distribution of the Mediterranean monk seal over the past 50 years using data from the Monk Seal Register (Beudels 1992), a large international database funded by the European Commission that brings together much of the published and unpublished information on the species and considers the possible causes of these distributional changes. We then assess the applicability of metapopulation models to this species and use the metapopulation framework to list the priorities for future research and conservation.

## Recent Changes in Distribution

Although anecdotal evidence suggests that Mediterranean monk seals used to rest and breed on sandbanks and open rocky shelves (see the references in

Marchessaux 1989*a*), the species is now seen almost only on beaches within caves on isolated and inaccessible coasts. This elusive behavior makes it difficult to determine population size and distribution reliably. The statistical basis for most published estimates of the size of local monk seal populations (see the caption to Figure 11.1) is unclear. Many are derived from interviews with local fishermen and naturalists, but the way in which these responses are converted into actual numbers is seldom documented. As a result one should not use published figures to determine trends in abundance, although such reports can provide qualitative information on changes in the species' distribution.

Figure 11.1 shows changes in the distribution of the Mediterranean monk seal since the 1950s. At the start of that decade the species occurred from Cap Blanc in Mauritania to the Black Sea, with small populations on the Spanish and French mainlands, in Corsica, and an almost continuous distribution along the North African coast in the western Mediterranean (Figure 11.1*a*). By the 1970s monk seals had disappeared from the French coast and much of the Spanish and Algerian coasts (Figure 11.1*b*). The most severe fragmentation of the distribution has occurred during the past two decades (Figure 11.1*c*). Monk seals have virtually disappeared from the western Mediterranean and the Black Sea. The extinction of these local populations was predicted on the basis of their known demographic characteristics by Durant and Harwood (1992). As a result of this fragmentation, seals in the Atlantic are separated by at least 1200 km from the nearest seals in the western Mediterranean and by at least 4500 km from the surviving populations in the eastern Mediterranean.

The reasons for the increased fragmentation of the monk seal's distribution are not entirely clear. Deliberate killing, mostly carried out by fishermen, and loss of suitable habitat for pupping have played the primary role in most parts of Europe. (For Greece see IUCN/SSC 1994; for Corsica see Graziani 1987.) But these factors do not explain the species' decline on the North African coast, where there is much less antipathy between fishermen and seals. Boudouresque and Lefevre (1992) have suggested that, in this region, reduced food availability as a result of overfishing may be the cause, but the evidence is circumstantial. Incidental loss of seals tangled in fishing gear may have played a role, but the extent of this has not been documented.

## Gene Flow Between the Two Populations

Other monachine seals (such as the southern elephant seal, *Mirounga leonina*) are known to travel thousands of kilometers over periods of a few months (McConnell et al. 1992), and Hawaiian monk seals are occasionally sighted

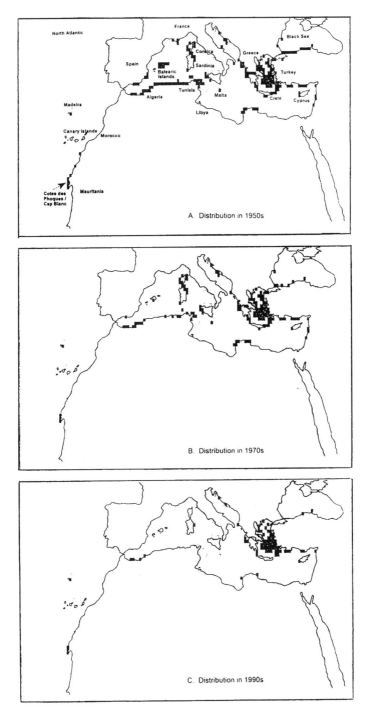

Figure 11.1. Changes in the distribution of the Mediterranean monk seal since 1950. Sources of data are listed in the Acknowledgments at the end of the chapter.

on the main islands of the Hawaiian chain, at least 1500 km from the nearest breeding site. But the 4500 km of apparently unsuitable habitat separating the eastern Mediterranean and the Atlantic breeding areas may have posed a barrier to gene flow between these two populations.

Monod (in Marchessaux 1989*a*) has documented that large numbers of seals were present in the Bay of Dakhla, on the Atlantic coast of Africa, when the area was first explored by Portuguese navigators between 1436 and 1441. Dakhla is approximately 300 km north of the short length of coastline where the species still occurs. Large numbers of seals were killed during these, and subsequent, visits. There are no reports of monk seals on this coast from the eighteenth century until the rediscovery of the current colonies in the 1920s, even though the area was visited by competent observers. Marchessaux (1989*a*) suggests that the species must have disappeared from the accessible parts of this coastline in the eighteenth century and survived only where there were suitable caves. If this is the case, the Atlantic population passed through a fairly severe bottleneck that lasted for at least ten generations. This bottleneck may have resulted in significant genetic differences between these seals and those in the Mediterranean as a result of founder effects and drift (see Chapter 3 in this volume). Van Bree (1979) has suggested that there are statistically significant differences in the frequency of dental aberrations in samples from the Atlantic and the western Mediterranean—evidence "that gene flow between the populations was limited"—but his brief abstract provides no information on sample sizes or analytic procedures.

We have begun to assess levels of intraspecific genetic differentiation by direct sequencing of part of the mitochondrial genome. Mitochondrial DNA (mtDNA) is maternally inherited, and mtDNA genotypes are believed to be selectively neutral. Due to their haploid nature, mtDNA genotypes can be lost or fixed in a population relatively rapidly (Avise et al. 1984). Because of this, they have been widely used in other threatened or endangered species to determine relationships between closely related populations (Moritz 1994). The most rapidly evolving region of the mitochondrial genome is the D-loop, or control region, and we have used variations in this region to examine geographic differences The cytochrome b gene evolves more slowly than the D-loop, and we have used sequence variation in this gene to examine phylogenetic relationships among the monachine seals.

Monk seal material for DNA extraction is in short supply and often in poor condition. We have been provided with samples from at least 27 different individuals: 7 from the Atlantic, 1 from the western Mediterranean, and 19 from the eastern Mediterranean. Some samples consisted of cultured lymphocytes, but the majority were of hair or skin, often decomposed. We used universal D-loop primers to generate monk-seal–specific sequence data from the highest-quality samples and then designed and synthesized monk-seal–specific primers to permit amplification of smaller fragments. This approach increased the efficiency of amplification using the polymerase chain reaction and was particularly necessary because of the poor quality of the DNA extracted from hair or rotting skin.

A 444 base-pair sequence from the D-loop region has now been obtained from seven individuals and shown to contain only two variable sites that define two genotypes. Both substitutions were transitions. Shorter sequences have been obtained from at least eight other individuals, but these have not been included in the subsequent analysis because they were incomplete. Sequence divergences were less than 1 percent; this is low but comparable with data obtained from other marine mammals (for example, Hoelzel and Dover 1991; Hoelzel et al. 1993). One of the two genotypes was recorded in all four samples from the eastern Mediterranean, but it was not observed in any individuals from the Atlantic or the western Mediterranean. These results suggest that there has been limited gene flow between the monk seal populations in the eastern Mediterranean and those in the west of the species' range.

Analysis of the cytochrome b gene sequence (H. Stanley, unpublished data) has shown that the difference between *M. monachus* and *M. schauinslandi* is as great as that between either species and any other member of the subfamily Monachinae. Considering the length of time that the two species have been separated (Repenning et al. 1979), this is not surprising. There is a great temptation to assume that *M. monachus* has reproductive and survival rates similar to those estimated for *M. schauinslandi* (Johanos et al. 1994; Gilmartin et al. 1993). Given the genetic differences between the two species, this temptation should be resisted unless there is convincing evidence to support the assumption.

## Metapopulation Structure and Monk Seal Conservation

The development of the metapopulation concept has been succinctly reviewed by Hanski (1991). Metapopulations are systems of local populations connected by dispersing individuals. Small populations are more vulnerable to extinction than large populations because of a variety of causes, primarily environmental stochasticity (Lande 1993), although this may be accentuated by genetic effects (for example, Smouse 1994). In the face of environmental stochasticity, the mean lifetime of a population increases with the logarithm of the number of individuals (Leigh 1981). Metapopulation theory, in its classic form, is concerned with the balance between extinction and recolonization in these local populations. Models derived from this theory have been primarily concerned with the proportion of suitable habitat patches that are occupied. A common framework for these models has been developed by Gotelli and Kelley (1993).

It is important to determine the geographic scale on which colonization and extinction events are taking place. Simply because a species occurs in local populations that are separated by regions of apparently unsuitable habitat does not mean that it exists as a metapopulation. Movement between local populations may be so great that they should be considered together as a single homogeneous unit. Yet dispersal rates may be so low that each local population has its own, unique dynamics. Hanski and Gilpin (1991) make a useful distinction between the local scale (at which individuals move and interact with one another in the course of their routine feeding and breeding activities) and the metapopulation scale (at which individuals infrequently move from one population to another, typically across habitat types that are not suitable for their feeding and breeding activities, often with substantial risk of failing to locate another suitable habitat patch in which to settle). Individual monk seals probably range over a wide geographic area when they are feeding, but they may be faithful to local and specific breeding sites. At present we know nothing about the colony fidelity and dispersal behavior of the Mediterranean monk seal, so it is impossible to determine the appropriate geographic scale for this species.

Metapopulation theory suggests that local populations exist on a knife edge of persistence. As Hanski (1991) has pointed out, metapopulation extinction may occur for a variety of reasons. As total population size declines, the rate of establishment of new local populations may fall lower than the rate of extinction. Many metapopulation models have two stable states, one of which corresponds to metapopulation extinction. Even a large population may become extinct if, by chance, it falls below a certain threshold level. Increasing habitat fragmentation can have these effects. Burgman et al. (1993) note that fragmentation results in a reduction in population size (because the extent of suitable habitat is reduced), a decrease in the average size of habitat patches, and an increase in the average distance between patches (which may lead to decreased dispersal or increased predation).

It is not surprising, therefore, that the metapopulation model has been seen as a paradigm for the management of rare and endangered species, which are often confined to virtually isolated patches of suitable habitat. Harrison (1991) found few examples that fit the classical metapopulation model. Instead, most populations fell into one of three categories:

1. Mainland–island or source/sink situations, where persistence depends on the existence of one or more extinction-resistant populations
2. Patch populations, where dispersal between patches or subpopulations is so high that the system is essentially a single extinction-resistant population

3. Nonequilibrium metapopulations, where local extinctions occur as part of a species' overall regional decline

Harrison concluded that local extinction is more an incidental than a central feature for most "metapopulations." In a later paper Harrison (1994) suggests that "any set of conspecific populations, possibly but not necessarily interconnected," should be considered a metapopulation. By this definition, metapopulation structure may have great relevance to a species' viability or very little relevance; this must be evaluated case by case. Harrison (1994) advocates an approach based on "spatially explicit computer simulation, in which real landscape geometry is integrated with demographic and dispersal data." Fahrig and Merriam (1994) have stressed that such models must consider the spatial relationship between habitat patches, the dispersal characteristics of the species of interest, and temporal changes in landscape features. Models of this kind have been developed for the northern spotted owl, *Strix occidentalis caurina* (Lamberson et al. 1992, 1994), but they require detailed information on the species' demography, which may not always be available.

It is clear that classical metapopulation models are of little direct relevance to the conservation of the Mediterranean monk seal. Durant and Harwood (1992) have shown that the risks of extinction for the fragmented populations in the western Mediterranean are very high, and Durant et al. (1992) have shown that realistic levels of dispersal are insufficient to counteract extinction. Over its entire geographic range, the Mediterranean monk seal is clearly an example of a nonequilibrium metapopulation (Harrison 1991), although the surviving populations in the eastern Mediterranean and the Atlantic may be viable.

Despite the problems with the concept noted by Harrison (1991), efforts to apply metapopulation theory to conservation problems have been important because they highlight the significance of spatial structure in the dynamics of fragmented populations. In the following section, we use information from the Monk Seal Register to elucidate the spatial structure of monk seal populations in the eastern Mediterranean and Atlantic and use the metapopulation framework, in the sense of Harrison (1994), to identify conservation priorities and research requirements.

## Monk Seal Metapopulations

The Mediterranean monk seal currently breeds at two sites in the Atlantic: the Côtes des Phoques (a 15-km stretch of coastline on the border between Morocco and Mauritania) and the Desertas Islands in Madeira. On the Côtes des Phoques the population has been estimated at 130 individuals (Francour et al. 1990), although the basis for this figure is not entirely clear. Marchessaux (1989*a*) summarizes the counts made at the Côtes des Phoques since 1945. Together with the count of Francour et al. (1990) and data in El Amrani et al. (1992) these show no significant trend with time. Taylor and Gerrodette (1993), however, have emphasized the importance of evaluating the statistical power of such time series to detect trends. No formal evaluation has been carried out for the Côtes des Phoques time series, but variation between counts in the same year is high, so it is likely that only large changes in abundance could be detected. Numbers of seals in the Desertas Islands of Madeira, where a natural reserve was established in 1990, have declined from an estimated 50 individuals in the 1970s (Sergeant et al. 1978) to around 12 individuals (Costa-Neves 1992). Freitas (1994) has used capture–recapture analysis of sightings of individually recognizable animals to estimate that there is a population of 10 individuals (with a 95 percent confidence interval of 8–15) in one part of the reserve.

The colonies on the Côtes des Phoques and in Madeira are separated by 1300 km. Seals are sighted occasionally in the Canary Islands, about halfway between the two sites, but there is no evidence of breeding. Recent reports (Gonzalez et al. 1994) suggest that more than 50 pups may be born in some years on the Côtes des Phoques, and this colony could act as a source for the recolonization of its former habitat. Thus the Atlantic population may have a mainland–island metapopulation structure (Harrison 1991). The Côtes des Phoques colony is particularly vulnerable to chance catastrophic events, however, because of the high levels of aggregation that have been observed. Such events include the introduction of an infectious disease (Osterhaus et al. 1992), an oilspill, or the collapse of the roof of one of the four caves currently in use. The vulnerability of the colony has led to proposals to remove seals from the Côtes des Phoques, either for captive breeding or for translocation to former pupping sites (such as the Canary Islands or the Bay of Dakhla). Griffith et al. (1989) have reviewed numerous attempts at translocation. Those involving threatened, endangered, or sensitive species were much less successful than those using native game species. Translocations involving carnivores, species with low reproductive rates, low numbers of individuals, or declining parent populations were particularly unsuccessful. Because monk seals have all of these characteristics, any proposals for translocation must be carefully evaluated and the potential benefits weighed against the impact of removals on the Côtes des Phoques colony.

Estimates of monk seal abundance in the eastern Mediterranean are more difficult to evaluate. Figures in Reijnders et al. (1993) suggest a population of 220 to 300. A recent comparison of results from studies in different regions of Greece using individual recognition and reports from fishermen gave a similar figure (IUCN/SSC 1994). These seals are spread over a very wide area, with local populations of 10 to 30 individuals. The extent of movement between them is not known, but the Monk Seal Register provides an objective methodology for comparing sightings of recognizable individuals in different parts of Greece. Reported levels of deliberate killing, usually by fishermen, in Greece are so high that extinction is almost certain within 100 years unless the population's underlying rate of increase is greater than 3 percent a year (IUCN/SSC 1994). As it seems unlikely that this figure will be achieved, clearly our highest priority is to eliminate, or reduce to very low levels, deliberate killing of this species.

It will also be necessary to develop methodologies for monitoring the monk seal population in Greece to evaluate the effectiveness of action to reduce levels of deliberate killing. Direct counts will not provide sufficiently precise estimates to allow detection of the relatively small annual changes that are likely to occur. Taylor and Gerrodette (1993) have shown that small populations of this kind can be monitored effectively through capture–recapture analysis of demographic rates. Deliberate tagging of large numbers of monk seals is not practical, but individual recognition, using natural marks, is feasible. Projects using this approach are already under way in parts of Greece. Techniques developed as part of the Monk Seal Register can be used to estimate demographic rates from these data. (See, for example, Barlow 1990 and Buckland 1990.) Estimates of these parameters will also be essential for the modeling studies described shortly.

The monk seal occurs at low densities throughout the eastern Mediterranean. Results of surveys carried out in Greece, compiled in the Monk Seal Register, indicate that caves suitable for pupping are patchily distributed throughout the area. Locally there may be quite high densities of suitable caves, but most are small and could hold only a few pups. The spatial structure of the monk seal population in the eastern Mediterranean, therefore, bears some resemblance to that of the northern spotted owl, where clusters of potential breeding sites occur within patches of suitable habitat, which are themselves embedded in a matrix of unsuitable habitat. Lamberson et al. (1992, 1994) have shown that the population dynamics of the northern spotted owl are determined to a large extent by the size of these clusters and the ability of dispersing individuals to find unoccupied breeding sites in other clusters.

This analogy indicates how a metapopulation approach can be used to prioritize data requirements and develop an appropriate conservation strategy for the eastern Mediterranean monk seal population. Survey data in the Monk Seal Register can be combined with observations at sites where monk seals still breed to map the distribution of suitable breeding caves. Care must be used in interpreting data on habitat suitability, however, because the habitat fragments currently used by a depleted species may not be optimal (Lawton 1993). The resulting maps can then be combined with demographic models of the dynamics of populations within existing clusters to predict future changes in distribution (Burgman et al. 1993).

To complete this process, more information is urgently required on the dispersal and survival of juvenile monk seals from existing colonies to estimate the probability that currently unoccupied sites will be recolonized. The resulting models then can be used to plan the development of a network of protected areas in the eastern Mediterranean. Not all of these areas will contain monk seals, but they should be protected in anticipation of future colonization. Although the highest conservation priority in the eastern Mediterranean must be the reduction—and preferably elimination—of deliberate killing, a suitable network of protected areas is a prerequisite for the recovery of this population.

## Acknowledgments

We thank Alex Aguilar, Angela Caltagirone, Albert Osterhaus, and the Hellenic Society for the Study and Protection of the Monk Seal for samples used in the genetic analysis and Mike Gilpin for pointing out the analogy between northern spotted owls and eastern Mediterranean monk seals.

Data for Figure 11.1 come from Ardizzone et al. (1992), Avella (1979), Avella and Gonzalez (1984a, 1984b), Bayed and Beaubrun (1987), Berkes et al. (1979), Bertram (1943), Boitani (1979), Borges et al. (1979), Boudouresque and Lefevre (1988), Bougazelli (1979), Boulva (1979), Boutiba (1990), Boutiba et al. (1988), Cebrian and Vlachoutsikou (1992), Costa-Neves (1992), Draganovic (1992), Duguy and Cheylan (1980), El Amrani et al. (1992), Francour et al. (1990), Gamulin-Brida (1979), Gamulin-Brida et al. (1965), Gomercic et al. (1984), Gonzalez and Avella (1989), Graziani (1987), Hadjichristophorou (1992), Harwood (1987), Harwood et al. (1984), Hernandez (1986), Kouroutos (1992), Ktari-Chakroun (1979), Lloze (1979), Lopez-Jurado (1980), Lozano-Cabo (1953), Maigret (1984), Maigret et al. (1976), Marchessaux (1987, 1989a, 1989b), Marchessaux and Duguy (1977), Marchessaux and Muller (1987), Mursaloglu (1992), Ozturk et al. (1992), Reiner and dos Santos (1984), Ronald (1984), Sergeant et al. (1978), Vamvakas et al. (1979), Verriopoulos and Kiortsis (1985), Vlachoutsikou and Cebrian (1992).

## REFERENCES

Anonymous. 1994. *Annual Report to Congress 1993*. Washington, D.C.: Marine Mammal Commission.

Ardizzone, G., R. Argano, and L. Boitani. 1992. Le déclin du phoque moine en Italie et sa survie dans un contexte méditerranéen. *Environmental Encounters* (Council of Europe Press) 13:30–31.

Avella, F. J. 1979. The status of the monk seal on the Spanish Mediterranean coast. Pages 95–98 in K. Ronald and R. Duguy, eds., *The Mediterranean Monk Seal*. Proceedings of the First International Conference, Rhodes, Greece, 2–5 May 1978. UNEP Technical Series. Oxford and New York: Pergamon Press.

Avella, F. J., and L. Gonzalez. 1984*a*. Monk seal (*Monachus monachus*) survey along the Mediterranean coast of Morocco. Pages 60–78 in K. Ronald and R. Duguy, eds., *Les Phoques Moines: Monk Seals*. Proceedings of the Second International Conference, La Rochelle, France, 5–6 October 1984. La Rochelle: Annales de la Société des Sciences Naturelles de la Charente-Maritime.

———. 1984*b*. Some data on the monk seal (*Monachus monachus*) in eastern Atlantic. Pages 56–59 in K. Ronald and R. Duguy, eds., *Les Phoques Moines: Monk Seals*. Proceedings of the Second International Conference, La Rochelle, France, 5–6 October 1984. La Rochelle: Annales de la Société des Sciences Naturelles de la Charente-Maritime.

Avise, J. C., J. E. Neigel, and J. Arnold. 1984. Demographic influences on mitochondrial DNA lineage survivorship in animal populations. *Journal of Molecular Evolution* 20:99–105.

Barlow, J. 1990. A birth-interval model for estimating cetacean reproductive rates from resighting data. *Reports of the International Whaling Commission* (Special Issue) 12:155–160.

Bayed, A., and P. C. Beaubrun. 1987. Les mammifères marins du Maroc: Inventaire préliminaire. *Mammalia* 51:437–446.

Berkes, F., H. Anat, M. Kislalioglu, and M. Esenel. 1979. Distribution and ecology of *Monachus monachus* on Turkish coasts. Pages 113–127 in K. Ronald and R. Duguy, eds., *The Mediterranean Monk Seal*. Proceedings of the First International Conference, Rhodes, Greece, 2–5 May 1978. UNEP Technical Series. Oxford and New York: Pergamon Press.

Bertram, G. C. L. 1943. Notes on the present states of the monk seal in Palestine. *Journal of the Society for the Preservation of the Fauna of the Empire* 47:20–21.

Beudels, R. 1992. The monk seal register: Its structure and organisation. *Environmental Encounters* (Council of Europe Press) 13:61–72.

Boitani, L. 1979. Monk seal *Monachus monachus* in Italy: Status and conservation perspectives in relation to the condition of the species in the Western Mediterranean. Pages 61–62 in K. Ronald and R. Duguy, eds., *The Mediterranean Monk Seal*. Proceedings of the First International Conference, Rhodes, Greece, 2–5 May 1978. UNEP Technical Series. Oxford and New York: Pergamon Press.

Borges, J. G., G. E. Maul, G. M. de Vasconcellos, and P. A. Zino. 1979. The monk seals of Madeira. Pages 63–64 in K. Ronald and R. Duguy, eds.,

The Mediterranean Monk Seal. Proceedings of the First International Conference, Rhodes, Greece, 2–5 May 1978. UNEP Technical Series. Oxford and New York: Pergamon Press.

Boudouresque, C., and J. Lefevre. 1988. Nouvelles données sur le status du phoque moine, Monachus monachus, dans la région de l'Oran (Algérie). Marseille: GIS Posidonie Publications.

———. 1992. Ressources alimentaires, phoque moine (Monachus monachus) et stratégie de protection. Environmental Encounters (Council of Europe Press) 13:73–78.

Bougazelli, N. 1979. Quelques données sur le phoque moine d'Algérie (Monachus monachus). Pages 175–178 in K. Ronald and R. Duguy, eds., The Mediterranean Monk Seal. Proceedings of the First International Conference, Rhodes, Greece, 2–5 May 1978. UNEP Technical Series. Oxford and New York: Pergamon Press.

Boulva, J. 1979. Perspective d'avenir du phoque moine de Méditerranée (Monachus monachus). Pages 85–94 in K. Ronald and R. Duguy, eds., The Mediterranean Monk Seal. Proceedings of the First International Conference, Rhodes, Greece, 2–5 May 1978. UNEP Technical Series. Oxford and New York: Pergamon Press.

Boutiba, Z. 1990. Observations récente de phoques-moines (Monachus monachus) sur le littoral Centre Algérien (région d'Alger). Mammalia 54:663–664.

Boutiba, Z., B. Squabria, and D. Robineau. 1988. Etat actuel de la population du phoques-moines (Monachus monachus) sur le littoral ouest algérien (région d'Oran). Mammalia 52:549–555.

Buckland, S. J. 1990. Estimation of survival rates from sightings of individually identifiable whales. Reports of the International Whaling Commission (Special Issue) 12:149–154.

Burgman, M. A., S. Ferson, and H. R. Akçakaya. 1993. Risk Assessment in Conservation Biology. London: Chapman & Hall.

Cebrian, D., and A. Vlachoutsikou. 1992. Recent data on the state of the population of Mediterranean monk seal (Monachus monachus) in Greece. Environmental Encounters (Council of Europe Press) 13:38–42.

Costa-Neves, H. 1992. The monk seal (Monachus monachus): Conservation and monitoring on the Desertas Islands (Madeira, Portugal). Environmental Encounters (Council of Europe Press) 13:21–24.

Draganovic, E. 1992. Distribution and legal protection of the monk seal along the eastern Adriatic coast of Yugoslavia. Environmental Encounters (Council of Europe Press) 13:32.

Duguy, R., and G. Cheylan. 1980. Les phoques des côtes de France 1: Le phoque moine Monachus monachus (Hermann, 1779). Mammalia 44: 203–209.

Durant, S. M., and J. Harwood. 1992. Assessment of monitoring and management strategies for local populations of the Mediterranean monk seal (*Monachus monachus*). *Biological Conservation* 61:81–92.

Durant, S. M., J. Harwood, and R. C. Beudels. 1992. Monitoring and management strategies for endangered populations of marine mammals and ungulates. Pages 252–261 in D. R. McCullough and R. H. Barrett, eds., *Wildlife 2001: Populations*. London: Elsevier Applied Science.

El Amrani, M., S. Hajib, P. Robert, and P. Escoubet. 1992. Observations sur la population de phoque moine *Monachus monachus* entre le Cap Barbas et la péninsule du Cap Blanc. *Environmental Encounters* (Council of Europe Press) 13:25–29.

Fahrig, L., and G. Merriam. 1994. Conservation of fragmented populations. *Conservation Biology* 8:50–59.

Francour, P., D. Marchessaux, A. Arigiolas, P. Campredon, and G. Vuignier. 1990. La population de phoque moine (*Monachus monachus*) de Mauritanie. *Revue d'Ecologie* (*La Terre et la Vie*) 45:55–64.

Freitas, L. 1994. Present status, conservation and future perspective of the Mediterranean monk seal (*Monachus monachus*) colony in Desertas Islands. Unpublished M.S. thesis, University of Aberdeen, Scotland.

Gamulin-Brida, H. 1979. Protection du phoque moine de l'Adriatique. Pages 163–166 in K. Ronald and R. Duguy, eds., *The Mediterranean Monk Seal*. Proceedings of the First International Conference, Rhodes, Greece, 2–5 May 1978. UNEP Technical Series. Oxford and New York: Pergamon Press.

Gamulin-Brida, H., M. Kamenarovic, and Z. Mikulic. 1965. Sur la distribution du phoque moine dans l'Adriatique. *Rapport et Procès-verbaux des Réunions Commission Internationale pour l'Exploration Scientifique de la Mer Méditerranée* 18:257–260.

Gilmartin, W. G., T. C. Johanos, and L. L. Eberhardt. 1993. Survival rates for the Hawaiian monk seal (*Monachus schauinslandi*). *Marine Mammal Science* 9:407–420.

Gomercic, H., D. Huber, and K. Ronald. 1984. A note on the presence of the Mediterranean monk seal (*Monachus monachus*) Hermann 1779 in the eastern part of the Adriatic Sea. Page 51 in K. Ronald and R. Duguy, eds., *Les Phoques Moines: Monk Seals*. Proceedings of the Second International Conference, La Rochelle, France, 5–6 October 1984. La Rochelle: Annales de la Société des Sciences Naturelles de la Charente-Maritime.

Gonzalez, L. M., and F. J. Avella. 1989. La extincion de la foca monje (*Monachus monachus*) en las costas Mediterraneas de la peninsula Iberica y propuesta de una estrategia de actuacion. *Ecología* 3:157–177.

Gonzalez, L. M., J. R. Gonzalez, M. San Felix, E. Grau, L. F. Lopez-Jurado, and A. Aguilar. 1994. Pupping season and annual productivity of the monk seal (*Monachus monachus*) in "Cabo Blanco" peninsula (Western Sahara–Mauritania). In *Proceedings of the Annual Conference of the European Cetacean Society,* 5–6 March 1994, Montpellier.

Gotelli, N. J., and W. G. Kelley. 1993. A general model of metapopulation dynamics. *Oikos* 68:36–44.

Graziani, J. J. 1987. History of the monk seal in Corsica and its disappearance. Pages 108–116 in *Coastal Seal Symposium.* 28–29 April, 1987, Oslo, Norway. Budapest: International Council for Game and Wildlife Conservation.

Griffith, B., J. M. Scott, J. W. Carpenter, and C. Reed. 1989. Translocation as a species conservation tool: Status and strategy. *Science* 245:477–480.

Hadjichristophorou, M. 1992. The Mediterranean monk seals in Cyprus. *Environmental Encounters* (Council of Europe Press) 13:33–34.

Hanski, I. 1991. Single-species metapopulation dynamics: Concepts, models, and observations. *Biological Journal of the Linnean Society* 42:17-38.

Hanski, I., and M. Gilpin. 1991. Metapopulation dynamics: Brief history and conceptual domain. *Biological Journal of the Linnean Society* 42:3-16.

Harrison, S. 1991. Local extinctions in a metapopulation context: An empirical evaluation. *Biological Journal of the Linnean Society* 42:73-88.

————. 1994. Metapopulations and conservation. Pages 111–128 in P. J. Edwards, R. M. May, and N. R. Webb, eds., *Large-Scale Ecology and Conservation Biology.* Oxford: Blackwell.

Harwood, J. 1987. Population biology of the Mediterranean monk seal in Greece. Report to the Commission of the European Community, Brussels.

Harwood, J., S. S. Andersen, and J. H. Prime. 1984. Special measures for the conservation of monk seals in the European Community. Report EUR 9228 EN. Brussels: Commission of the European Community.

Hernandez, E. 1986. Le phoque moine dans les îles Canaries: Données historiques et notes relatives à sa réintroduction. Conseil de l'Europe, Convention relative à la conservation de la vie sauvage et du milieu naturel de l'Europe, 1ère réunion du groupe d'experts sur le phoque moine de Méditerranée, Strasbourg, 15–16 Septembre 1986.

Hoelzel, A. R., and G. A. Dover. 1991. Genetic differentiation between sympatric killer whale populations. *Heredity* 66:191–195.

Hoelzel, A. R., J. Halley, S. J. O'Brien, C. Campagna, T. Arnbom, B. J. LeBoeuf, K. Ralls, and G. A. Dover. 1993. Elephant seal genetic variation and the use of simulation models to investigate historical population bottlenecks. *Journal of Heredity* 84:443–449.

IUCN/SSC 1994. Population and habitat viability assessment for the Greek population of the Mediterranean monk seal (*Monachus monachus*).

Gland: International Union for the Conservation of Nature and Natural Resources.

Johanos, T. C., B. L. Becker, and T. J. Ragen. 1994. Annual reproductive cycle of the female Hawaiian monk seal (*Monachus schauinslandi*). *Marine Mammal Science* 10:13–30.

King, J. E. 1956. The monk seals (genus *Monachus*). Bulletin of the British Museum (Natural History). *Zoology* 3:201–256.

Kouroutos, V. 1992. Distribution, monitoring and conservation projects for monk seal in Greece. *Environmental Encounters* (Council of Europe Press) 13:36–37.

Ktari-Chakroun, F. 1979. Le phoque moine, *Monachus monachus* (Hermann, 1779) en Tunisie. Pages 179–180 in K. Ronald and R. Duguy, eds., *The Mediterranean Monk Seal*. Proceedings of the First International Conference, Rhodes, Greece, 2–5 May 1978. UNEP Technical Series. Oxford and New York: Pergamon Press.

Lamberson, R. H., R. McKelvey, B. R. Noon, and C. Voss. 1992. A dynamic analysis of northern spotted owl viability in a fragmented forest landscape. *Conservation Biology* 6:505–512.

Lamberson, R. H., B. R. Noon, C. Voss, and K. S. McKelvey. 1994. Reserve design for territorial species: The effects of patch size and spacing on the viability of the northern spotted owl. *Conservation Biology* 8:185–195.

Lande, R. 1993. Risks of population extinction from demographic and environmental stochasticity and random catastrophes. *American Naturalist* 142:911–927.

Lawton, J. H. 1993. Range, population abundance and conservation. *Trends in Ecology and Evolution* 8:409–413.

Leigh, E. G. 1981. The average lifetime of a population in a varying environment. *Journal of Theoretical Biology* 90:231–239.

Lloze, R. 1979. Répartition et biologie du *Monachus monachus* (Hermann, 1779), phoque moine, sur la côte oranienne. Pages 101–112 in K. Ronald and R. Duguy, eds. *The Mediterranean Monk Seal*. Proceedings of the First International Conference, Rhodes, Greece, 2–5 May 1978. UNEP Technical Series. Oxford and New York: Pergamon Press.

Lopez-Jurado, L. F. 1980. Observaciones de foça monje (*Monachus monachus* Hermann) en las costas del sureste de la peninsula Iberica. *Donana Acta Vertebrata Sevilla* (Spain) 7:91–93.

Lozano-Cabo, F. 1953. Nota sobre la presencia de un ejemplar de *Monachus monachus* (Hermann) en las costas de Alicante. *Boletín de la real Sociedad Espanola de Historia Natural* 51:135–138.

Maigret, J. 1984. The monk seal (*Monachus monachus*) on the Saharian coast: Present status of the colony. Pages 52–55 in K. Ronald and R. Duguy, eds., *Les Phoques Moines: Monk Seals.* Proceedings of the Second International Conference, La Rochelle, France, 5–6 October 1984. La Rochelle: Annales de la Société des Sciences Naturelles de la Charente-Maritime.

Maigret, J., J. Trotignon, and R. Duguy. 1976. Le phoque moine *Monachus monachus* Hermann 1779, sur les côtes méridionales du Sahara. *Mammalia* 39:413–422.

Marchessaux, D. 1987. The Mediterranean monk seal in Turkey: A survey. *Scientific Reports of Port-Cros National Park, France* 13:13–23.

———. 1989*a*. Recherches sur la biologie, l'ecologie et le statut du phoque moine, *Monachus monachus.* Marseille: GIS Posidonie Publications.

———. 1989*b*. Distribution et statut des populations du phoque moine *Monachus monachus* (Hermann, 1779). *Mammalia* 53:621–642.

Marchessaux, D., and R. Duguy. 1977. Le phoque moine, *Monachus monachus* (Hermann, 1779), en Grèce. *Mammalia* 41:419–439.

Marchessaux, D., and N. Muller. 1987. Le phoque moine, *Monachus monachus:* Distribution, statut et biologie sur la côte Saharienne. Park National Port-cros Publication 1–68.

McConnell, B. J., C. Chambers, and M. A. Fedak. 1992. Foraging ecology of southern elephant seals in relation to the bathymetry and productivity of the Southern Ocean. *Antarctic Science* 4:393–398.

Moritz, C. 1994. Applications of mitochondrial DNA analysis in conservation: A critical review. *Molecular Ecology* 3:401–411.

Mursaloglu, B. 1992. Biology and distribution of the Mediterranean monk seal *Monachus monachus* on Turkish coasts. *Environmental Encounters* (Council of Europe Press) 13:54–57.

Osterhaus, A. D. M. E., I. K. G. Visser, R. L. DeSwart, M. F. Van Bressem, M. W. G. Van de Bildt, C. Örvell, T. Barrett, and J. A. Raga. 1992. Morbillivirus threat to Mediterranean monk seal? *Veterinary Record* 130:141–142.

Ozturk B., A. Candan, and H. Erk. 1992. Cruise results covering the period from 1987 to 1991 on the Mediterranean monk seal occurring along the Turkish coastline. *Environmental Encounters* (Council of Europe Press) 13:43–44.

Reijnders, P., S. Brasseur, J. van der Toorn, P. van der Wolf, I. Boyd, J. Harwood, D. Lavigne, and L. Lowry. 1993. *Seals, Fur Seals, Sea Lions, and Walrus.* Gland: International Union for the Conservation of Nature and Natural Resources.

Reiner, F., and M. dos Santos. 1984. L'extinction imminente du phoque moine à Madère. Pages 79–87 in K. Ronald and R. Duguy, eds., *Les Phoques Moines: Monk Seals.* Proceedings of the Second International Conference, La Rochelle, France, 5–6 October 1984. La Rochelle: Annales de la Société des Sciences Naturelles de la Charente-Maritime.

Repenning, C. A., C. E. Ray, and D. Grigorescu. 1979. Pinniped biogeography. Pages 357–369 in J. Gray and A. J. Boucot, eds., *Historical Biogeography, Plate Tectonics, and the Changing Environment*. Corvallis: Oregon State University Press.

Ronald, K. 1984. The recent status of the monk seal in Yugoslavia. Pages 48–50 in K. Ronald and R. Duguy, eds., *Les Phoques Moines, Monk Seals*. Proceedings of the Second International Conference, La Rochelle, France, 5–6 October 1984. City: Annales de la Société des Sciences Naturelles de la Charente-Maritime.

Sergeant, D., K. Ronald, J. Boulva, and F. Berkes. 1978. The recent status of *Monachus monachus*, the Mediterranean monk seal. *Biological Conservation* 14:259–287.

Smouse, P. E. 1994. Demographic consequences of inbreeding in remnant populations. *American Naturalist* 144:412–431.

Taylor, B., and T. Gerrodette. 1993. The uses of statistical power in conservation biology: The vaquito and the northern spotted owl. *Conservation Biology* 7:489–500.

Vamvakas, C. E., N. Tsimenidis, and H. Kainadas. 1979. Contribution to the knowledge of the distribution patterns of the monk seals *Monachus monachus* in Greek seas: Conservation plan by the establishment of Marine Parks. Pages 147–150 in K. Ronald and R. Duguy, eds., *The Mediterranean Monk Seal*. Proceedings of the First International Conference, Rhodes, Greece, 2–5 May 1978. UNEP Technical Series. Oxford and New York: Pergamon Press

van Bree, P. J. H. 1979. Notes on the differences between monk seals from the Atlantic and the Western Mediterranean. Page 99 in K. Ronald and R. Duguy, eds., *The Mediterranean Monk Seal*. Proceedings of the First International Conference, Rhodes, Greece, 2–5 May 1978. UNEP Technical Series. Oxford and New York: Pergamon Press.

Verriopoulos, G., and V. Kiortsis. 1985. Fréquence et répartition du phoque moine (*Monachus monachus*) en Grèce: Résultats d'une enquête (1982–1984). *Rapport et Procès-verbaux, Commission International pour l'Exploration Scientifique de la Méditerranée* 29:169–170.

Vlachoutsikou, A., and D. Cebrian. 1992. Population status, habitat use, interaction with fisheries and biology study of the Mediterranean monk seals on Zakynthos Island, Greece. WWF Project 3871: Monk seal conservation in the eastern Mediterranean. Athens: World Wildlife Fund for Nature.

# 12

# An Analysis of the Steller Sea Lion Metapopulation in Alaska

*Anne E. York, Richard L. Merrick, and Thomas R. Loughlin*

Steller sea lions (*Eumetopias jubatus*) are the largest otariid and, with males two to three times larger than females, show a marked sexual dimorphism (Figure 12.1). Adult Steller sea lion males are about the size of a Kodiak bear; they average 566 kg (maximum approaching 1120 kg) and grow to 2.8–3.3 m long. Adult females average 263 kg (350 kg maximum) and are approximately 2.3 m long (Fiscus 1961; Calkins and Pitcher 1982; Loughlin and Nelson 1986). The fur of both sexes is light buff to reddish brown and is darker on the chest and abdomen; pups are a deep chocolate brown at birth and molt to a lighter brown after about 6 months (Loughlin et al. 1987). Pups weigh 16–23 kg and are about 1 m long at birth (Calkins and Pitcher 1982). Survival rates of adult females average 80 to 85 percent a year; survival from birth to age 3 is about 50 percent (Pitcher and Calkins 1981; York 1994). Males are sexually mature by 3 to 7 years of age, but they generally are not physically large enough to establish and maintain a territory until 9 to 11 years of age. By 13 or 14 years of age, males are too old and battered to maintain a territory (Pitcher and Calkins 1981; Gisiner 1985). Females reach sexual maturity between 3 and 6 years of age and may produce young into their 20s (Pitcher and Calkins 1981; Calkins and Goodwin 1988); an estimated 62 percent of mature females give birth each year (Pitcher and Calkins 1981). Males establish territories in early May on sites traditionally used by females for giving birth (Gentry 1970; Gisiner 1985). Pups are born from late May to early July; most are born in mid-June (Pitcher and Calkins 1981). Females give birth to a single pup and nurse their pups during the following 4 to 24 months; however, most pups are weaned just prior to the females' next breeding season.

Figure 12.1. A mixed group of Steller sea lions at a breeding rookery on Marmot Island in the Gulf of Alaska. Note the strong sexual dimorphism in size of the large male in the center and females in the background.

## Population Decline and Present Status

The Steller sea lion breeding range extends across the north Pacific Ocean rim from the Kuril Islands and Okhotsk Sea, through the Aleutian Islands, south to central California (Kenyon and Rice 1961; Loughlin et al. 1984; Figure 12.2). The number of sea lions throughout this range has declined by over 50 percent since the 1960s, from an estimated 250,000 to 300,000 to about 116,000 in 1989 (Loughlin et al. 1992; Table 12.1). Most of the decline occurred in Alaska. The declines in Alaska began in the eastern Aleutians in the early 1970s (Braham et al. 1980). This was the center of abundance of the species, and as late as the 1960s some rookeries contained over 15,000 adults and juveniles (collectively called nonpups) (Figure 12.3). By 1985, population declines had spread throughout the Aleutians and eastward into the Gulf of Alaska, at least to the Kenai Peninsula (Merrick et al. 1987). Between 1985 and 1989 the rate of decline increased, and after 1989 population declines were observed in Prince William Sound and the eastern gulf area. A 1994 rangewide survey indicates that the decline is continuing (NMFS 1995). Presently, the area from southeastern Alaska through Oregon is the only region where the number of animals is stable or slightly increasing. Historically, the centers of abundance and distribution were the Gulf of Alaska and the Aleutians, but as the population declined the centers shifted eastward.

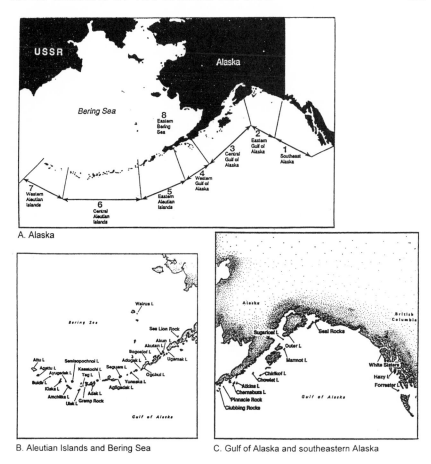

Figure 12.2. Alaska subareas and individual Steller sea lion rookery locations. Kenai-Kiska comprises areas 3–6.

The causes of the decline are unknown. York (1994) has analyzed rates of declines and associated changes in the age structure of the Marmot Island population for 1975–1985 and shown that the simplest explanation for both phenomena was a 20 percent decline in the annual survival of juvenile females. Pascual and Adkison (1994) have analyzed trends at various rookeries and reached a similar conclusion. Consistent with these analyses, Merrick et al. (1988) and Chumbley et al. (1995) have observed that the fraction of juveniles counted in surveys has declined in recent years. Past research efforts have eliminated redistribution, pollution, predation, subsistence use, commercial harvest, disease, and natural fluctuations as principal causes for the decline. (See the review in NMFS 1992.) The major effort to identify causes now focuses on environmental changes and the effects of commercial fishing

## TABLE 12.1.

Estimates of the number of nonpup Steller sea lions by year, geographic area, and stock for the 1960s through 1994

| Area | 1960s | % | 1970s | % | 1985 | % | 1989 | % | 1994 | % |
|---|---|---|---|---|---|---|---|---|---|---|
| Oregon and California | 8,000 | 4 | 5,000 | 3 | 5,200 | 4 | 5,300 | 8 | 7,200 | 14 |
| Southeast Alaska | 7,000 | 4 | 8,000 | 5 | 8,000 | 7 | 12,300 | 18 | 11,400 | 22 |
| Gulf of Alaska | 69,000 | 36 | 55,000 | 34 | 38,000 | 33 | 31,600 | 46 | 17,100 | 33 |
| Aleutians | 99,000 | 51 | 90,000 | 56 | 61,000 | 53 | 19,000 | 28 | 14,800 | 28 |
| Bering Sea | 9,000 | 5 | 4,000 | 2 | 3,000 | 3 | 900 | 1 | 1,700 | 3 |
| Eastern U.S. stock[a] | 15,000 | 8 | 13,000 | 8 | 13,200 | 11 | 17,600 | 25 | 18,600 | 36 |
| Western U.S. stock[b] | 177,000 | 92 | 149,000 | 92 | 102,000 | 89 | 51,500 | 75 | 33,600 | 64 |
| All U.S. | 192,000 | | 162,000 | | 115,200 | | 69,100 | | 52,200 | |

Note: Numbers are adjusted to account for missed sites and animals-at-sea by the methods of Loughlin et al. (1992). The percentage that each area represents of the U.S. population is also shown. Estimates from 1960 through 1989 are from Loughlin et al. (1992), with Bering Sea, Oregon, and California numbers based on new data.

[a] Southeastern Alaska, Oregon, and California.
[b] Gulf of Alaska and westward.

Figure 12.3. View of a portion of the Ugamak Island (eastern Aleutian Islands) rookery, once the largest Steller sea lion rookery in the world. This population has declined from an estimated 13,553 individuals in 1969 to 937 individuals in 1994.

on sea lion survival. Environmental changes influence sea lions indirectly, principally by affecting their prey. Commercial fisheries affect sea lions directly by incidental take, shooting, entanglement in derelict debris or indirectly through competition for prey, disturbance, or disruption of prey schools (Loughlin and Merrick 1989; Ferrero and Fritz 1994). Studies using satellite-linked telemetry on sea lion distribution (Merrick et al. 1994) with concomitant surveys on prey distribution and abundance will be correlated with commercial fishing information to establish the nature and magnitude of overlap between the two.

## Evidence for Metapopulations

A metapopulation is a "population of populations" (Levins 1970). For a particular species, specifying the units of the metapopulation requires specifying the spatial and temporal scales over which the population operates. Hanski and Gilpin (1991) have described the local scale (interaction of individuals in the course of their routine feeding and breeding), the metapopulation scale (like the local scale but with less frequent interaction), and the geographic scale (interaction of individuals over the range of the species). It is not always clear how to apply these definitions to marine mammals, however, given that females with pups routinely leave them on the rookery to feed at sea, and a particular feeding area may draw animals from a large area. For Steller sea lions, and probably for most pinnipeds, we consider the local population to be the "rookery." Thus, in this context, a metapopulation is either a rookery or a group of rookeries, ranging from small to larger geographic clusters. In some contexts, however, it might be reasonable to consider a local population to be smaller—for example, a small population within a rookery—but we have few population data at this level of detail. Genetic evidence for the existence of (at least) two separate stocks of Steller sea lions in Alaska (Bickham et al. 1996) suggests that western Alaska and southeastern Alaska are the limits for the geographic scale.

### Dispersal and Migration Patterns

Steller sea lions are not known to migrate, but they do disperse widely at times of the year other than the breeding season. Animals marked as pups in the Kuril Islands (Russia) have been sighted near Yokohama, Japan (over 1600 km away), and in China's Yellow Sea (over 3700 km away), and pups marked near Kodiak, Alaska, have been sighted near Ketchikan in southeastern Alaska (Loughlin in press). As they approach breeding age, Steller sea lions have a propensity to stay in the general vicinity of the breeding islands and, as a general rule, return to their island of birth to breed as adults. Despite the wide-ranging movements of juveniles and adult males, migration between areas or rookeries by breeding adult females (other than between adjoining rookeries) appears relatively low. Calkins and Pitcher (1982) reported that of 15 recognizable branded females giving birth at Marmot Island (Figure 12.2) and Sugarloaf Island (32 km north of Marmot Island), only one located on a rookery different from the one where she was born. Twenty-one of 424 females branded as pups in 1987 and 1988 at Marmot Island were resighted as adults; only 2 were observed to give birth at a site other than Marmot Island, and those were at nearby Sugarloaf Island (Loughlin in press). Similar information for males is not available.

Fidelity of adult females to specific rookeries is further supported by radio-tracking studies conducted since 1990 (Merrick et al. 1994; Merrick 1995). Satellite-telemetry studies conducted in summer and winter found that adult females forage from a central place (such as a rookery) and return to that place at the end of a trip, which may last from hours to months. The winter range is greater, but most adult females return to the rookery of tagging at the end of the trip, even during the winter.

### Genetic Structure

Bickham et al. (1996) examined a 450-nucleotide base-pair segment of the mitochondrial DNA (mtDNA) control region from skin tissues of 224 Steller sea lions from Oregon, Alaska, and Russia. Their study indicates nucleotide variability at 29 sites that defined 52 haplotypes, which were further grouped into eight maternal lineages. There was no common haplotype that predominated throughout the range; however, there were many haplotypes of relatively low frequency. Based on a phylogeographic analysis (Dizon et al. 1992) utilizing information on distribution, population dynamics, phenetics, and genetics, Loughlin (in press) suggests that these lineages may be subdivided into two genetically differentiated populations, including an eastern population (Oregon and southeastern Alaska) and a western population (Prince William Sound and west). The average sequence divergence among the 52 haplotypes was 1.7 percent (range 0.4–3.8 percent), which is relatively high for the most divergent haplotypes (3.8 percent), indicating that the haplotypes have persisted for a long time. Nucleon diversity estimates for six of the eight populations combined was $h = 0.927$, suggesting high levels of genetic diversity. Furthermore, Bickham et al. (1996) calculated gene flow estimates ($N_c m$), which provided an estimate of about 9.5 females per generation dispersing to another rookery.

Ono (1993) conducted similar mtDNA analysis on skin samples from 11 Steller sea lions from Año Nuevo Island, California, resulting in seven haplotypes, six of which were identical to those identified from southeastern Alaska and Oregon by Bickham et al. (1996). One haplotype was unique to Año Nuevo Island.

# Analysis of Metapopulation Structure

Using survey data, we developed models of adult female population dynamics on three spatial scales: local, cluster or metapopulation, and geographic.

## Survey Methods and Data

Aerial surveys conducted throughout coastal Alaska during mid-June (peak of the breeding season) resulted in a single estimate of adults and juveniles on land for each site surveyed. There are several hundred rookeries and haul-out sites of nonbreeding animals in the survey area; many sites contained few animals and were surveyed on an irregular basis. A subset of these sites (the "trend" sites), which contained most of the population, was surveyed more regularly and was used to produce an index of abundance of the survey area. For example, 27,860 nonpups were counted in 1990 on 152 sites between the Kenai Peninsula and Kiska Island in the western Aleutians; the 77 trend sites contained 22,754 animals, 82 percent of the total (Merrick et al. 1991). Aerial photographic surveys were conducted daily from the second week of June through the end of June between the hours of 1000 and 1600 local time (Withrow 1982). Aircraft were flown at slow speeds (<180 kph), at low altitude (<125 m), and close to shore (<1.6 km). Overlapping 35-mm transparencies were projected onto white paper, and adults and juveniles (non-pups) were counted.

Surveys have been conducted in Alaska on an irregular basis since the late 1950s, but methods were not standardized until the mid-1970s. Moreover, complete simultaneous areawide surveys were not conducted until 1985. Consequently, we restricted our analysis to the period from 1976 through 1994 with an emphasis on the period 1985–1994. During this last period, surveys were conducted six times over the past decade in 1985, 1989, 1990, 1991, 1992, and 1994 (Figure 12.4). Data are available from each of these

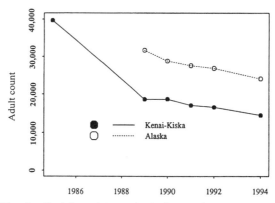

Figure 12.4. Trends of adult and juvenile Steller sea lions counted at Kenai-Kiska ($n$ = 26) and all Alaska ($n$ = 32) trend rookeries during 1985–1992. No count is available for all of Alaska for 1985.

surveys for almost all major rookeries and haul-out sites in Alaska. Areas out-side Alaska were excluded from this analysis because the data were incomplete or had been collected by methods different from the current survey protocol.

Our analysis is restricted to the area from the Kenai Peninsula to Kiska Is-land (areas 3–7 in Figure 12.2) for three reasons: historically, this area has contained the most Steller sea lions (even in 1992 the area contained about 60 percent of the Alaskan population); mtDNA research (Bickham et al. 1996) suggests that the Kenai-Kiska area contains a genetically distinct sub-population of the Alaska Steller sea lion population; and the Kenai-Kiska area has the most complete, longest-running, and consistently collected count database.

## *Model Development*

We developed three models corresponding to the three spatial models in the definition of metapopulations (Hanski and Gilpin 1991). The first was a rookery model (local scale), the second was a cluster of rookeries model (metapopulation scale) in which the clusters were derived from a cluster analysis based on geographic proximity and similarities in the chronology of their patterns of decrease, and the third was a Kenai-Kiska aggregate model (geographic scale). Because we were interested in investigating the long-term viability of the Steller sea lion population in the Kenai-Kiska area, we used the stochastic model of exponential growth described by Dennis et al. (1991) to model counts of adult female sea lions. The model is flexible and requires only population counts as input (Thompson 1991). Thus, the primary data-base used for our analysis consisted of counts on rookeries at the height of the breeding season. We model the female population because of the strong sexual dimorphism of the species. It is easy to incorporate covariates (such as area or time period) into the model and to explore how various factors affect the rates of decline. The model's suitability can be assessed by using the gen-eral methods of linear models.

Let $N_1, N_2, \ldots, N_k$ be counts of numbers of adult female Steller sea lions at times $t_0, t_1, \ldots, t_k$, for a given population (which may be defined by a rookery, a cluster of rookeries, or the entire area), and let $X_i$ be the natural logarithm of $N_i$. The model (Dennis et al. 1991) assumes that $X(t)$ is ap-proximated by a Wiener process with drift with parameters mean $\mu$ and variance $\sigma^2$. This means that the elements of the Leslie matrix governing the population change over time are a multivariate stationary time series. It is not necessary that the times $(t_i)$ are evenly spaced, which in our application was rarely the case. Estimates of the rate of increase and its variance result from a regression analysis on a transformed time series: $Y$ is fitted as a linear function (with no intercept) of $\tau$ where $\tau_i = (t_i - t_{i-1})^{0.5}$ and $y_i = (X_i - X_{i-1})/\tau_i$ —that is, $\tau$ is the square root of the length of time between surveys and $y$ is a average rate of decline over that time period. Then $\mu$ is the slope of the regression line (through zero), and $\sigma^2$ is estimated from squares of the regression residuals. The rate of increase of the population, $r$, is estimated as $\mu + \sigma^2/2$, and $\lambda$, the principal eigenvalue of the underlying Leslie matrix, is estimated as $\exp(r)$. (Note that $N_t = \lambda \, N_{t-1}$.) Formulas for the variances of these quantities and for the likelihood of reaching a particular threshold population size—say $n_{EX}$, (for extinction, $n_{EX} = 1$), and the median and mean time to reach $n_{EX}$ (conditional on the population reaching $n_{EX}$)—are found in Dennis et al. (1991).

Estimates of rates of decline for the Kenai-Kiska and the cluster models were derived from the regression model described above using all data collected from rookeries (but not haul-out sites) after 1975. We noted earlier that the 1975–1985, 1985–1990 and 1990–1994 periods appeared to exhibit different rates of decline, and a formal analysis of that observation was conducted with the regression model. Numbers of surveys on individual rookeries were roughly the same during each time period, so to gain precision in the estimate of the variance of the rates of decline, we used the individual data points from each available rookery observation rather than a sum of all the data over subareas. The residuals of the regression models were examined with normal quantile plots to verify distributional assumptions.

For all models, parameters were estimated using the linear model form of the preceding equations with rookeries, clusters, and time periods coded as nested factors. The distribution of future population size for 200 years was simulated using parameters from the Dennis model. For example, let $\mu_i$ and $\sigma_i^2$ be the estimated parameters for rookery or cluster $i$; then if $x_1, x_2, \ldots, x_{200}$ are a sample of 200 normal $(\mu_i, \sigma_i^2)$ random variables, then one realization of the estimated population size for population $i$ was

$$\left\{ \exp\!\left(n_{it}\right) \right\}_{t=1}^{200}$$

where

$$\ln\!\left(n_{it}\right) = \ln\!\left(n_{i0}\right) + \sum_{t=1}^{200} x_i$$

$n_{i0}$ is the estimated size of the adult female population on rookery $i$ in 1994 and $t$ runs from 1 to 200 years from the present. One thousand simulations for each rookery and cluster were obtained by repeating this process. For each year, the number of animals remaining at each rookery or cluster was totaled to estimate the annual Kenai-Kiska population; the rookery totals were summed within each year to derive the distribution of total population size at time $t$ ($0 < t < 200$). Similarly, we maintained statistics on individual rookery or cluster extinctions ($n_{EX} < 1$). This approach allowed use of all the data available for each rookery and permitted an analysis using the trends at each site. For the rookery model only, the annual population growth rates were truncated at 15 percent on the upper end to ensure that growth remained biologically reasonable for a large mammal. If separate estimates of $\mu_i$ and $\sigma_i^2$ were appropriate for various time periods, then the population was modeled using a mixture of $\mu$ with the mixing proportions corresponding to the proportion of years pertaining to the particular values of $\mu$. Modeling and statistical analyses were performed using Splus (Becker et al. 1988) and SPSS (Norušis 1992).

The raw data used in this analysis were counts of adult animals on rookeries from aerial surveys conducted by NMFS and Alaska Department of Fish and Game (ADFG) during June and July 1976 to 1994. To account for females not present during the surveys on the rookeries, the total number of adult females in the initial population was estimated by multiplying the number of counted nonpups by a correction factor developed as follows. Let

$P$ = number of pups born

$F_{tot}$ = total number of adult females in the population

$b$ = birthrate (average number of pups born to each adult female)

$A_{tot}$ = total number of nonpups counted on rookeries

$m$ = ratio of pups to nonpups on rookeries ($P/A_{tot}$)

Then in equation form, $P = bF_{tot}$ and $P = mA_{tot}$. Therefore, the number of adult females, $F_{tot}$, equals $A_{tot}\ m/b$, and the ratio of numbers of adult females in the population to counted numbers of nonpups on the rookeries, $F_{tot}/A_{tot}$, equals $m/b$. The correction factor, $m/b$, was estimated from independent estimates of $m$ and $b$; $m$ was derived from the slope of a regression line of pup counts on nonpup counts (Figure 12.5) at 0.722 (SE = 0.027); $b$ was derived from data in Pitcher and Calkins (1981) and Calkins and Goodwin (1988) as a binomial (86, 0.627)/86 random variable. This resulted in an estimate of $m/b$ of about 1.159 (SE = 0.106; 95 percent confidence interval 0.976–1.392). In the simulations of the population trajectories, the starting population for 1994 (for each rookery) was estimated by multiplying the observed count of nonpups by a number drawn in a random fashion from the distribution of correction factors ($m/b$) simulated as a normal random variable (mean = 0.72; SD = 0.027) divided by the mean of a binomial random variable (binomial (86, 0.627)/86).

## Cluster Analysis of Rookeries

Visual inspection of population trends at separate rookeries suggested that adjoining rookeries had similar trends. We formally analyzed this relationship using cluster analysis (Everitt 1993). Clusters of rookeries were formed by

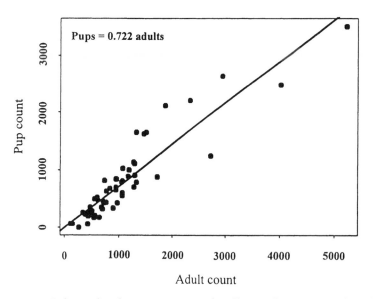

Figure 12.5. Relationship between counts of Steller sea lion pups and adults on rookeries.

grouping individual rookeries into larger and larger clusters until all were members of a single cluster (agglomerative hierarchical clustering). Rookeries were grouped based on the similarities of two variables: distance of a rookery in nautical miles from Outer Island and the slope of a rookery's population trend for some time period (Table 12.2). The former variable was used to maintain some spatial order in the clustering. Slopes of the population trend were calculated using simple regression of the natural log of a rookery's counts over time. Trends were calculated for the periods 1959–1975, 1959–1994, 1976–1985, 1976–1994, 1985–1989, 1985–1994, and 1989–1994 using all available counts. All values were standardized by subtracting their mean and dividing by their standard deviation prior to their entry into the analysis.

Similarities or distances between rookeries or clusters were calculated using the squared Euclidian distance, which was the sum of the squared differences in $Z$ scores for each variable. Ward's (1963) method was used to compare cluster distances. By this approach, the union of all possible pairs of clusters was calculated at each step in the analysis, and the two clusters whose union resulted in the smallest increase in the error sum-of-squares were combined. This was repeated until all rookeries were grouped together as the final cluster. Ward's method was used because it works well with variable data (Everitt 1993), and it was appropriate in this analysis because of the multiple causes of fluctuations in rookery counts.

The cluster analysis was conducted using various combinations of the slope and distance measures. Results were interpreted from dendrograms (visual displays of the clustering) and a matrix of the distances between clusters as produced by the software (CLUSTER from SPSS; Norušis 1992). This information suggested where the logical spatial breaks in the clusters occurred (for example, between the central and western Gulf of Alaska).

## Model Results

We used the three spatial models to assess the probability of population persistance through time, with the assumption that the population continues to behave as it has during the past 25 years.

### Individual Rookery Model

Rates of decline were estimated individually for each rookery using the method of Dennis et al. (1991). The simple rookery model fit the data well except for those rookeries in the western Gulf of Alaska where populations increased in the early 1990s. The use of additional parameters to permit varying rates of decline at different time periods for those rookeries in the western gulf improved the overall fit; the model was also improved by combining the counts from two rookeries that lie very close to each other in the Shumagin Islands (Chernabura and Atkins islands).

When based on the individual rookery model using 1976 to 1994 data, the population simulation predicted that within the Kenai-Kiska area the median number of adult females would decline to fewer than 50 animals and 80 percent of the rookeries would disappear within 100 years (Figures 12.6 and 12.7). Although the female population would be reduced to fewer than 5000 animals during the next 20 years, few rookeries would be completely vacated. At about 40 years from the present, the rate of rookery extinctions would increase sharply. After 60 years, almost half the rookeries would be vacant; after 100 years, only a few rookeries would remain (Figure 12.6). The median time to reach $N_{EX}$ = 100 animals was 101 years compared with 120 years to reach $N_{EX}$ = 50 animals (Figure 12.7).

The relatively high probabilities of overall population persistence under the rookery model are due to positive growth rates at five small rookeries (Akutan Island, Clubbing Rocks, Ugamak Island, Sea Lion Rocks, and Akun Island). These sites have had increasing populations since 1989 (Table 12.2) and relatively low variance in their counts. As a result, the rookery model predicts that some or all of these sites (the specific sites vary between simulations) would persist beyond 100 years, despite the extinction of all other rookeries.

TABLE 12.2.

Distance from Outer Island and rookery population trends (regression correlation coefficient $r$) for six time periods

| Rookery | Distance (nm) | Annual trend | | | | | |
|---|---|---|---|---|---|---|---|
| | | 1959–1975 | 1976–1985 | 1985–1989 | 1989–1994 | 1976–1994 | 1985–1994 |
| Outer | 0 | 0.0158 | -0.1354 | | -0.2070 | -0.1457 | -0.2070 |
| Sugarloaf | 64 | -0.0436 | -0.0620 | -0.1186 | -0.1129 | -0.0978 | -0.1292 |
| Marmot | 83 | 0.0493 | -0.0759 | -0.1899 | -0.1358 | -0.1266 | -0.1685 |
| Chirikof | 283 | 0.0181 | -0.0021 | -0.1519 | -0.2117 | -0.0867 | -0.1796 |
| Chowiet | 298 | -0.0579 | 0.0032 | -0.2568 | -0.0523 | -0.0734 | -0.1298 |
| Atkins | 405 | -0.0303 | -0.0696 | -0.1818 | -0.0446 | -0.0968 | -0.1047 |
| Chernabura | 420 | -0.0635 | -0.1202 | 0.0277 | 0.0463 | -0.0489 | 0.0273 |
| Pinnacle Rock | 484 | -0.0310 | -0.0105 | -0.0376 | -0.0682 | -0.0325 | -0.0573 |
| Clubbing Rocks | 506 | -0.0129 | 0.0031 | -0.0949 | 0.0074 | -0.0181 | -0.0294 |
| Sea Lion Rocks | 506 | -0.0127 | -0.1776 | -0.1118 | 0.0774 | -0.1100 | -0.0256 |
| Ugamak | 590 | -0.0494 | -0.1601 | -0.3015 | 0.1096 | -0.1094 | -0.0301 |
| Akun | 610 | -0.0077 | -0.1232 | -0.2662 | 0.1234 | -0.1154 | -0.0614 |
| Akutan | 632 | -0.0347 | -0.1062 | -0.1966 | 0.0909 | -0.0757 | -0.0208 |
| Bogoslof | 698 | 0.0043 | -0.0741 | -0.1588 | -0.1084 | -0.1031 | -0.1240 |

| | | | | | | |
|---|---|---|---|---|---|---|
| Ogchul | 737 | −0.0508 | −0.0907 | −0.2311 | −0.0117 | −0.1088 | −0.1014 |
| Adugak | 762 | 0.0184 | −0.0821 | −0.2226 | −0.0435 | −0.1154 | −0.1236 |
| Yunaska | 812 | 0.1329 | −0.1236 | −0.2080 | −0.0098 | −0.1284 | −0.1080 |
| Seguam | 882 | 0.0499 | −0.1319 | −0.3966 | −0.0034 | −0.1735 | −0.1610 |
| Agligadak | 900 | 0.0070 | −0.1098 | −0.3340 | −0.5936 | −0.2391 | −0.3695 |
| Kasatochi | 978 | 0.0047 | −0.1026 | −0.1435 | −0.1801 | −0.1349 | −0.1591 |
| Adak/Lake Pt. | 1039 | −0.0496 | 0.0069 | −0.2780 | 0.0955 | −0.0500 | −0.0486 |
| Gramp Rock | 1086 | 0.0445 | −0.0465 | −0.1366 | −0.0650 | −0.0777 | −0.0903 |
| Tag | 1092 | 0.0735 | −0.1019 | −0.1175 | −0.1259 | −0.1180 | −0.1273 |
| Ulak | 1109 | 0.0185 | 0.0382 | −0.2220 | −0.0647 | −0.0717 | −0.1234 |
| Ayugadak | 1179 | 0.0446 | −0.1224 | −0.1476 | −0.0706 | −0.1148 | −0.1025 |
| Lief Cove | 1213 | 0.2498 | −0.1768 | −0.3032 | −0.0884 | −0.1899 | −0.1768 |
| Cape St. Stephen | 1218 | 0.0723 | −0.0814 | −0.2672 | −0.1773 | −0.1614 | −0.2031 |
| Buldir | 1251 | 0.0349 | | | −0.2196 | −0.1803 | −0.2378 |
| Agattu | 1324 | 0.0033 | −0.1379 | −0.0777 | −0.0766 | −0.0988 | −0.0757 |
| Attu | 1354 | 0.0037 | −0.0600 | −0.2239 | −0.0114 | −0.1106 | −0.1164 |

*Note:* A positive value of *r* indicates an increasing slope and a negative value a decreasing slope.

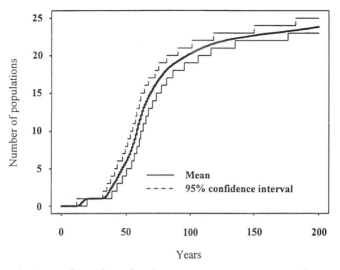

Figure 12.6. Expected number of rookery extinctions over time in the Kenai-Kiska area as predicted by the rookery simulation model estimates of rates of decline for each rookery. Slopes were estimated over 1976–1994 for all data collected outside the western Gulf of Alaska. Within the western gulf, separate slopes were calculated for 1975–1990 and 1990–1994.

## *Metapopulation Model*

Most of the cluster analyses combining multiple correlations (for example, a 1959–1994 analysis using the 1959–1975, 1976–1985, 1985–1989, and 1989–1994 rates of decline) produced unreasonable clusters in which sites that were up to several hundred kilometers apart were included in the same cluster but intermediate sites were assigned to another cluster. This unreasonable clustering was likely due to the reliance on just two counts for the earlier period made before methods were standardized. Clusters determined from the 1959–1975, 1976–1985, 1985–1989, 1989–1994, 1976–1994, and 1985–1994 single-slope analyses produced clusters that were cohesive and without excessive chaining. Patterns of change from the first four clusters (Table 12.3) in combination provide a reasonable description of the course of decline during the 1976–1994, 1985–1994, and 1989–1994 time periods and provide cohesive clusters of rookeries and a synthesis of trends that were used for the cluster-based model (Table 12.4).

The estimated rates of decline among the clusters of rookeries have varied over both space and time (Figure 12.8 and Table 12.5). The patterns in all clusters except the western Gulf of Alaska are similar. Rates of decline varied from 4 to 13 percent a year during 1975–1985 and 1990–1994, with an

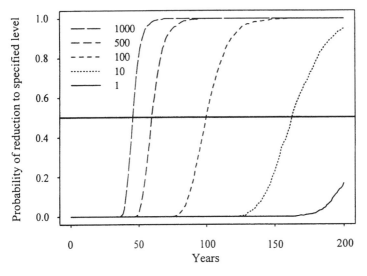

A. Probability of reduction to various levels

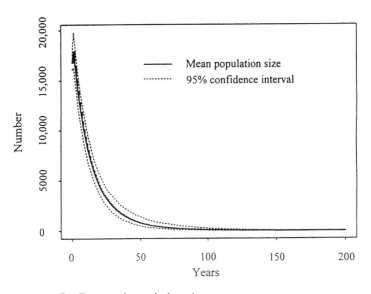

B. Expected population size

Figure 12.7. **A.** Probabilities of reduction of the Kenai-Kiska Steller sea lion population to levels of $N_{\text{EX}}$ = 1, 10, 100, 500, and 1000 predicted by the rookery-simulation model using the mixed individual rookery trend model. **B.** Future population size (mean and 95 percent confidence interval) predicted by the same model.

TABLE 12.3.

Spatial and temporal pattern of the decline of rookeries of
Steller sea lions in the Kenai-Kiska area: 1959–1994

| Period and area | Description |
|---|---|
| ***1959–1975*** | |
| Chowiet westward to Ogchul | Declining slowly |
| Elsewhere | Stable or increasing |
| ***1976–1985*** | |
| Outer Island–Attu | Declining moderately |
| Sea Lion Rocks–Akutan | Declining sharply |
| Yunaska–Kasatochi | Declining sharply |
| Ayugadak–Kiska | Declining sharply |
| Elsewhere | Declining moderately |
| ***1985–1989*** | |
| Chernabura–Sea Lion Rocks | Declining slowly |
| Elsewhere | Declining sharply |
| ***1989–1994*** | |
| Chernabura–Akutan | Stable or increasing |
| Chowiet–Atkins | Declining slowly |
| Ogchul–Seguam | Declining slowly |
| Elsewhere | Declining moderately |

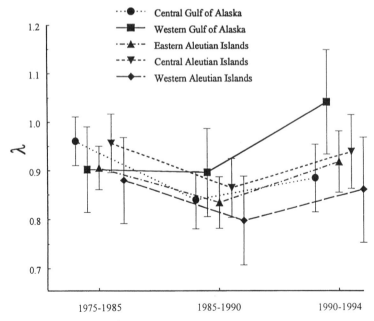

Figure 12.8. Estimates of $\lambda$ (natural logarithm of the rate of increase of the population) for five clusters of rookeries (see text for details) over three time periods (that is, $N_t = \lambda\, N_{t-1}$).

TABLE 12.4.

Groupings of Steller sea lion rookeries in the Kenai-Kiska area determined by a cluster analysis derived from their location and similarities in rates of decline over different periods of time

| Cluster | Area | Rookeries |
| --- | --- | --- |
| 1 | Central Gulf of Alaska | Outer, Sugarloaf, Marmot, Chirikof, Chowiet, Atkins |
| 2 | Western Gulf of Alaska | Chernabura, Pinnacle–Clubbing (before 1985) |
| 3 | Eastern Aleutian Islands | Pinnacle–Clubbing (after 1985), Sea Lion Rocks, Ugamak, Akun, Akutan |
| 4 | Central Aleutian Islands | Yunaska–Seguam, Agligadak–Kasatochi, Adak, Gramp, Tag, Ulak, Ayugadak, Kiska, Buldir |
| 5 | Western Aleutian Islands | Agattu, Attu |

additional 5 to 12 percent decline during 1985–1990. In the western gulf, the population declined at about 10 percent a year during the 1975–1985 and 1985–1990 periods, but since 1990 the population has increased at over 4 percent a year.

When based on the cluster model using the 1976–1994 data, the population simulation predicted that within the Kenai-Kiska area the mean number of adult females would decline to a few dozen animals and the mean number of clusters to one ($\pm 1$) within the next 80 to 100 years (Figures 12.9 and 12.10). The expected female population would be reduced to fewer than 400 females after 50 years, and the median time to reduce the population to 100 females was about 70 years (Figure 12.9b); during the next 20 to 50 years, one cluster of rookeries would probably disappear. About 50 years, from the present, the rate of cluster extinctions would increase sharply. After 70 years, over half the clusters would probably be vacant; after 120 years, the model predicts that only the one cluster would remain. The relatively high probabilities of persistence of the population under the cluster model are due to positive growth rates in the western Gulf of Alaska cluster.

### Kenai-Kiska Geographic Model

The rate of decline in the Kenai-Kiska area has varied over time. During 1975–1985, the annual rate of decline was 5.9 percent (SE = 1.5 percent); during 1990–1994, it was 4.5 percent (SE = 2.0 percent) (Table 12.6). These rates were not significantly different ($P = 0.35$), and the combined annual rate of decline for the two periods is estimated at 6.9 percent (SE = 1.2 percent). The combined rate of decline is higher than the individual rates as a consequence of the lognormal assumption of population size. The estimate of the rate of increase of the population is $r = \mu + \sigma^2/2$; when the estimates are pooled, usually resulting in smaller values of $\sigma^2$, the pooled estimate of $r$ may also be smaller. The population declined at a significantly ($P < 0.001$) higher rate (15.6 percent a year, SE = 1.8 percent) during 1985–1990. Thus, since

TABLE 12.5.
Parameter estimates for the cluster model

| Area and period | μ | SE (μ) | r | SE (r) | λ | SE (λ) | LCB | UCB |
|---|---|---|---|---|---|---|---|---|
| *Central Gulf of Alaska* | | | | | | | | |
| 1 | -0.0406 | 0.0265 | -0.0403 | 0.0265 | 0.9605 | 0.0255 | 0.9102 | 1.0108 |
| 2 | -0.1755 | 0.0354 | -0.1748 | 0.0355 | 0.8396 | 0.0298 | 0.7807 | 0.8985 |
| 3 | -0.1232 | 0.0397 | -0.1224 | 0.0399 | 0.8848 | 0.0353 | 0.8151 | 0.9545 |
| *Western Gulf of Alaska* | | | | | | | | |
| 1 | -0.1033 | 0.0237 | -0.1030 | 0.0237 | 0.9021 | 0.0214 | 0.8598 | 0.9444 |
| 2 | -0.1101 | 0.0319 | -0.1096 | 0.0320 | 0.8962 | 0.0287 | 0.8396 | 0.9529 |
| 3 | 0.0407 | 0.0341 | 0.0413 | 0.0343 | 1.0422 | 0.0357 | 0.9716 | 1.1128 |
| *Eastern Aleutian Islands* | | | | | | | | |
| 1 | -0.0996 | 0.0253 | -0.0993 | 0.0254 | 0.9054 | 0.0230 | 0.8601 | 0.9508 |
| 2 | -0.1815 | 0.0318 | -0.1810 | 0.0319 | 0.8345 | 0.0266 | 0.7819 | 0.8870 |
| 3 | -0.0856 | 0.0347 | -0.0850 | 0.0350 | 0.9185 | 0.0321 | 0.8551 | 0.9820 |
| *Central Aleutian Islands* | | | | | | | | |
| 1 | -0.0453 | 0.0321 | -0.0448 | 0.0321 | 0.9562 | 0.0307 | 0.8956 | 1.0169 |
| 2 | -0.1467 | 0.0353 | -0.1461 | 0.0355 | 0.8641 | 0.0306 | 0.8036 | 0.9246 |
| 3 | -0.0634 | 0.0411 | -0.0626 | 0.0413 | 0.9393 | 0.0388 | 0.8627 | 1.0160 |
| *Western Aleutian Islands* | | | | | | | | |
| 1 | -0.1291 | 0.0507 | -0.1278 | 0.0508 | 0.8800 | 0.0447 | 0.7917 | 0.9684 |
| 2 | -0.2277 | 0.0572 | -0.2261 | 0.0575 | 0.7977 | 0.0458 | 0.7071 | 0.8882 |
| 3 | -0.1513 | 0.0630 | -0.1493 | 0.0634 | 0.8613 | 0.0546 | 0.7534 | 0.9692 |

*Note:* Time period 1 = 1975–1985, 2 = 1985–1990, 3 = 1990–1994; $\mu$ and $\sigma^2$ are the parameters of the underlying Wiener process, $r$ is the instantaneous rate of increase, and $\lambda$ is the rate of increase of the population; LCB and UCB are approximate 95% confidence intervals for $\lambda$.

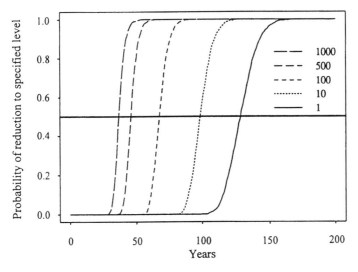

A. Probability of reduction to various levels

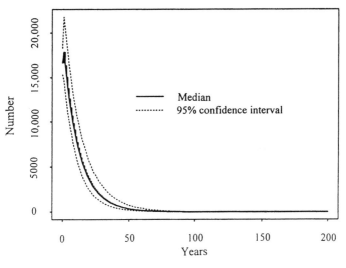

B. Expected population size

Figure 12.9. **A.** Probabilities of reduction of the Kenai-Kiska Steller sea lion population to levels of $N_{EX}$ = 1, 10, 100, 500, and 1000 predicted by the cluster model. **B.** Expected number of female Steller sea lions remaining in the Kenai-Kiska area as predicted by the cluster model.

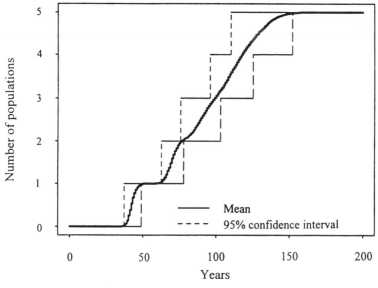

Figure 12.10. Expected number of extinct rookery clusters over time in the Kenai-Kiska area as predicted by the cluster simulation model.

1975 there has been a steady rate of decline of 6.9 percent a year, with an additional drop of about 8.7 percent a year during the late 1980s.

If we model the future population based on the pattern of decline of the past 20 years, then about 75 percent of the time the population was declining at 6.9 percent and 25 percent of the time it was declining at an additional 8.7 percent, or 15.6 percent a year. Clearly this model predicts a continued decline for the sea lion population. Compared to a 1994 female population estimate of 16,781 (Merrick et al. 1994) with correction factor applied, the probability of a reduction to a population of fewer than 500 females is 0.5 within 40 years and 1.0 within 50 years; the probability of extinction (no females) is 0.5 after 106 years and 0.75 after 110 years. The projected median population 20 years from the present is fewer than 3000 females (95 percent CI = 1845–4780); 50 years from the present it is 192 (95 percent CI = 91–352) (Figures 12.11 and 12.12).

The relative influence of the two types of decline (the background rate of decline and the additional drop) on the expected future population size was analyzed by comparing the expected population trajectories assuming, first, a background rate of decline of 6.9 percent and, second, an improvement to no decline (0 percent) coupled with additional declines of 8.7 percent a year occurring every 4 years (25 percent of the time), every 20 years (5 percent of the time), or every 100 years (1 percent of the time) (Figure 12.11). The observed

TABLE 12.6.

Parameter estimates for the Kenai-Kiska area model

| Period | μ | SE (μ) | $\sigma^2$ | r | SE (r) | λ | SE (λ) | LCB | UCB |
|---|---|---|---|---|---|---|---|---|---|
| 1975–1985 | -0.0786 | 0.0144 | 0.0391 | -0.059 | 0.015 | 0.943 | 0.015 | 0.914 | 0.971 |
| 1985–1990 | -0.1600 | 0.0176 | 0.0282 | -0.146 | 0.018 | 0.864 | 0.015 | 0.834 | 0.895 |
| 1990–1994 | -0.0560 | 0.0198 | 0.0229 | -0.045 | 0.020 | 0.956 | 0.019 | 0.919 | 0.994 |
| 1975–1985/ | | | | | | | | | |
| 1990–1994[a] | -0.0708 | 0.0116 | 0.0034 | -0.069 | 0.012 | 0.933 | 0.011 | 0.912 | 0.954 |
| 1985–1990[a] | -0.1600 | 0.0175 | 0.0078 | -0.156 | 0.018 | 0.855 | 0.015 | 0.826 | 0.885 |

*Note:* μ and $\sigma^2$ are the estimates of the mean and variance of the underlying Wiener process; r is the instantaneous rate of increase; λ is the rate of increase of the population; and LCB and UCB are estimated 95% confidence intervals for λ. The rates of increase during 1975–1985 and 1990–1994 were not significantly different ($P = 0.35$), so a combined estimate for these periods is also provided.

[a] Combined rate of decline.

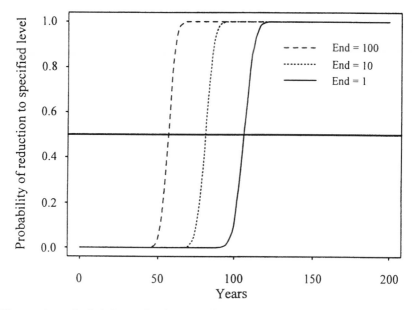

Figure 12.11. Probabilities of reduction of the Kenai-Kiska Steller sea lion popula-tion to levels of $N_{EX}$ = 1, 10, and 100 predicted by the Kenai-Kiska model with a background rate of decline of 7 percent and probability of an additional 8 percent decline 25 percent of the time.

frequency of the larger decline was every 4 years over the past 20 years. Using the background rate of decline of 6.9 percent, the probability of extinction within 120 years was greater than 0.9 in all cases. The median projected pop-ulation after 60 years was 133 females (95 percent CI = 72–461) if additional declines occurred every 4 years, 232 females (95 percent CI = 95–555) if they occurred every 20 years, and 382 females (95 percent CI = 284–621) if they occurred every 100 years. With the background rate of decline assumed equal to zero, the population did not go extinct before 120 years in any of the three cases. If the additional declines occurred 25 percent of the time, how-ever, the median projected population was reduced to 3654 females (95 per-cent CI = 1219–10,020) after 50 years and to 529 females (95 percent CI = 116–2089) after 100 years.

## Discussion

Results of the models provide important insight into the future of the Alaskan Steller sea lion population. If recent trends persist in the Kenai-Kiska area, all three models suggest that the next 20 to 30 years appear crucial to

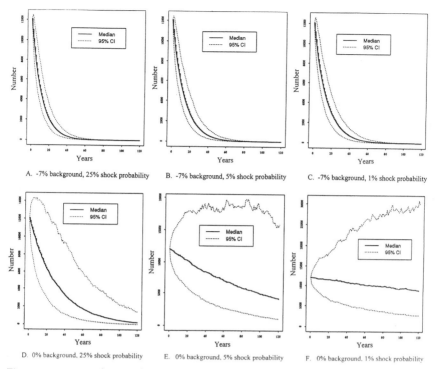

Figure 12.12. Median and 95 percent confidence interval of numbers of female Steller sea lion populations predicted by the Kenai-Kiska model. **A–C:** background rate of decline of 7 percent and probability of an additional 8 percent decline with frequency 25, 5, and 1 percent. **D–F:** background rate of decline of 0 percent and probability of an additional 8 percent decline with frequency 25, 5, and 1 percent.

survival of the southwestern Alaska sea lion population. Though there is no indication that the population is likely to go extinct in 30 years (Figures 12.7, 12.9, and 12.11), populations on some rookeries will probably be reduced to low levels (mean size less than 200 adult females). After 40 years, the extinction of some rookeries is expected. At about the time most rookeries have been vacated, the probability of extinction of the entire Kenai-Kiska geographic population increases rapidly. This trend culminates with all models predicting that the median Kenai-Kiska geographic population will be reduced to fewer than 150 adult females within 100 years. The pattern of the order of rookery extinction indicates that the sea lion population within the Kenai-Kiska area is expected to fragment first into small clusters of sites. As these rookeries disappear, the population will contract to the core of the Kenai-Kiska area in the western Gulf of Alaska and eastern Aleutians (160° to 166° W longitude). The rookery model predicts the longest mean persistence time for the Kenai-Kiska population, the geographic model the shortest, and the cluster model in between, similar to the rookery model in the near term and to the geographic model in the long term (Figure 12.11). The rookery model delays the projected time to reach given population sizes about 12 to 25 years relative to the cluster model and about 25 years relative to the geographic model (Kenai-Kiska, 7 percent background decline with sharper declines 25 percent of the time). The relative importance of the two types of decline apparent in the geographic model with respect to the expected future population size is such that even if the background rate of decline decreases to zero but the sharper decline operates with the same frequency as observed during the past 25 years, the population would still be in serious jeopardy.

Using simple modeling, York (1994) has shown that a 17 to 20 percent decrease in the annual rate of survival from birth to age at first reproduction is the simplest explanation for observed population declines from 1975 to 1985 and a decrease in the average age of adult females. It is mathematically possible that a further decrease in juvenile survival resulted in the 15.6 percent a year decline observed from 1985 to 1990, but it is more likely that an additional increase in the mortality of adults or age of first reproduction occurred. Consider the following simple model of the sea lion population: let $N_t$ be the number of adult females in the population at the end of the pupping season in year $t$. If all females become sexually mature at age $a$ years, then the number of adult females at $N_{t+1}$ is the sum of those that survived from the previous year plus the surviving female pups born a year before. If $s$ is the average annual survival of adult females, $s_j$ is the average annual survival of juvenile females, $b$ is the average number of female pups born to adult females, and $r = \ln(\lambda)$ is the rate of increase of the population, then $N_{t+1} = s\,N_t + s_j^a\,b\,N_{t+1-a}$ and $N_{t+1} = \lambda\,N_t$. Combining these two equations implies $1 = s/\lambda + b\,(s_j/\lambda)^a$. Similar to the life tables reported in York (1994) and Pitcher and Calkins (1981), if $\lambda = 1$, $a = 3$, $b = 0.315$, and $s = 0.85$, then $s_j$ is 0.781. If $\lambda$ decreases to 0.932 (7 percent a year decline), then $s_j$ decreases to 0.609 (assuming no change in $a$, $s$, or $b$), roughly a 22 percent decrease a year in juvenile survival (consistent with York 1994). If $\lambda$ decreases further to 0.855 (observed during 1985–1990—background rate of decline plus the sharper decrease), then under the same assumptions, $s_j$ would decrease to 0.226 (a 71 percent annual decrease in juvenile survival); even if fecundity were lowered by 50 percent, $s_j$ would decrease to 0.285 (a 69 percent annual decrease). Juvenile survival would not have to decrease so sharply if adult survival decreased: a 5 percent decline in $s$ with $\lambda = 0.855$ implies that $s_j$ would decrease to 0.480 (a 39 percent annual decrease), and a 10 percent decline in $s$ implies that $s_j$ would decrease to 0.553 (a 24 percent annual decrease). Assuming no change in adult survival or fecundity or age at first reproduction, the largest rate of decline that could be explained by a decrease in juvenile survival is the natural logarithm of the estimated adult survival rate, 15.9 percent a year. (This is a slightly stronger decline than the 15.6 percent estimated for the Kenai-Kiska area during 1985–1990 and certainly a much smaller decline than in the lower part of the 95 percent confidence interval of the estimated rate of decline during the mid-1980s [Table 12.6].) If the age at first reproduction increased, then the required decreases in juvenile survival would not be so drastic. Assuming no change in adult survival, then in the case of 7 percent annual decline, a 13 percent decrease in $s_j$ is required if $a$ increases to 4 years and an 8 percent decline in $s_j$ if $a$ increases to 5 years; in the case of the 15.6 percent annual decline, a 60 percent decrease in $s_j$ is required if $a$ increases to 4 years, a 51 percent decrease if $a$ increases to 5 years, and a 44 percent decrease if $a$ increases to 6 years.

Although the eastern Gulf of Alaska and the western Aleutians were not included in the Kenai-Kiska analyses, the similarity in recent trends between these areas and the Kenai-Kiska region (Sease et al. 1993) indicates that Steller sea lions could also disappear there. Thus, in 100 years most of the Steller sea lions remaining in North America would most likely be in the area between southeastern Alaska and northern California.

## Caveats

Several caveats are necessary with these analyses. The strongest assumption built into most of the models is that the population will behave as it has since the mid-1970s. We cannot predict the future rate of increase of the population, but if the western Alaska population of Steller sea lions continues to decline as it has during the past 20 years, it will surely go extinct. Capping the rate of increase at 0.15 in the rookery-simulation model biases the results somewhat. This cap was used to constrain interannual variability to a biologically reasonable level and was included because of the low rate of decline with high variance at a few sites. The strong consistency among the models, however, suggests that this bias is likely small. The models do not account for immigration or emigration, and it is possible, though unlikely, that fluctuations in counts in some areas (for example, the western gulf) are due to movement of animals among rookeries.

The choice of model (rookery, cluster, or geographic) has only a marginal effect on the results. The geographic Kenai-Kiska model predicts sharper decreases than either the cluster or the rookery models. Using a mixture of decline rates acknowledges that both moderate declines (as in 1990–1994) and severe declines (as in 1985–1989) can be experienced by the population. The past history of declines in the eastern Aleutians (Merrick et al. 1987), the best-surveyed area, has shown that Alaska's Steller sea lion population decline has not followed a constant trajectory. Periods of apparently moderate decline seem to have been interspersed with periods of acute decline. Whether 1990–1994 is another of these periods of moderation or is truly an indication that the declines are abating remains unknown. Consequently, we use all available data taken since 1975 as the basis for our viability analyses.

All three models have drawbacks, but taken together they seem to provide a reasonable picture of the population's probability of persistence. The Kenai-Kiska geographic model oversimplifies the analysis to some extent because it uses only one trend for the entire area and ignores differences among rookeries. But by virtue of the large area covered, there would be relatively little movement by sea lions into or out of the area. The rookery model (local scale), which explicitly recognizes that each rookery has its own probability of persistence, provides results that may be misleading because it ignores the possibility of dispersal among rookeries. The rookery model also provides detail on the pattern of fragmentation that could be expected as the population declines at varying rates in different areas. The Kenai-Kiska geographic model provides a more conservative estimate of the population trends than the individual rookery model. The cluster model (metapopulation model) is an attempt to bridge the gap between the coarse and fine scales—it recognizes the spatial pattern of differential rates of decline among the clusters, but it increases the probability that the clusters are closed to dispersal. The population and local-scale models can be considered as boundaries and the metapopulation model as a median for the prediction of future persistence of the population. The near-term metapopulation trajectories (0–25 years from the present) are similar to the rookery model; the mid-term trajectories (25–50 years from present) are about equidistant between the local and geographic models; and the long-term trajectories (50–200 years from the present) lie closer to the geographic-scale model (Figure 12.13).

Inspection of the rates of decline from Table 12.2 indicates that population trajectories changed abruptly between certain rookeries. These results are summarized for the 1959–1975, 1976–1985, 1985–1989, and 1989–1994 time periods in Table 12.3. Some of these breakpoints are common to more than one period and may indicate a persistent environmental change to which the decline rates are linked. Exclusion of animals on haul-out sites will not significantly affect the analysis because most were either juveniles or nonbreeding adult males. Ultimately, the fate of the population rests on the persistence of the breeding female population, which is why we focus on rookeries.

The models may overestimate the time to reach a given population size because the only variability explicitly considered in the model is environmental

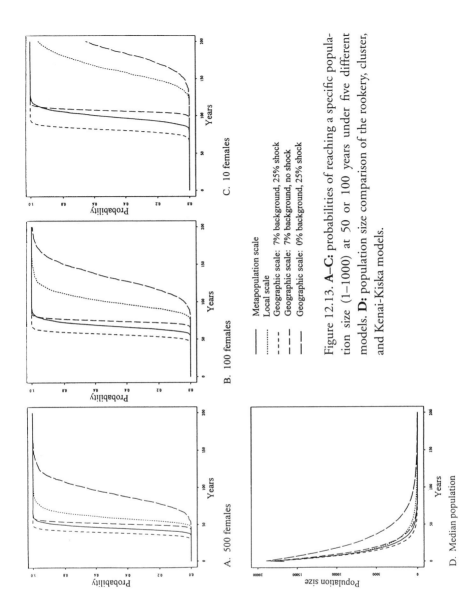

A. 500 females

B. 100 females

C. 10 females

D. Median population

Metapopulation scale
Local scale
Geographic scale: 7% background, 25% shock
Geographic scale: 7% background, no shock
Geographic scale: 0% background, 25% shock

Figure 12.13. **A–C:** probabilities of reaching a specific population size (1–1000) at 50 or 100 years under five different models. **D:** population size comparison of the rookery, cluster, and Kenai-Kiska models.

stochasticity (May 1974; Shaffer 1981). Other factors that can affect population dynamics—and as a result the probability of the population's persistence—include demographic and genetic stochasticity (Shaffer 1981), natural catastrophes (May 1974), and immigration stochasticity (Hanski 1991). The significance of the first two factors to Steller sea lions is unclear. Natural catastrophes (such as the epizootic that recently struck North Sea phocid seals) and catastrophes of human origin would significantly increase the probabilities of extinction. At least one such event may have struck Alaska's Steller sea lion population from 1985 to 1989. As events of the frequency and scale of the 1985 to 1989 decline were incorporated into all three models, some measure of the effect of catastrophic changes is included in this analysis. But an additional event (like the one that occurred in the eastern Aleutians during the early 1970s when the population there declined by over 58 percent; Braham et al. 1980) could overwhelm an already depleted population, driving it more rapidly to extinction.

Immigration could be a positive factor for the sea lion population if significant numbers of females disperse from their natal rookeries with increasing populations to breed on other rookeries. Available data (from both tagging and genetic studies) indicate that sea lion female dispersal rates have been low. Hanski (1991), citing Roff (1974), Hamilton and May (1977), and others, argues that the dispersal rate results from natural selection and evolves to overcome local extinctions. If the present measures of Steller sea lion immigration rates are indicative of its potential to disperse, this factor cannot possibly overcome the local extinction rates predicted by the present estimates of decline. To what extent the apparent lack of dispersal is a consequence of the population decline—or if dispersal would increase if the population were increasing—is unclear. The Steller sea lion, as a species, appears to have evolved to exploit and reproduce across a wide range, but as individuals they exploit fairly small areas. Perhaps the sea lion's apparent lack of dispersal ability is a result of an evolutionary past in which there was little local variability and major environment changes were sufficiently widespread that long-term trends of either increase or decrease were a regular pattern. It would be comforting to believe that sea lions are at present caught in one of these long-term declines and that in the future they will recover (as they must have recovered from any long-term decline in the past). But the ecosystem that sea lions now occupy contains an intense competitor—technological humans—that was not present during their evolutionary history.

If the downward trend in the Steller sea lion population continues during the next 20 or 25 years, the population will reach a critical state. Are there last-ditch rescue measures that could be used to save the population, such as artificial relocation (see Chapter 16 in this volume)? To our knowledge, there has never been a successful large-scale transplant of an otariid species. Sea lions breed in colonies, and their behavior is such that conspecific attraction is an important component of their behavior (Gentry 1970). Due to immigration–extinction stochasticity, species that exhibit strong conspecific attraction may be at greater risk of extinction than those that disperse randomly following habitat fragmentation (Ray et al. 1991; Hanski 1991). As a protective measure since 1990, fisheries managers have prohibited fishing within 10 nautical miles of all rookeries and within 20 nautical miles of some rookeries for part of each year. We believe that the main cause of the sea lion decline is related to juvenile survival, and given that juvenile sea lions are not completely surveyed, we will not know whether this measure is having a positive effect until substantial numbers of females born since 1990 have been recruited into the breeding population.

## Acknowledgments

We wish to thank James Lerczak, Dale McCullough, Grant Thompson, and Brian Walker for critical reviews of this chapter.

## REFERENCES

Becker, R. A., J. M. Chambers, and A. R. Wilks. 1988. *The New S Language.* Pacific Grove, Calif.: Wadsworth and Brooks/Cole.

Bickham, J. W., J. C. Patton, and T. R. Loughlin. 1996. High variability for control-region sequences in a marine mammal: Implications for conservation and biogeography of the Steller sea lion (*Eumetopias jubatus*). *Journal of Mammalogy* 77:95–108.

Braham, H. W., R. D. Everitt, and D. J. Rugh. 1980. Northern sea lion population decline in the eastern Aleutian Islands. *Journal of Wildlife Management* 44:25–33.

Calkins, D. G., and E. Goodwin. 1988. Investigation of the declining sea lion population in the Gulf of Alaska. Unpublished report. Alaska Department of Fish and Game, Anchorage.

Calkins, D. G., and K. W. Pitcher. 1982. Population assessment, ecology and trophic relationships of Steller sea lions in the Gulf of Alaska. Pages 447–546 in *Environmental Assessment of the Alaskan Continental Shelf.* Final report. Washington, D.C.: U.S. Department of Commerce and U.S. Department of the Interior.

Chumbley, M. K., J. Sease, M. Strick, and R. Towell. 1995. Field studies of northern sea lions at Marmot Island, Alaska, during 1979–94. National Marine Mammal Laboratory, Seattle.

Dennis, B., P. L. Munholland, and J. M. Scott. 1991. Estimation of growth and extinction parameters for endangered species. *Ecological Monographs* 61:115–143.

Dizon, A. E., C. Lockyer, W. F. Perrin, D. P. DeMaster, and J. Sisson. 1992. Rethinking the stock concept: A phylogeographic approach. *Conservation Biology* 6: 24–36.

Everitt, B. S. 1993. *Cluster Analysis.* London: Edward Arnold.

Ferrero, R., and L. W. Fritz. 1994. Comparisons of the walleye pollock, *Theragra chalcogramma,* harvest to Steller sea lion, *Eumetopias jubatus,* abundance in the Bering Sea and Gulf of Alaska. NOAA Technical Memo NMFS-AFSC-43. Seattle: U.S. Department of Commerce.

Fiscus, C. H. 1961. Growth in the Steller sea lion. *Journal of Mammalogy* 42:195–200.

Gentry, R. L. 1970. Social behavior of the Steller sea lion. Ph.D. thesis, University of California, Santa Cruz.

Gisiner, R. C. 1985. Male territorial and reproductive behavior in the Steller sea lion, *Eumetopias jubatus.* Ph.D. thesis, University of California, Santa Cruz.

Hamilton, W. D., and R. M. May. 1977. Dispersal in stable habitats. *Nature* 269:578–581.

Hanski, I. 1991. Single-species metapopulation dynamics: Concepts, models and observations. *Biological Journal of the Linnean Society* 42:17–38.

Hanski, I., and M. Gilpin. 1991. Metapopulation dynamics: Brief history and conceptual domain. *Biological Journal of the Linnean Society* 42:3–13.

Kenyon, K. W., and D. W. Rice. 1961. Abundance and distribution of the Steller sea lion. *Journal of Mammalogy* 42:223–234.

Levins, R. 1970. Extinction. Pages 77–107 in M. Gerstenhaber, ed., *Some Mathematical Questions in Biology.* Providence, R.I.: American Mathematical Society.

Loughlin, T. R. In press. Using the phylogeographic method to identify Steller sea lion stocks. In A. Dizon and W. Perrin, eds., *Use of Genetic Data to Diagnose Management Units.* NOAA Technical Report. Seattle: U. S. Department of Commerce.

Loughlin, T. R., and R. L. Merrick. 1989. Comparison of commercial harvest of walleye pollock and northern sea lion abundance in the Bering Sea and Gulf of Alaska. Pages 679–700 in *Proceedings of the International*

*Symposium on the Biology and Management of Walleye Pollock.* Sea Grant Report AK-SG-89-01. Anchorage: University of Alaska.

Loughlin, T. R., and R. Nelson. 1986. Incidental mortality of northern sea lions in Shelikof Strait, Alaska. *Marine Mammal Science* 2:14–33.

Loughlin, T. R., M. A. Perez, and R. L. Merrick. 1987. *Eumetopias jubatus. Mammalian Species* 283:1–7.

Loughlin, T. R., A. S. Perlov, and V. A. Vladimirov. 1992. Range-wide survey and estimation of total abundance of Steller sea lions in 1989. *Marine Mammal Science* 8:220–239.

Loughlin, T. R., D. J. Rugh, and C. H. Fiscus. 1984. Northern sea lion distribution and abundance: 1956–1980. *Journal of Wildlife Management* 48:729–740.

May, R. M. 1974. *Complexity and Stability in Model Ecosystems.* 2nd ed. Princeton: Princeton University Press.

Merrick, R. L. 1995. The relationship of the foraging ecology of Steller sea lions (*Eumetopias jubatus*) to their population decline in Alaska. Ph.D. thesis, University of Washington, Seattle.

Merrick, R. L., L. M. Ferm, R. D. Everitt, R. R. Ream, and L. A. Lessard. 1991. Aerial and ship-based surveys of northern sea lions (*Eumetopias jubatus*) in the Gulf of Alaska and Aleutian Islands during June and July 1990. NOAA Technical Memo NMFS-F/NWC-196. Seattle: U.S. Department of Commerce.

Merrick, R. L., P. Gearin, S. Osmek, and D. Withrow. 1988. Field studies of northern sea lions at Ugamak Island, Alaska, during the 1985 and 1986 breeding seasons. NOAA Technical Memo NMFS-F/NWC-143. Seattle: U.S. Department of Commerce.

Merrick, R. L., T. R. Loughlin, and D. G. Calkins. 1987. Decline in abundance of the northern sea lion, *Eumetopias jubatus,* in Alaska, 1956–86. *Fisheries Bulletin* (U.S.) 85:351–365.

Merrick, R. L., T. R. Loughlin, G. A. Antonelis, and R. Hill. 1994. Use of satellite-linked telemetry to study Steller sea lion and northern fur seal foraging. *Polar Research* 13:105–114.

NMFS. 1992. Recovery plan for the Steller sea lion (*Eumetopias jubatus*). Steller Sea Lion Recovery Team for NMFS, Silver Spring, Maryland.

———. 1995. Status review of the United States Steller sea lion (*Eumetopias jubatus*) population. Unpublished manuscript. National Marine Mammal Laboratory, Alaska Fisheries Science Center, Seattle.

Norušis, M. J. 1992. SPSS for Windows: Professional statistics, release 5. Chicago: SPSS.

Ono, K. A. 1993. Steller sea lion research at Año Nuevo Island, California, during the 1992 breeding season. Report 40JGNF210351. La Jolla, Calif.: Southwest Fisheries Science Center.

Pascual, M. A., and M. D. Adkison. 1994. The decline of northern sea lion populations: Demography, harvest, or environment? *Ecological Applications* 4:393–403.

Pitcher, K. W., and D. G. Calkins. 1981. Reproductive biology of Steller sea lions in the Gulf of Alaska. *Journal of Mammalogy* 62:599–605.

Ray, C., M. Gilpin, and A. T. Smith. 1991. The effect of conspecific attraction on metapopulation dynamics. *Biological Journal of the Linnean Society* 42:123–134.

Roff, D. A. 1974. The analysis of a population model demonstrating the importance of dispersal in a heterogeneous environment. *Oecologia* 15:259–275.

Sease, J. L., J. P. Lewis, D. C. McAllister, R. L. Merrick, and S. M. Mello. 1993. Aerial and ship-based surveys of Steller sea lions (*Eumetopias jubatus*) in Southeast Alaska, the Gulf of Alaska, and Aleutian Islands during June and July 1992. NOAA Technical Memo NMFS-AFSC-17. Seattle: U.S. Department of Commerce.

Shaffer, M. L. 1981. Minimum population sizes for species conservation. *BioScience* 31:131–134.

Thompson, G. G. 1991. Determining minimum viable populations under the Endangered Species Act. NOAA Technical Memo NMFS F/NWC-198. Seattle: U.S. Department of Commerce.

Ward, J. H. 1963. Hierarchical grouping to optimize an objective function. *Journal of the American Statistical Association* 58:236–244.

Withrow, D. E. 1982. Using aerial surveys, ground truth methodology, and haul out behavior to census Steller sea lions, *Eumetopias jubatus*. M.S. thesis, University of Washington, Seattle.

York, A. E. 1994. The population dynamics of the northern sea lions, 1975–85. *Marine Mammal Science* 10:38–51.

# 13

## Metapopulation Models, Tenacious Tracking, and Cougar Conservation

*Paul Beier*

In 1987, the California Department of Fish and Game (CDFG) and the University of California began a 5½ year study to document the home range, density, movements, food habits, reproduction, survival, and related parameters for cougars (*Felis concolor*) in the southern third of the Santa Ana Mountains of southern California. When he hired me to lead this study in early 1988, Dr. Reginald Barrett ruefully quipped that urban growth in the study area was so rapid that I would "document the demise of this cougar population." Over the next few months, this remark seemed like a prophecy. After the death of the territorial male in February 1988, there was no reproductive activity among the seven radio-tagged females and no evidence of a breeding male in the study area for 12 months (Padley 1990). These females were visiting one another at intervals that seemed to coincide with estrus cycles (Padley 1990), provoking among the field crew a spate of ribald humor that gradually segued into anxiety. Increasingly concerned, I expanded the study in early 1989 to include the entire mountain range and was relieved to note reproduction among the cougars to the north. Also at this time two young males settled into the original study area and the females promptly stopped visiting one another and started bearing cubs.

The optimism created by the 1989 birth pulse was soon tempered by the high mortality observed in the new cohort and, moreover, by an awareness that in less than 5 years urban sprawl could completely isolate the population and internally fragment at least one large piece of habitat from the core area. I believed that maintaining connectivity within and beyond the Santa Ana Mountains would allow this population to persist and that unless habitat areas functioned spatially as a metapopulation, this population was doomed.

But 2 years of data provided only biological anecdotes to support this view, and I did not know if my hunch was correct or, if it was correct, which meta-population configurations would persist or fail. If those with power over land use decisions were to be persuaded to limit urban growth, I would need scientific evidence that was both rigorous and easily understood. In particular, I would need:

1. A population viability analysis (PVA) that would predict the fate of a cougar population under various scenarios, highlighting two elements at the heart of metapopulation dynamics—patch size and patch connectivity. The emphasis on these elements was crucial because these are the two factors controlled by land use decisions. Patch area is controlled by restricting human development; connectivity is controlled by protecting corridors for wildlife movement to adjacent populations. Although the predictions of my model would also depend on age-specific survival rates, carrying capacity, subroutines for various types of stochasticity, and the functions relating vital rates to density, planning commissioners might be distracted by these aspects of population ecology. To get their attention and keep it, I needed to highlight the extent to which the population's fate depended on land use decisions.

2. A PVA model that simulated cougar population dynamics realistically. Erroneous model predictions can arise from two interacting sources: uncertainty in parameter estimates (which I address via sensitivity analyses—making predictions under both high and low estimates for poorly known parameters) and oversimplification of a species' life history (which I minimize by species-specific algorithms). Because huge profits for developers are at stake, biologists representing development interests could rightly argue that planners should not rely on a model that ignores certain aspects of cougar life history, such as the cougar's ability to shorten the interbirth interval from about 24 months (when cubs survive to dispersal) to as little as 4 months (if cubs die early). Indeed a generic model that was too pessimistic might not only limit development unfairly but also cause a cougar population to be written off as a "hopeless case." Generic models could also be too optimistic; by accounting only for females, for instance, a model would fail to predict the lack of reproduction observed when males were absent in 1988. Because the consequences of error were so severe, I wanted the model to be as objective and realistic as possible.

3. Data on corridor use. My data on the local population and published values were available to estimate most vital rates, but I had no data

on cougar immigration. In this landscape, dispersing cougars would immigrate via narrow habitat corridors or not at all. I expected the model to show that immigration was critically important to the metapopulation. Thus, it would be crucial to know whether dispersing cougars would immigrate via such corridors into semi-isolated habitat patches and to know where corridors should be retained in the landscape.

Meeting these three needs became the focus of the past 3 years of the research effort. This chapter recapitulates my published descriptions of the population viability analysis (Beier 1993) and cougar use of corridors (Beier 1995). In addition, I partition stochastic variation in observed survival rates into environmental and demographic components, apply the model to another cougar metapopulation (in and near the Santa Monica Mountains), and discuss the successes and failures of this metapopulation analysis and fieldwork in helping to conserve the Santa Ana Mountains metapopulation.


# Methods

I modeled population trajectories under various levels of habitat area and immigration. Also, I used radio tracking to estimate population parameters, and document movement between habitat patches and identify travel routes and potential habitat corridors for the Santa Ana Mountain cougar metapopulation.

## Simulation Model

I did not directly model the dynamics of an entire cougar metapopulation. Instead, the model simulated the trajectory of a subpopulation linked to adjacent populations via several levels of immigration. Linked Leslie matrices for males and females were used to model the Allee effect, make vital rates depend on the density of same sex individuals, and allow for sex-biased immigration. Additional subroutines introduced demographic and catastrophic stochasticity into survival and fecundity rates.

For each combination of input conditions, I simulated a 100-year population trajectory 100 times, recording the number of runs on which the population went extinct, mean population size in year 100, and other summary statistics. I set the initial number of adults (animals $\geq 2$ years of age) at carrying capacity (with equal numbers in each year class) and initial numbers of 0-year-olds and 1-year-olds at 50 and 25 percent, respectively, of the total number of adult females. This initial age distribution represents a population that had stable reproductive and survival rates for the decade before the start of the simulations. Because all models are simplifications and thus prone to error, and because I ran only 100 simulations per combination of input conditions, I considered any extinction risk of 2 percent or greater to be significant and unacceptable.

The factors controlled by land use decisions were area of habitat and level of immigration. I ran simulations starting with 200 km$^2$ of habitat and in increments of 200 km$^2$ until the extinction risk declined to zero. No estimates of immigration rates for cougars were available. Therefore, in addition to simulating no immigration (no corridor), I simulated the three lowest levels of immigration that would qualify the area as part of a metapopulation (that is, interacting with adjacent groups of conspecifics): one male immigrant per decade, two males per decade, and three males plus one female per decade. If these levels had no influence on population persistence, I planned to simulate higher immigration rates. The sex bias in these immigration rates reflects the fact that males are more likely than females to disperse out of their natal mountain range (Ashman et al. 1983; Anderson et al. 1992).

For each combination of habitat area and level of immigration I simulated the population dynamics under many combinations of estimates for life history and environmental attributes (Table 13.1). Because male and female equilibrium densities and juvenile survival rates are hard to estimate, may vary geographically, and have a profound influence on the results, the simulations used 36 permutations of values for these parameters. I used separate estimates of carrying capacity for adult females and males because social intolerance among adult females (calibrated to prey abundance and influenced by vegetation and topography) is thought to regulate their density, whereas territoriality among males (competing for access to females) regulates male density (Seidensticker et al. 1973).

The model included subroutines to simulate density dependence (Table 13.2), including an Allee effect, inhibition of reproduction for the youngest females when the population exceeded carrying capacity, enhancement of survival rates at low density, and decline in survival rates (especially for juveniles and dispersers) at high density. Lacking empirical data, I chose the survival rate functions for their computational simplicity (Beier 1993). However, I tested various alternative functions for density dependence in survival rates and found that neither risk of extinction nor ending population size varied among them (Beier 1993). Extinction risk was markedly higher in preliminary analyses with density-independent survival rates, but this approach also produced ending population sizes (for populations that survived) that far exceeded carrying capacity (Beier 1993).

Most studies report adult sex ratios skewed toward females; at some age, therefore, survival rates of males must be lower than those of females.

TABLE 13.1.

Values for biological parameters used in simulating cougar population dynamics

| Parameter | Values used |
| --- | --- |
| Mean litter size | 2.8[a] |
| Juvenile survival rate[b] | 0.65 for females, 0.60 for males<br>0.75 for females, 0.70 for males |
| Adult survival rate[c] | 0.75<br>0.85 |
| Probability that a resident female bears a litter in a given year | For 0- and 1-year-old females: 0%<br>For 2-year-old females: <40%<br>For older females: 0% if litter from previous year survived 1 year; 100% if litter from previous year died |
| Maximum life span | 12 years |
| Carrying capacity (breeding adults per 100 km² | Sex ratio of 2 adult females per adult male:<br>0.4 females, 0.2 males<br>0.6 females, 0.3 males<br>0.8 females, 0.4 males<br>1.0 females, 0.5 males<br>1.2 females, 0.6 males<br><br>Sex ratio of 3–4 adult females per adult male:<br>0.8 females, 0.2 males<br>1.2 females, 0.4 males<br><br>Sex ratio of ~1 adult female per adult male:<br>0.4 females, 0.4 males<br>0.8 females, 0.6 males |

*Source:* Anderson (1983), Anderson et al. (1989), Ashman et al. (1983), Beier and Barrett (1993), Currier et al. (1977), Eaton and Velander (1977), Hemker et al. (1984), Hemker et al. (1986), Hopkins (1981), Hopkins (1989), Hornocker (1970), Lindzey et al. (1988), Logan et al. (1986), M. Jalkotzy and I. Ross, (Calgary, Alberta, unpublished data), Murphy (1983), Neal et al. (1987), Quigley et al. (1989), Robinette et al. (1961), Robinette et al. (1977), Seidensticker et al. (1973), Shaw (1977), Sitton and Wallen (1976), Young (1946), with justifications for specific values given by Beier (1993).

*Note:* Survival and breeding probabilities are for a population at carrying capacity and were modified as indicated in Table 13.2.

[a] A value of 2.4 produced population sizes much smaller than carrying capacity, even when used together with optimistic estimates for other parameters. A value of 3.2 produced about the same extinction risk as a value of 2.8.

[b] Both sexes 0 and 1 year old and males 2 years old.

[c] Females ≥ 2 years old and males ≥ 3 years old.

TABLE 13.2.
Density-dependent relationships used in simulating cougar population dynamics

| Vital rate | Relationship to density | Rationale |
|---|---|---|
| Litter size | Density independent | Low cost of gestation (92 days, 500-g neonate mass). |
| Probability that a resident female will breed | When NF > KF, 20% of resident females in excess of K breed; youngest females assigned to nonbreeding status. | Young females reproduce only after home range establishment (Seidensticker et al. 1973). |
| Probability that a resident female will breed | Allee effect:[a] When NM < KM (some male territories are vacant), % of NF breeding is multiplied by: $1 - [(KM - NM \cdot 1.15^{(KM-NM)}/KM]$ | To reflect our 1989 observations. Because territory of each adult male increases by 15% for each "deficit male," the effect is very mild except at very low $N$. |
| Survival rate, 0-year-olds (M, F) and 1–2-year-old F | $= S(KF/NF)^{0.5}$, truncated to values between 0.3 and 0.9 | To reflect competition among mothers and dispersing females. |
| Survival rate, 1-year-old males | $=$ minimum of $S(KF/NF)^{0.5}$ or $S(KF/NF)^{0.5}(KM/NF)^{0.5}$, truncated to values between 0.3 and 0.9 | To reflect competition among mothers for food resources as well as mortality due to adult males during dispersal. |
| Survival rate, 2-year-old males | $= S(KM/NM)^{0.5}$, truncated to values between 0.3 and 0.9 | To reflect mortality due to adult males during dispersal. |
| Survival rate, females >2 year old | $= S(KF/NF)^{0.25}$, truncated to values between 0.5 and 0.95 | To reflect competition among mothers and dispersing females. |
| Survival rate, males >2 years old | $= S(KM/NM)^{0.25}$, truncated to values between 0.5 and 0.95 | To reflect competition among males for territories and access to females. |

Note: S = annual survival rate at carrying capacity, KF and KM = carrying capacity for breeding males and females, respectively, and NF and NM = number of ≥2-year-old females and males, respectively.

[a] Formula was incorrectly typeset in Beier (1993).

Robinette et al. (1977), Ashman et al. (1983), Lindzey et al. (1988), and Anderson et al. (1989) did not report sex differences in adult survival rates, however, nor did I document them in this population (Beier and Barrett 1993). Therefore, the model included a small difference between the sexes in juvenile survival rates, with the sex-specific subroutines for density dependence (Table 13.2) creating further sex differences in survival rates (especially for juveniles) to maintain the specified adult sex ratio.

I included two of the three types of stochasticity commonly included in simulation models (Table 13.3). Catastrophic stochasticity was included to reflect low-frequency, high-magnitude events such as prolonged droughts or severe epidemics that might cause large reductions in prey numbers. I modeled this variation by decreasing carrying capacity by 20 percent for a 3-year period starting every 25 years. Because demographic stochasticity arises from binomial processes (such as surviving or dying, breeding or not breeding, being born male or female), I modeled it by applying the appropriate probability to each simulated cougar (Table 13.3). If the survival rate for yearling

TABLE 13.3.

Stochasticity in population parameters used to simulate cougar population dynamics

| Population parameter | How stochasticity was modeled | Type of stochasticity |
|---|---|---|
| Survival rate | Each animal in an age class survived with probability = density-dependent survival rate. | Demographic |
| Primary sex ratio | Each newborn had 50% chance of being male. | Demographic |
| Litter size | Each litter had 2, 3, or 4 cubs, with probabilities appropriate to the specified mean litter size. | Demographic |
| Immigration rate | Each year 1 male or female immigrated with probability = 0.1 · specified number of immigrants per decade, and immigrants were assigned to the 1-year, 2-year, or 3-year age class with probability = 0.3, 0.5, and 0.2, respectively. | Demographic |
| Probability that a resident female bred | Each female bred with probability = density-dependent probability of breeding for that age class. | Demographic |
| Carrying capacity | 20% decrease in carrying capacity in years 25–27, 50–53, 75–77.[a] | Catastrophic |

*Note:* Demographic stochasticity was achieved by applying the appropriate age- and sex-specific binomial probabilities to each animal.

[a] Extinction risk did not decrease when a 0% decline was simulated; nor did it increase when a 40% decline was used.

males was 0.60 and there were two yearling males in a given year, for example, all outcomes (2, 1, or 0 survivors) were possible (with binomial probabilities 0.36. 0.48, and 0.16, respectively) in a biologically realistic manner.

Environmental stochasticity, a third type of variation relevant to PVA, reflects low-magnitude, year-to-year variation in survival rates for all individuals resulting from fluctuation in food crops, subepidemic diseases, poaching, and similar factors. I attempted to estimate the magnitude of environmental stochasticity from the observed month-to-month variation in survival rates of radio-tagged cougars. In the absence of a catastrophe, this variation can come from only two sources: variation due to sampling a small population with a given probability (that is, demographic stochasticity) and variation in that probability of survival (that is, environmental stochasticity) (Lacy 1993). Therefore, I estimated environmental stochasticity by comparing observed variation to the variation due to demographic stochasticity. The variance of 64 monthly survival rates for adults (mean = 97.51 percent) was 25.6 (SD = 5.06 percent). The variance expected if the observed variation were based solely on demographic stochasticity was 30.5, yielding an estimate of zero (rounding −4.9 to the nearest feasible value) for variance due to environmental stochasticity. To further explore this issue I examined the frequency distribution of adult survival rates under the assumption that environmental stochasticity contributed to the observed month-to-month variation (Table 13.4). This Monte Carlo sample contained significantly fewer outcomes near the true mode and significantly more outcomes <0.80 (a monthly rate never observed in the field). Thus, incorporating environmental stochasticity into my PVA would have reduced the realism of the model and predicted higher extinction risks.

### Fieldwork in the Santa Ana Mountains

*Study Area.*   The cougar metapopulation in the Santa Ana Mountains of southern California consists of about 20 adults on about 2070 km² of habitat with a tenuous linkage to a larger population in the Palomar Range (Figure 13.1) (Beier and Barrett 1993). The surrounding urban areas do not offer even marginal cougar habitat. About 1270 km² of this habitat (61 percent) is protected from urban uses, primarily within lands owned by the U.S. Forest Service and U.S. Navy (Beier 1993). Of the protected land, about 1114 km² forms a contiguous block (the "protected core area"); if all private lands were developed, the other 154 km² of protected land would be isolated into fragments unusable by cougars. The terrain is rugged and elevation varies from zero to 1680 m. Vegetation includes chaparral, coastal scrub, oak (*Quercus agrifolia* and *Q. engelmannii*) woodlands, annual grasslands, and small areas of coniferous forest on high-elevation, north-facing slopes. Few drainages have

TABLE 13.4.

Relative frequency (%) of monthly survival rates for adult cougars in the Santa Ana Mountains (1987–1992) compared to that expected if environmental stochasticity (ES) is absent or accounts for half of the observed monthly variation

| Monthly survival rate | Observed frequency | Expected frequency with demographic stochasticity | |
|---|---|---|---|
| | | No ES[a] | ES accounting for 50% of observed variance[b] |
| 0.0–0.8 | 0/64 = 0.0% | 1.7 | 5.5 |
| 0.801–0.85 | 1/64 = 1.6% | 2.8 | 3.2 |
| 0.851–0.9 | 10/64 = 15.6% | 11.8 | 14.6 |
| 0.901–0.95 | 2/64 = 3.1% | 1.4 | 1.9 |
| 0.951–1.0 | 51/64 = 79.7% | 82.3 | 74.8 |

*Note:* Survival rate was computed as the number of cougars alive at the end of the month divided by the number alive at the start of the month (mean 7.9; range 5 to 12 animals at start of month). Expected frequencies were generated by repeated sampling, applying to each animal alive at the start of a month either a constant survival probability of 0.9751 ("no ES") or a probability from a normal distribution with mean 0.9751 and SD = 0.0358 ("ES accounting for 50% of observed variance").

[a] Did not differ from observed frequency distribution, two-sample Kolmogorov-Smirnov test, $P = 0.06$. Standard deviation of 0.0552 was nearly the same as the observed SD of 0.0506.

[b] Differed from observed frequency distribution, two-sample Kolmogorov-Smirnov test, $P < 0.0005$. Standard deviation of 0.732 was larger than the observed SD of 0.0506.

perennial aboveground water flow, but seeps and springs are well distributed throughout the range.

Because mule deer (*Odocoileus hemionus*) accounted for 81 percent of the prey biomass consumed (Beier and Barrett 1993), deer density probably was the ultimate determinant of carrying capacity. There were 2.3 to 4.6 deer per square kilometer in the Santa Ana Mountains (Beier and Barrett 1993). Hunting (legal and illegal combined) took less than 5 percent of the adult deer a year in one study area in the northeastern corner of the range (Environmental Science Associates 1992).

The human population of the eastern half of Orange County and the western sixth of Riverside County is projected to grow from 1.15 million in 1987 to 2.09 million by 2010 (Anonymous 1989). Most growth is expected to occur, not within existing cities, but via "planned communities" built in

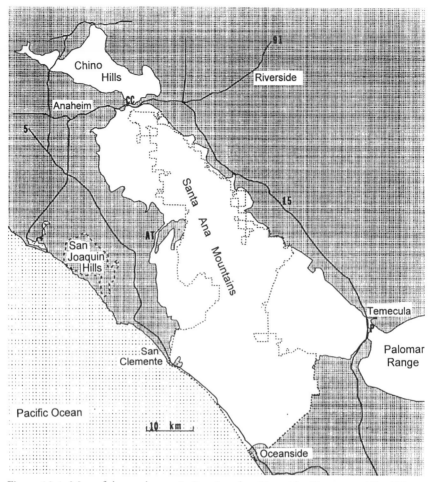

Figure 13.1. Map of the study area in Los Angeles, Riverside, Orange and San Diego counties, California. All highways (numbers) shown are six- to ten-lane freeways. The unshaded area includes 2070 km² of cougar habitat in the Santa Ana Mountains (including the Chino Hills) and a portion of the larger habitat area in the adjacent Palomar Range (east of Interstate 15). The center of the Santa Ana mountain range is about 65 miles southeast of Los Angeles and about 60 miles north of San Diego. The dotted line encloses the protected core area (seven interconnected protected parcels totaling 1114 km²). Urban (nonhabitat) areas are indicated by dense stippling with the dashed line enclosing the 75 km² of suitable habitat in the San Joaquin Hills (cougars recently extinct). Three habitat corridors are designated CC (Coal Canyon), AT (Arroyo Trabuco), and P (Pechanga).

privately owned open-space areas, including some of the best cougar habitat. Such rapid growth causes the outright loss of several square kilometers of habitat a year in Orange and Riverside counties. In addition, some wildlands are lost to the cougar population because they become isolated by freeways and other development. After urbanization isolated a 75-km$^2$ fragment of cougar habitat (Figure 13.1: San Joaquin Hills) in the late 1970s, for example, cougars became extinct there between March 1987 and June 1990 (Beier and Barrett 1990).

*Radio Telemetry.* I captured and radio-tagged 21 adult and 11 juvenile cougars from 1988 to 1992 (Beier 1995; Beier et al. 1995; Figure 13.2). To document travel paths, I monitored animals overnight, determining the focal animal's location by ground triangulation every 15 minutes from before sunset until after sunrise on each of 178 tracking sessions (Beier et al. 1995). Cougars traveled an average of 5.4 km per night (Beier et al. 1995). I radio-tracked adults and juveniles to document the routes by which cougars traveled from the protected core area to smaller protected parcels surrounded by private land. These smaller parcels (15,448 ha total) included six regional parks, one state park, three private reserves, and a university field station. To document use of corridors for dispersal, I followed the dispersal movements of the nine radio-tagged juveniles (all that survived to dispersal age) in relation to three potential corridors and several habitat peninsulas (Beier 1995). I identified the three potential corridors (Figure 13.1: Coal Canyon, Pechanga, and Arroyo Trabuco) before radio-tracking the dispersers. These areas were not designed as wildlife movement corridors, but were simply habitats made linear by urban growth.

The Coal Canyon Corridor (1.5 km long) provided the only potential habitat link between the Chino Hills and the rest of the Santa Ana Mountains (Figure 13.3). Most of the corridor was occupied by two shrubless golf courses and a horse stable devoid of understory. It was crossed by an eight-lane freeway (State Route 91) with heavy night traffic that precluded any at-grade road crossing. The only potential freeway undercrossings were a vehicle underpass (not used at night but devegetated, brightly lit, and noisy at all hours) and a 2.6 × 3.3 m concrete box culvert 200 m long.

The Pechanga Corridor (4 km long) was the last potential habitat link between the Santa Ana Mountains and the Palomar Range. It was crossed by a six-lane freeway (Interstate 15) and contained two golf courses. Although a bridge over the Santa Margarita River provided a quiet, well-vegetated, and relatively dark freeway undercrossing, a cougar using this underpass must also cross more than 400 m of golf course and skirt two fences along a golf driving range. The corridor was bordered by tract homes and contained an

Figure 13.2. Female cougar F6 at capture in May 1989.

abandoned rock quarry, a two-lane paved road, several residences, and a concrete embankment on one side of the main watercourse.

The Arroyo Trabuco Corridor (6 km long) was 400 to 600 m wide and lined with tract homes. Steep bluffs (20 to 70 m deep) minimized urban vistas, noise, and light pollution in the arroyo, and the dominant vegetation was an oak-sycamore (*Platanus racemosa*) riparian forest. Unlike the other two corridors, the arroyo was not the sole link between the large habitat area at either end (Figure 13.1).

I estimated survival rates for radio-tagged adults (≥2-year-old animals with stable home ranges) and juveniles (dependent young and dispersers lacking a stable home range) by using 1-month time intervals and the product limit procedure with staggered entry (Pollock et al. 1989). The product of 12 monthly rates yielded an estimate of annual survival rate, and the average of these running products yielded a single point estimate. I computed population density by applying mark–recapture estimators to the numbers of cougars tagged and the numbers of tagged and untagged animals killed on depredation or by vehicles in three half-year periods. Because road surveys for

Figure 13.3. The Coal Canyon Corridor, looking southward from the Chino Hills. From the mouth of Coal Canyon (top of photo), northbound cougars had to walk under State Route 91, through an equestrian center (bare areas), and then across the Santa Ana River (hidden in the riparian forest running from left center to lower right) and a golf course (lower left) to reach the southern edge of the Chino Hills (just below bottom of photo).

cougar tracks suggested that approximately 10 percent of suitable habitat lacked resident females, and because female home ranges overlap broadly, I estimated carrying capacity for females as 20 percent higher than the maximum calculated density.

## Results and Discussion

Both the model results and the field data emphasize the importance of movement corridors to maintenance of the Santa Ana and Santa Monica cougar metapopulations. I describe my efforts to bring this information to bear on the land use decisions of the different jurisdictions within this area.

### Population Model

One does not need a model to make the point that patch size and connectivity are important for keeping cougars on the landscape. These landscape features are inherent in the metapopulation concept. The value of the model is that it permits us to estimate which combinations of patch size and connectivity will conserve a cougar population, given our best estimates for that population's carrying capacity, survival rates, and other parameters. In this model, as expected, both area of habitat and the presence (or quality) of an immigration corridor influenced the probability of extinction (Figure 13.4; for other results see Beier 1993). Despite variation in biological parameters, at least 98 percent of simulated populations persisted for 100 years when there was more than 2200 km² of habitat available, except under the most pessimistic estimates of biological parameters (carrying capacity of no more than 0.4 adult female and 0.2 adult male per 100 km² in concert with adult survivorship $\leq 0.75$). With only 1000 km² of habitat and no immigration, simulated populations had 98 percent persistence under only the most optimistic estimates of biological parameters (carrying capacity greater than 1.0 adult female and 0.5 adult male per 100 km² in concert with adult survivorship $\geq 0.85$ and juvenile survivorship $\geq 0.65$). Thus in the absence of an immigration corridor, the critically small habitat area lies between 1000 and 2200 km². Within this range, the critical size depended on demographic parameters.

Immigration improved the probability of survival at surprisingly low levels—as low as one male per decade. For any combination of biological parameter estimates, the critical habitat area was 200 to 600 km² smaller with an immigration corridor than without. I simulated only three low levels of immigration, and certainly a corridor allowing more immigration would make even smaller areas viable. Thus in areas where isolation or fragmentation of a cougar population appears imminent, protecting and enhancing any remaining corridor is a valuable measure.

These minimum habitat areas (1000–2200 km², depending on carrying capacity and vital rates) would hold 15 to 20 adult cougars—far fewer than necessary to preserve genetic variability over several centuries (Franklin 1980). Lacking a quantitative way to model how inbreeding would increase extinction risk, I relied on the generalization that small populations will succumb to demographic events before inbreeding becomes a problem (Lande 1988), and my model ignored this risk. Mills and Smouse (1994), however, argue convincingly that moderate levels of inbreeding depression can influence population persistence in the first few generations after habitat fragmentation and isolation. I, therefore, stress that the minimum areas suggested by this model will not guarantee survival of a cougar population. In cases where no immigration corridor is provided, populations confined to such small areas will require monitoring and perhaps periodic introduction of new genetic material by translocation.

These minimum areas (and the number of cougars they contain) are

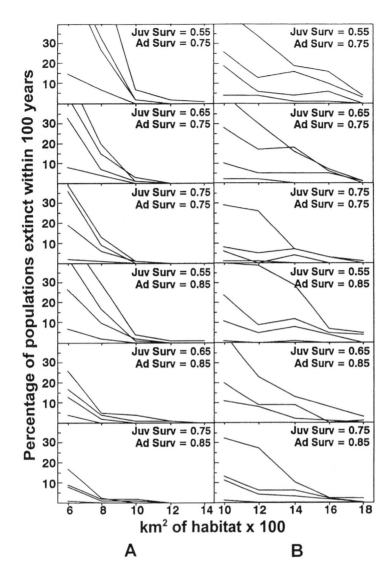

Figure 13.4. Effect of habitat area and immigration on cougar population persistence given (**A**) a carrying capacity of 0.8 breeding adult female and 0.6 breeding adult males per 100 km² or (**B**) a carrying capacity of 0.8 breeding adult female and 0.2 breeding adult male per 100 km². In each graph the top through bottom lines indicate the percentage of simulated populations that went extinct when the numbers of immigrants per decade were 0, 1 male, 2 males, or 3 males and 1 female, respectively. Juv Surv (juvenile survival rate) and Ad Surv (adult survival rate) are defined in Table 13.1.

comparable to the minimum area and number suggested by Shaffer (1983) for grizzly bears. Both my model and Shaffer's are species-specific and incorporate density dependence. Moreover, both produce minimum areas and populations much smaller than predicted by analytic models (for example, Belovsky 1987), simulation models lacking density dependence (for example, Ginzburg et al. 1990), or models that incorporate fewer subroutines to mimic cougar life history (Seal and Lacy 1989, although their model also models inbreeding depression and uses lower survival and reproductive rate estimates). It is tempting to favor models that call for protecting larger areas on the grounds that they are more "conservative," but such models have two drawbacks. First, in the long run they undermine the credibility of conservation biology and PVA especially when populations persist under conditions that such models predict will lead to rapid extinction. Second, to the extent that such analyses misclassify viable populations as "hopelessly" small, they can be a *less* conservative approach.

The modeling exercise yielded two results relevant to how stochasticity is incorporated into PVA models. First, natural catastrophes of moderate severity (up to 40 percent loss of carrying capacity), frequency (every 25 years), and duration (3 years) had little impact on extinction risk and were not apparent in graphs of population trajectory (Beier 1993). Shaffer (1983) similarly concluded that catastrophes are unimportant to the population dynamics of grizzly bears. Second, observed variation in survival rates was attributable entirely to demographic stochasticity; thus environmental stochasticity (ES) appeared trivial or absent in this population. Although ES, if it exists, can greatly increase extinction risk (Shaffer 1987; Lande 1993) and many PVAs therefore include subroutines for it, I am unaware of previous attempts to empirically measure ES for a PVA or (except for Lacy 1993) to partition observed variance between demographic stochasticity and environmental stochasticity. Although ES is probably important for populations of small-bodied or herbivorous animals (more at the mercy of weather and other transient phenomena) and is obviously important for exploited populations, it should not be incorporated blindly into all PVA models.

### Applying the Model to the Santa Ana Mountains

I estimated the carrying capacity for the Santa Ana Mountains as 0.80 breeding female and 0.30 breeding male per 100 km². The mean annual survival rate was 0.74 for adults (0.9751 per month, SD = 0.0502, $n = 64$ months) and 0.48 for juveniles (0.9403 per month, SD = 0.1045, $n = 34$ months). Using these estimates, the model predicted that the cougar population in the Santa Ana Mountains is clearly endangered. Although there is low (less than two percent) risk of extinction in the next 100 years with the current 2070 km² of habitat and no immigration, every parcel of habitat lost increases the risk of extinction (Figure 13.5). Consistent with our field observations in 1988, the model also predicted frequent local shortages of breeding males and an unstable sex and age structure under current conditions. If the population is confined to the 1114-km² block of contiguous protected lands, extinction risk rises to approximately 65 percent; an immigration corridor, necessarily including some lands now in private ownership, greatly improves the prognosis.

### Applying the Model to the Santa Monica Mountains

Urban sprawl has created a second cougar metapopulation to the north of the study area in the Santa Monica Mountains (660 km²), Simi Hills (130 km²), and Santa Susana Mountains (contiguous with more than 5000 km² of cougar habitat) (Figure 13.6). Using the most optimistic published estimates for vital rates and carrying capacity, the model predicts that cougars would rapidly become extinct in both the Santa Monicas and the Simi Hills if these areas were isolated from the Santa Susanas. Edelman (1990) identified several potential movement corridors linking these areas (Figure 13.6), all on private land, and the Santa Monica Mountains Conservancy is actively working to

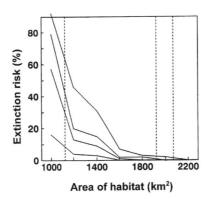

Figure 13.5. Extinction risk for the cougar population in the Santa Ana Mountains. The top through bottom lines give the percentage of simulated populations that went extinct within 100 years when the numbers of immigrants per decade were 0, 1 male, 2 males, or 3 males and 1 female, respectively. From right to left, the vertical lines indicate the total available habitat in 1992, the total habitat available if the Chino Hills (150 km²) is lost, and the total area of the protected and interconnected habitat block. Simulations were run with a carrying capacity of 0.80 breeding female and 0.30 breeding male per 100 km², an adult survival rate of 0.74, and a juvenile survival rate of 0.48.

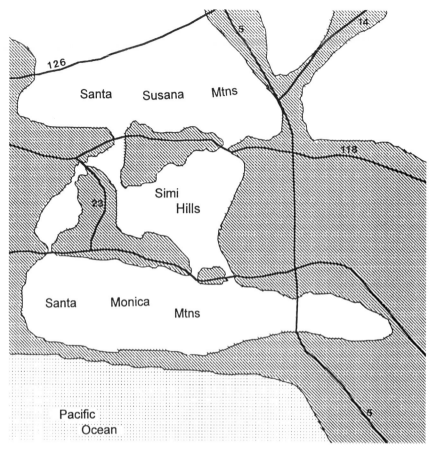

Figure 13.6. Map of the mountain ranges that lie just northwest of Los Angeles; all highways shown (numbers) are 4–12 lane freeways. All potential connections are threatened by urban development. The unshaded areas indicate cougar habitat in the Santa Monica Mountains (660 km²), Simi Hills (130 km²), Santa Susana Mountains (>2000 km², including contiguous areas to the north), and potential connections among these areas, in relation to urban areas (cross-hatched).

preserve these links. Because the average home range of an adult female cougar in southern California is about 110 km² (Beier and Barrett 1993), the Simi Hills are unlikely to support more than two adult female cougars and often may be unoccupied for several years after the death of resident females. Nonetheless, the offspring of resident females in the Simi Hills are the most likely immigrants to the Santa Monicas, even if the Simi Hills support breeding females less than 50 percent of the time. Thus, even if a formal survey for cougars classifies the Simi Hills as "suitable but unoccupied" habitat in any given year, from the perspective of the metapopulation this habitat is as critically important as any occupied habitat.

### Use of Corridors by Dispersers

I assessed corridor use from two perspectives. Both dramatically showed that cougars will disperse via habitat corridors in a landscape fragmented by urbanization and, moreover, that some dispersers will use corridors containing unnatural features such as golf courses and major freeways. First, from the perspective of the dispersers, five of the nine juveniles used corridors and one of them used two corridors (Table 13.5). Second, all three potential corridors were used. Three dispersers successfully used the Arroyo Trabuco. Two radio-tagged dispersers successfully used the Coal Canyon Corridor, a third was hit by a vehicle there, and tracks indicate that an uncollared juvenile also used the corridor. One disperser used the Coal Canyon Corridor more than 22 times over 19 months to establish a home range that included both the Chino Hills and the northern part of the Santa Ana Mountains (Beier 1995). Of the three dispersers encountering the Pechanga Corridor, one apparently did not enter it, another traversed it in a single night, and a third strayed from the corridor into a residential area where he was captured by animal control officers. Between October 1990 and December 1993, five unradioed cougars were killed by vehicles where I-15 crosses the Pechanga Corridor. Although my early observations (Beier 1993) led me to be pessimistic about the utility of this corridor, the five road-kills there suggest that successful crossings also are likely.

In addition, seven of the nine dispersers explored to the tips of habitat peninsulas, making a total of ten such forays averaging 2.9 km in one-way length (Table 13.5). All peninsulas extend into dense urban areas, and most were heavily used during the day by humans. Despite intense monitoring of 20 radio-tagged adults, there was no evidence that adult cougars visited any of the ten habitat peninsulas.

### Travel Paths Likely to Become Corridors

In the Santa Ana Mountains, overnight radio monitoring identified 18 routes by which cougars traveled from the 1114-km$^2$ protected core area to the 11 regional parks and reserves surrounded by private land. Although only four of these routes had significant human intrusions in 1992, 17 (including these four) were threatened by proposed highway projects and anticipated urban development. The only secure connection was provided by the Santa Margarita River, which links the University of California Field Station to the protected core area.

TABLE 13.5.

Use of habitat corridors and peninsulas by dispersing juvenile cougars radio-monitored in southern California: 1990–1992

| Disperser[a] | Corridor use (successful unless otherwise noted) | Length (km) | Width (m) | Peninsula use |
| | | | | Description of movement |
| --- | --- | --- | --- | --- |
| F17 | None | 1.1 | 50–250 | Moved between horse stables, church, freeway (SR 91), and a 75-ha newly graded area to daybed at peninsula tip, <50 m from new housing tract in City of Anaheim. |
| M3 | None | 3.0 | 1000 | Moved through mix of orchards and scrub into edge of residential area in City of Irvine.[b] |
| M5 | 1 km into Pechanga Corridor, detoured into peninsula | 1.4 | 100–300 | 1. Partway into Pechanga Corridor, detoured into Temecula Creek habitat peninsula, walking 1.4 km to its tip and then into residential area in City of Temecula.[b] |
| | | 4.5 | 50–400 | 2. Explored Santiago Creek in Cities of Orange and Anaheim; daybedded in peninsula tip (10-ha riparian forest at Katella St.). |
| | | 1.5 | 50–100 | 3. Crossed 1 km of open terrain on three occasions to reach 50-ha riparian forest at Peters Canyon Reservoir and on one occasion continued an additional 500 m down canyon. |
| M6 | Coal Canyon (≥22 crossings) | | | None. |

| | | | | |
|---|---|---|---|---|
| M7 | Encountered but did not use Pechanga Corridor | | | None. |
| M8 | 1. Arroyo Trabuco<br>2. Coal Canyon | 1.0<br><br>2.3 | 30–400<br><br>1000 | 1. Crossed under I-5 and spent 14 days in 350-ha riparian forest ("Trestles Beach") in estuary of San Mateo Creek between City of San Clemente, freeway, and Pacific Ocean.<br>2. Crossed grassland at edge of Chino Hills and was killed at tip of peninsula trying to cross SR 60. |
| M10 | 1. Struck by vehicle in Coal Canyon Corridor<br>2. Arroyo Trabuco | 0.8 | 400 | Used a small habitat peninsula west of Robbers' Peak in the City of Anaheim; peninsula contained high-voltage power lines, water tanks, fire trails, and was heavily used by joggers. |
| M11 | Arroyo Trabuco | >7 | 100–650 | Followed San Luis Rey River (bordered by urban areas) under I-5 into City of Oceanside; at 0230 he was sighted three blocks from ocean, pursued, and shot by police. |
| M12 | Pechanga Corridor | 6.5 | 500–2000 | During overnight monitoring, traveled the open grassy ridge separating cities of San Juan Capistrano and San Clemente, daybedding <30 m from I-5. |

*Note:* All habitat peninsulas were bordered by freeways or densely populated urban areas. Despite intense monitoring, there was no evidence that any adult cougar used any of these habitat peninsulas.

[a] F, M = female and male, respectively.

[b] Treed by domestic dogs, captured, and released back to wild.

*Bringing the Findings to Decision Makers*

The model clearly showed that preserving the Pechanga Corridor is critically important to conserving the cougar population in the Santa Ana Mountains, and field data show that the corridor is usable. Although the five jurisdictions (City of Temecula, BLM, Pechanga Indian Reservation, and San Diego and Riverside counties) with authority over land use in the corridor do not coordinate their plans, Temecula has authority over the part of the corridor most threatened by urban encroachment. In 1992 I provided written reports and personally presented my findings to the city early in the process of drawing up its first land use plan. The city ignored this input, zoning almost all land into various urban uses, and has since approved every development proposal in its jurisdiction. Because it is possible to create a corridor outside the city limits, Temecula's response was legal under state environmental law. But now the only hope for this corridor lies in an effort by The Nature Conservancy to protect key parcels via purchase or easements.

The model convincingly predicted that loss of the Coal Canyon Corridor would guarantee the extinction of cougars from the 150 km² of habitat in the Chino Hills (Figure 13.1), reducing by 7.5 percent the total habitat available to the population and pushing the population toward the steeply rising part of the risk curve (Figure 13.5). The fieldwork showed that in fact the corridor was used. Because the local press covered these findings prominently in 1990 and 1991, local awareness was much greater than in Temecula (which is not served by its own print or broadcast media). Although less than 2 km long, this corridor is within the jurisdiction of three counties and two incorporated cities and is crossed by a state freeway. The City of Yorba Linda passed a resolution to preserve the part of the corridor in its jurisdiction so long as other jurisdictions kept it intact. But early in 1991, a developer proposed to build 1500 homes on a 150-ha parcel in the mouth of Coal Canyon in the city of Anaheim—a project that would sever the corridor. I provided information, attended meetings, and made three presentations to the city urging a scaled-back project. Anaheim delayed its decision several times to allow the developer to present alternatives to the Coal Canyon Corridor. In each case, I demonstrated that the alternative corridors were infeasible. Ultimately, Anaheim formally acknowledged that the project would destroy the corridor and "result in the loss of potential for a cougar population to occur in the Chino Hills." When the city hinted it would approve the project anyway because other jurisdictions could destroy other parts of the corridor, supporters of the corridor urged Anaheim to enter into a Coordinated Resources Management Plan (CRMP), a voluntary legal mechanism for interjurisdictional cooperation, with the other parties. On 3 March 1992, the Anaheim City Council perfunctorily declined to initiate a CRMP and then, citing their lack of authority over the entire corridor, unanimously approved the development project. Several groups, including the CDFG, sued Anaheim under the California Environmental Quality Act. Although the suit eventually failed, it delayed the project until a downturn occurred in the housing market. The developer is now willing to sell the parcel to the state if funding can be obtained.

Our field data may mitigate the impact of a proposed tollroad in southern Orange County that would slice through a pristine area with no human residents along its 21-km length (Anonymous 1990; Figure 13.7). All-night radio-tracking revealed the routes by which cougars traveled between the protected core area (east of the proposed road) and five smaller areas of dedicated open space west of the road. Although these routes now traverse pristine open space, they will become corridors (at best) as highway-induced growth removes the adjacent habitat. The regional transportation agency responded to this information by proposing bridged undercrossings at the most important crossing points (Figure 13.7), but preserving a corridor is not as simple as building a bridge at one point along the corridor. The agency has acknowledged that the tollroad, by providing "critical infrastructure to large expanses of open space," will induce massive urban growth (Anonymous 1990). (Under current policy, freeway access is a prerequisite for issuance of development permits.) Such growth could sever all of the wildlife corridors, rendering the underpasses pointless. Although state law requires mitigating the impacts of induced growth, the agency refused requests (from myself and CDFG) to purchase easements to the three most important corridors as mitigation for this induced growth. Because it was impossible to estimate the amount and location of induced growth, the agency argued that it was "speculative and infeasible" to mitigate for it and said that mitigation should occur in association with each proposed development project. Having repeatedly seen developers argue that it was "speculative and infeasible" to require them to preserve part of a corridor when the next project might destroy another section, I initiated a lawsuit on this issue with other plaintiffs in 1991. (The lawsuit was dropped when I moved out of the state.) The tollroad still lacks some of its environmental permits, which may present opportunities to deny or mitigate the project or use it as a means for obtaining comprehensive regional planning. I also provided findings on travel routes between the protected core area and the other six protected parcels to the appropriate agencies. This inforation caused the U.S. Forest Service to cancel a planned land swap and greatly improved the land use plan for one rural area (Anonymous 1991).

In early 1992, the Mountain Lion Foundation, relying on the quarterly reports from my study, filed a petition to list the cougar metapopulation in the

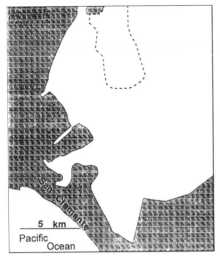 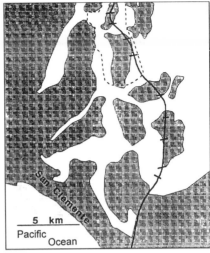

A. Potential reserve areas in 1994        B. Proposed reserve design

Figure 13.7. Maps of the reserve plan proposed for southern Orange County under the first attempt at a regional conservation plan. **A.** Unshaded area indicates distribution of potential reserve areas (including disturbed habitats such as annual grasslands and a small amount of agricultural land) in relation to urban areas (cross-hatched). **B.** The proposed NCCP reserve design. Additional cross-hatching indicates proposed urban developments. Heavy line indicates route of a proposed freeway; transverse bars indicate underpasses were planned to allow for cougar movements between the protected core area (east of the freeway alignment) and five smaller protected parcels to the west. The largest known concentration of the California gnatcatcher and the San Diego cactus wren (two primary target species of the NCCP) is indicated by the dashed line. Source maps: Habitat, Target Species, and Reserve Concept Maps, Dudek and Associates, Encinitas, California, dated 3 Oc-

Santa Ana Mountains and Chino Hills as endangered under the Endangered Species Act. The California Department of Fish and Game, which disagreed with the Mountain Lion Foundation on many issues, opposed the petition. After the initial headlines, the petition received little media attention, and the U.S. Fish and Wildlife Service delayed action for 30 months beyond the deadline for response. Then, in December 1994, the service rejected the petition.

In July 1991, the California gnatcatcher (*Polioptila californica californica*) was about to be listed as threatened under the Endangered Species Act, and the coastal cactus wren (*Campylorhynchus brunneicapillus sandiegensis*), the orange-throated whiptail lizard (*Cnemidophorus hyperythrus*), and the Santa

Ana Mountains cougar population were under consideration for listing. At this point the State of California initiated the Natural Communities Conservation Planning (NCCP) program (Anonymous 1993). In March 1993, the U.S. Fish and Wildlife Service accepted this effort as the vehicle for designating critical habitat for the gnatcatcher and for developing a biological reserve system for southern California that would prevent the need for listing other species.

The NCCP planning areas include the entire range of the Santa Ana Mountains cougar population. As part of the NCCP effort, a scientific review panel (SRP) was convened, composed of conservation biologists whose expertise included metapopulation dynamics. In August 1991 and again in January 1992, I pleaded with the SRP to make an explicit map of the reserve areas. I used as an example the three detailed maps provided by the northern spotted owl recovery team (Thomas et al. 1990), which realized that broad guidelines alone would be subject to interpretation unfavorable to the conservation of the species. The SRP responded that producing a map was not part of its mandate and assured me that the guidelines would be so explicit that CDFG, the lead agency, would be forced to map an effective reserve system. In November 1993, CDFG incorporated the SRP report, including seven "basic tenets of reserve design," into its NCCP Conservation Guidelines.

Subsequent events show that the SRP's faith in the implementation process was unjustified. California Department of Fish and Game has allowed landowners and developers to draw the map of the proposed reserve for southern Orange County. In fact, the map (Figure 13.7a) violates five of the seven basic tenets as follows (retaining the original numbering and wording):

"2. *Larger reserves are better*" and "4. *Keep habitat contiguous.*" Two areas proposed for urban developments would fragment the largest known populations of gnatcatchers and cactus wrens in southern California. From a metapopulation perspective, these are clearly the main source populations for these two endangered species. Indeed, it would be difficult to draw a map that would inflict more discontinuity on these source populations.

"5. *Link reserves via corridors.*" The proposed urban developments would prevent nonvolant animals from reaching over half of the underpasses proposed as mitigations for impacts to animal movement caused by the planned Foothill Tollroad (Figure 13.7b). The locations of these underpasses had been determined on the basis of my data on cougars traveling from the protected core area to five protected parcels west of the tollroad.

"6. *Maintain reserve units that are biologically diverse ... [including]* other habitat types that occur in a mosaic with CSS." Virtually all rare grassland vegetation that occurs in a mosaic with coastal sage scrub is designated for urban development.

"7. *Protect reserves from encroachment. Blocks of habitat that lack roads or are otherwise inaccessible to human disturbance better serve target species. . . . [The] greatest potential for encroachment is from urban edges. . . . [which provide access for] weeds, cats, dogs, children, . . . [and] wildfire.*" The reserve design as mapped creates excessive edge, human disturbance, and the need for more roads to connect the developed areas.

## Lessons

A metapopulation model in conjunction with site-specific data can influence decisions on land use. Model predictions alone could not be implemented without data on the location of travel routes and would not be persuasive without data showing that the target species will use corridors.

My attempts to bring these findings to decision makers had decidedly mixed results. The main obstacle preventing the results of metapopulation analyses from being implemented is the lack of regional planning authority. Under the current mechanism for implementation, concerned citizens must detect and force mitigations on each proposed project that threatens the metapopulation. For the cougar population in the Santa Ana Mountains, this requires monitoring and being prepared to litigate decisions made by five county governments, seventeen municipal governments, two transportation authorities, and the world's largest water district. The linkages among meta-population centers are most vulnerable. Because a corridor is only as strong as its weakest link, a single oversight on the part of conservationist volunteers is sufficient to lose a linkage and imperil the metapopulation.

A land use plan will fail to conserve a metapopulation if scientific advice for that plan is limited to general rules ("preserve large areas and connect them") and developers and politicians are trusted to draw the map from those rules. To make their advice useful, conservation biologists should develop maps that present biologically optimal designs. Although nonbiological considerations will influence the ultimate map, biologists should provide formal and independent review of the final map for its consistency with stated conservation objectives.

Biologists can be effective advocates for incorporating metapopulation concepts into land use planning, but only if they are committed to a sustained effort: writing detailed comments, preparing testimony for public hearings, and meeting frequently with persons who may not want biological expertise. Scientific advocates must be scrupulously careful with the facts, acknowledge the limits of the data or models they use, and admit to mistakes. Biologists must resist well-meaning pressure from citizen activists to overstate the biological evidence; overstatement is a sure way to lose credibility.

Although a reserve designed for a cougar metapopulation would not suffice for habitat specialists (such as the California gnatcatcher), an area-sensitive species such as the cougar is an appropriate umbrella species (Noss 1991) for a regional conservation plan because its low density renders it most sensitive to habitat fragmentation. Cougars also are more likely than less mobile species to yield data on habitat linkages essential to metapopulation function. Using telemetered cougars to identify movement corridors is certainly a big improvement over the prevalent practice in southern California, which is to label leftover shards of habitat, and passages under bridges built for hydrologic or geologic reasons, as "wildlife corridors."

There is a lack of data on whether most species use corridors (Chapter 5 in this volume), and a few biologists have speculated that connectivity might have biological drawbacks in addition to benefits (Simberloff et al. 1992). Clearly we need more data and better data. But biologists should not let developers misrepresent this spirit of inquiry as a disagreement among the experts on the *value* of habitat connectivity. By such misrepresentation, developers have persuaded planning agencies that habitat fragmentation should proceed unhindered and, moreover, that conservationists should bear the burden of proof for preserving each remaining corridor. Without abandoning our scientific endeavors, we need to make decision makers hear our united voice on the value of connectivity, thus shifting the burden of proof to those who would destroy the last remnants of natural connectivity.

## Acknowledgments

Fieldwork was supported by California Agricultural Experiment Station Project 4326-MS, California Department of Fish and Game, the County of Orange, and the U.S. Marine Corps. Alan Brody, David Choate, Donna Krucki, Doug Padley, and Duggins Wroe helped with fieldwork. Reginald H. Barrett (University of California) provided crucial support throughout the project. I thank the Endangered Habitats League, Friends of Tecate Cypress, Mountain Lion Foundation, The Nature Conservancy, Rural Canyons Conservation Fund, Santa Monica Mountains Conservancy, Save Our Forests and Rangelands, and the Sierra Club for their efforts to maintain connectivity in southern California. This chapter is dedicated to Dan Silver, Ray Chandos,

Connie Spenger, Gordon Ruser, and Duncan McFetridge; the courage and dedication of these full-time unpaid leaders repeatedly shamed me out of my excuse that I was "too busy to help." I thank my wife Maryann van Drielen and daughters Kathy, Celia, and Michelle for tolerating my work habits. Bruce Ackerman, Reg Barrett, David Ehrenfeld, Todd Fuller, T. P. Hemker, Bob Lacy, David Maehr, Reed Noss, and two anonymous referees constructively reviewed the chapter or previous papers summarized here. This study was conducted under Animal Use Protocol R139-0394 issued by the Animal Care and Use Committee at the University of California, Berkeley.

## REFERENCES

Anderson, A. E. 1983. A critical review of literature on the puma *Felis concolor*. Special Report 54. Denver: Colorado Division of Wildlife.

Anderson, A. E., D. C. Bowden, and D. M. Kattner. 1989. Survival in an unhunted mountain lion population in southwestern Colorado. *Proceedings of the Mountain Lion Workshop* 3:57.

———. 1992. The puma in the Uncompahgre Plateau, Colorado. Technical Publication 40. Denver: Colorado Division of Wildlife.

Anonymous. 1989. Growth management plan. Los Angeles: Southern California Association of Governments.

———. 1990. Draft environmental impact report TCA EIR 3. Costa Mesa, Calif.: Foothill-Eastern Transportation Corridor Agency.

———. 1991. Foothill-Trabuco specific plan, environmental impact report 531. Orange County, Santa Ana, Calif.

———. 1993. Southern California coastal sage scrub natural community conservation planning conservation guidelines. Sacramento: California Department of Fish and Game.

Ashman, D. L., G. C. Christensen, M. L. Hess, G. K. Tsukamoto, and M. S. Wickersham. 1983. *The Mountain Lion in Nevada*. Carson City: Nevada Department of Wildlife.

Beier, P. 1993. Determining minimum habitat areas and habitat corridors for cougars. *Conservation Biology* 7:94–108.

———. 1995. Dispersal of juvenile cougars in fragmented habitat. *Journal of Wildlife Management* 59:228–237.

Beier, P., and R. H. Barrett. 1990. Cougar surveys in the San Joaquin Hills and Chino Hills. Orange County Cooperative Mountain Lion Study. Berkeley: Department of Forestry and Resource Management, University of California.

————. 1993. The cougar in the Santa Ana Mountain Range. Final report, Orange County Cooperative Mountain Lion Study. Berkeley: Department of Forestry and Resource Management, University of California.

Beier, P., D. Choate, and R. H. Barrett. 1995. Activity patterns of cougars during different behaviors. *Journal of Mammalogy* 76:1056–1070.

Belovsky, G. E. 1987. Extinction models and mammalian persistence. Pages 35–58 in M. E. Soulé, ed., *Viable Populations for Conservation.* Cambridge: Cambridge University Press.

Currier, M. J. P., S. L. Sheriff, and K. R. Russell. 1977. Mountain lion population and harvest near Cañon City, Colorado, 1974–1977. Special Report 42. Denver: Colorado Division of Wildlife.

Eaton, R. L., and K. A. Velander. 1977. Reproduction in the puma: Biology, behavior, and ontogeny. *The World's Cats* 3(3):45–70.

Edelman, P. 1990. Critical wildlife corridor/habitat linkage areas between the Santa Susana Mountains, Simi Hills, and Santa Monica Mountains. Malibu: Santa Monica Mountains Conservancy.

Environmental Science Associates. 1992. Eastern Transportation Corridor deer telemetry study. Final report. San Francisco: Environmental Science Associates.

Franklin, I. R. 1980. Evolutionary change in small populations. Pages 135-150 in M. E. Soulé and B. A. Wilcox, eds., *Conservation Biology: An Evolutionary-Ecological Perspective.* Sunderland, Mass.: Sinauer Associates.

Ginzburg, L. R., S. Ferson, and H. R. Akçakaya. 1990. Reconstructibility of density dependence and the conservative assessment of extinction risks. *Conservation Biology* 4:63–70.

Hemker, T. P., F. G. Lindzey, and B. B. Ackerman. 1984. Population characteristics and movement patterns of cougars in southern Utah. *Journal of Wildlife Management* 48:1275–1284.

Hemker, T. P., F. G. Lindzey, B. B. Ackerman, and A. J. Button. 1986. Survival of cougar cubs in a non-hunted population. Pages 327–332 in S. D. Miller and D. D. Everett, eds., *Cats of the World.* Washington, D.C.: National Wildlife Federation.

Hopkins, R. A. 1981. The density and home range characteristics of mountain lions in the Diablo Range of California. M.S. thesis, California State University, San Jose.

————. 1989. The ecology of the puma in the Diablo Range, California. Ph.D. thesis, University of California, Berkeley.

Hornocker, M. G. 1970. An analysis of mountain lion predation upon mule deer and elk in the Idaho Primitive Area. *Wildlife Monographs* 21:1–39.

Lacy, R. C. 1993. Vortex: A computer simulation model for population viability analysis. *Wildlife Research* 20:45–65.

Lande, R. 1988. Genetics and demography in biological conservation. *Science* 241:1455–1460.

————. 1993. Risks of population extinction from demographic and environmental stochasticity and random catastrophes. *American Naturalist* 142:911–927.

Lindzey, F. G., B. B. Ackerman, D. Barnhurst, and T. P. Hemker. 1988. Survival rates of mountain lions in southern Utah. *Journal of Wildlife Management* 52:664–667.

Logan, K. A., L. I. Irwin, and R. Skinner. 1986. Characteristics of a hunted mountain lion population in Wyoming. *Journal of Wildlife Management* 50:648–654.

Mills, L. S., and P. E. Smouse. 1994. Demographic consequences of inbreeding in remnant populations. *American Naturalist* 144:412–431.

Murphy, K. M. 1983. Relationships between a mountain lion population and hunting pressure in western Montana. M.S. thesis, University of Montana, Missoula.

Neal, D. L., G. N. Steger, and R. C. Bertram. 1987. Mountain lions: Preliminary findings on home-range use and density in the central Sierra Nevada. Research Note PSW-392. Berkeley: Pacific Southwest Forest and Range Experiment Station.

Noss, R. F. 1991. From endangered species to biodiversity. Pages 227–246 in K. A. Kohm, ed., *Balancing on the Brink of Extinction*. Washington, D.C.: Island Press.

Padley, W. D. 1990. Home ranges and social interactions of mountain lions in the Santa Ana Mountains, California. M.S. thesis, California State Polytechnic University, Pomona.

Pollock, K. H., S. R. Winterstein, C. M. Bunck, and P. D. Curtis. 1989. Survival analysis in telemetry studies: The staggered entry design. *Journal of Wildlife Management* 53:7–15.

Quigley, H. B., G. M. Koehler, and M. G. Hornocker. 1989. Dynamics of a mountain lion population in central Idaho over a 20-year period. *Proceedings of the Mountain Lion Workshop* 3:54.

Robinette, W. L., J. S. Gashwiler, and O. W. Morris. 1961. Notes on cougar productivity and life history. *Journal of Mammalogy* 42:204–217.

Robinette, W. L., N. V. Hancock, and D. A. Jones. 1977. The Oak Creek mule deer herd in Utah. Salt Lake City: Utah Division of Wildlife Resources.

Seal, U. S., and R. C. Lacy. 1989. Florida panther viability analysis and species survival plan. Apple Valley, Minn.: Captive Breeding Specialist Group.

Seidensticker, J. C., Jr., M. G. Hornocker, W. V. Wiles, and J. P. Messick. 1973. Mountain lion social organization in the Idaho Primitive Area. *Wildlife Monographs* 35:1–60.

Shaffer, M. L. 1983. Determining minimum population sizes for the grizzly bear. *International Conference on Bear Research and Management* 5:133–139.

———. 1987. Minimum viable populations: Coping with uncertainty. Pages 69–86 in M. E. Soulé, ed., *Viable Populations for Conservation*. Cambridge: Cambridge University Press.

Shaw, H. G. 1977. Impact of mountain lion on mule deer and cattle in northwestern Arizona. Pages 17-32 in R. L. Phillips and C. Jonkel, eds., *Proceedings of the 1975 Predator Symposium*. Missoula: University of Montana.

Simberloff, D., J. A. Farr, J. Cox, and D. W. Mehlman. 1992. Movement corridors: Conservation bargains or poor investments? *Conservation Biology* 4:493–504.

Sitton, L. W., and S. Wallen. 1976. *California Mountain Lion Study*. Sacramento: California Department of Fish and Game.

Thomas, J. W., E. D. Forsman, J. B. Lint, E. C. Meslow, B. R. Noon, and J. Verner. 1990. Habitat conservation area maps. In *A Conservation Strategy for the Northern Spotted Owl*. Report of the Interagency Scientific Committee to address the conservation of the northern spotted owl. Portland: USDA Forest Service; USDI Bureau of Land Management/Fish and Wildlife Service/National Park Service.

Young, S. P. 1946. History, life habits, economic status, and control. Pages 1–173 in S. P. Young and E. A. Goldman, *The Puma: Mysterious American Cat*. Washington, D.C.: American Wildlife Institute.

# 14

# Brown/Grizzly Bear Metapopulations

*F. Lance Craighead and Ernest R. Vyse*

In this chapter we examine the genetic and demographic dynamics of brown bear (*Ursus arctos*) populations in light of current metapopulation theory. Interior and arctic brown bears are often referred to as grizzly bears (Figure 14.1). Our research has centered on demographic and genetic studies of one such arctic brown/grizzly bear population. The characteristics of this population are typical of most historic brown bear populations: the bears occupy continuous habitat with frequent dispersal of individuals (usually males) between demes of more sedentary females. Brown bears are omnivorous, intelligent, and highly adaptable and can also be found in geographic situations that approximate classic metapopulation structures. Moreover, as large areas of occupied habitat have become fragmented due to human development, the "typical" structure has been replaced with smaller, isolated populations that may or may not interact as a metapopulation.

We begin by considering possible natural metapopulations within island systems and reviewing current mitochondrial DNA (mtDNA) and behavioral data. We then consider populations in continuous, mainland habitat, using our microsatellite DNA analyses and those of D. Paetkau and C. Strobeck, and discuss whether these systems can properly be termed metapopulations. We conclude by looking at habitat islands that have been fragmented by human activity and considering their metapopulation characteristics.

Brown bears may have evolved in Asia, most likely in forested habitat, or a tundra/forest mosaic. They dispersed throughout similar habitat, crossing the Bering land bridge into the New World about 35,000 (Talbot and Shields in press) to 40,000 years before the present (P. Matheus, University of Alaska, personal communication, 1995) or even earlier (Churcher and Morgan

Figure 14.1. Subadult grizzly bears on the tundra, Brooks Range, Alaska. Photo by H. V. Reynolds.

1976). Mitochondrial DNA sequence-divergence data (Waits et al. 1996*a*, 1996*b;* Talbot and Shields in press), when viewed in the light of fossil and molecular dating, indicate that there were at least two brown bear migration episodes from the Old World to the New World. These resulted in separate mtDNA lineages (or clades) in the New World. At the limits of their ancestral habitat, brown bears have adapted to new habitat types.

Brown bear populations probably reached the greatest extent of dispersal around the 1500s, before the widespread use of firearms. At that time there were two large continental populations in North America and Eurasia. The Eurasian distribution extended as far south as North Africa and the Middle East, while the North American distribution extended south into Mexico (Cowan 1972; Servheen 1990).

## Natural Island Metapopulations

During the last ice ages, as brown bear distribution reached the Russian Far East, populations were established on the Kamchatka peninsula and on the islands of Sakhalin, the Kuriles, and Hokkaido. These colonization events probably took place at about the same time as the first migrants colonized the New World, approximately 40,000 years before the present.

As they became established in Pleistocene Alaska, brown bears colonized the ABC Islands (Admiralty, Baranof, and Chichagof) in southeastern Alaska, the islands of Kodiak and Afognak, and the islands of Prince William Sound. Both mtDNA restriction fragment data (Cronin et al. 1991) and sequence data (Shields and Kocher 1991; Waits et al. 1996*a*, 1996*b;* Talbot and Shields in press) indicate that the ABC lineage is more closely related to polar bears than are other brown bear lineages. This is probably the most ancient lineage in North America.

Very little is known about the population dynamics or the population genetics of brown bears on most of these islands. In the following section we speculate on likely metapopulation dynamics. As genetic data are collected and analyzed from these island populations, they should provide insights into the functioning of a wide variety of metapopulation structures. There appears to be a size limitation to islands that currently support self-sustaining bear populations. In many cases this limit appears to be related to the area necessary in a given climatic regime to develop salmon spawning streams large enough to provide a consistent food source. But area alone, without considering food availability, does not appear to be a good predictor of brown bear population size or, by extension, population persistence over time: there are several large islands adjacent to occupied bear habitat that do not support bears.

Few data are available on brown bear dispersal distances across water. Bears are powerful swimmers and have been observed many kilometers from land. A subadult brown bear released on an island in Prince William Sound swam perhaps as far as 15.1 km to return to his point of capture 93 km away within 28 days (Miller and Ballard 1982). Other anecdotal reports are discussed in the following sections. Although it appears that expanses of open water greater than about 5 km can act as a barrier (or a filter) blocking or restricting migration, such barriers are less effective against males, subadult males in particular. Of course, to create gene flow an individual must not only migrate but also successfully reproduce after crossing a barrier.

Many natural island metapopulations have been isolated to some degree and have persisted over long periods of time without the benefit of increased gene flow from neighboring or transient males. The population sizes and spatial structure of these metapopulations should provide some insight into the minimal configuration of areas and population sizes needed to maintain populations in human-fragmented habitat.

### Sakhalin, Hokkaido, and the Kuriles

Sakhalin and Hokkaido were connected to mainland Asia several times during the Pliocene and Pleistocene as sea levels varied (Figure 14.2). It is likely that both these islands were colonized around the same time. Because of their large numbers, it is unlikely that brown bear populations have ever gone extinct on either island. Sakhalin, with an estimated population of 1400 (Servheen 1990) in a 77,000-km$^2$ area, is less than 10 km from the mainland of the Russian Far East at the closest point and migrant individuals are probably exchanged occasionally. Hokkaido, about 40 km from Sakhalin, is more isolated and may have received no immigrants since the

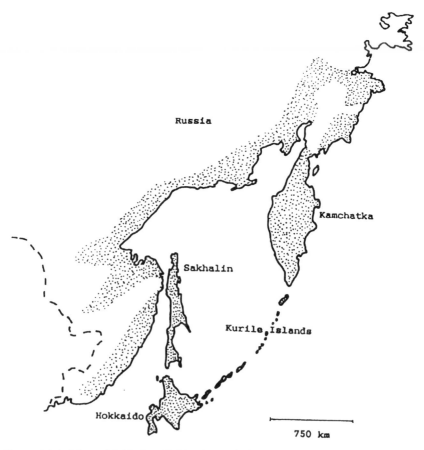

Figure 14.2. Map of Sakhalin, Hokkaido, and the Kurile Islands showing historic brown bear distribution. Stippled areas depict densities greater than 70 bears per 1000 km². After M. Kretchmar, Institute of Biological Problems of the North, Magadan, Russia (personal communication).

last, Wisconsin, glaciation. Historically, however, Hokkaido may have supported as many brown bears as Sakhalin. Up to 3000 bears have been reported in recent times (Domico 1988), but Servheen (1990) considered the population size as unknown in 1989. The population is much reduced from historic levels due to increasing human alteration of habitat, and it appears to be fragmented into three subpopulations by human development (Servheen 1990). Hunting of brown bears on Hokkaido is currently allowed. The Asiatic black bear (*Ursus thibetanus*) is found on Hokkaido and on Honshu, Kyushu, and Shikoku islands farther south in Japan (Hanai 1985), so it is possible that the brown bear also used to occur farther south but has been extirpated.

Between Hokkaido and the Kamchatka peninsula the Kurile Islands form a classic stepping-stone array of smaller intermediate islands. Both Kamchatka, with 12,000 to 14,000 brown bears (Dunishenko 1987) in a 472,000-km² area (Revenko 1996), and Hokkaido have historically had large brown bear populations. Recent legal and illegal hunting abetted by the struggling Russian economy and the demand for bear gallbladders has markedly reduced the Kamchatka brown bear population (Revenko 1996). The larger Kurile Islands adjacent to either of these "mainlands" have resident bear populations that may periodically go extinct and then become recolonized. These larger islands are separated from each other and from the "mainlands" by about 25 km. A total of 700 brown bears is estimated on the larger Kurile Islands (Dunishenko 1987). Current bear populations are probably restricted to islands that are large enough to support salmon populations that spawn in freshwater streams (Y. Zhuravlev, Russian Academy of Science, Vladivostok, personal communication, 1994). The smaller islands in the center of the chain do not support resident bear populations.

Although it is possible that bears visit the nearer of these small islands, it is unlikely that migrant bears would travel the length of the chain. During glacial episodes, however, when these islands were larger and less distant, or connected as a peninsula, brown bears from Kamchatka could well have reached Hokkaido. There is a slight possibility of exchange in recent times between Sakhalin and Hokkaido. Recent research using minisatellite DNA fingerprinting of brown bears on Hokkaido reveals little genetic diversity (Tsuruga et al. 1994a, 1994b), but the interpretation of these results is difficult because, using a variety of minisatellite markers, grizzly bears exhibit very little variability (Craighead 1994). Additional genetic studies of these islands should reveal an interesting history and the current dynamics of this metapopulation.

### Kodiak and Afognak

Kodiak and Afognak islands both support viable brown bear populations and are near enough that there is a frequent exchange of individuals between them and between Raspberry Island, Shuyak Island, and other islands in the archipelago (Figure 14.3). Population density was estimated at about 208 bears per 1000 km², or 2732 brown bears (1968–3538) in 13,191 km² (ADFG 1993). Excluding Afognak and Shuyak islands (with very different habitats), there are 2584 brown bears estimated in 9934 km², or a density of 380 bears per 1000 km² (Barnes and Smith 1996). The archipelago was probably colonized during one or more glacial episodes when the islands were connected with the mainland Alaska peninsula or separated by a narrow barrier of water. These bears are sufficiently distinct morphologically to be considered the only current subspecies of brown bear, *Ursus arctos middendorffi*.

In recent times there have been occasional incidental sightings of brown bears swimming in the waters of Shelikov Strait between Kodiak and the Alaska peninsula, a distance of about 45 km. A mtDNA haplotype found in all Kodiak bears (Cronin et al. 1991) is also found among bears on the Alaska peninsula mainland. Moreover, mtDNA sequence data (Talbot and Shields 1995) indicate that there are distinct lineages on Kodiak and other distinct lineages in nearby Katmai on the Alaska peninsula. Additional sequences that are shared by both populations are also found throughout western Alaska and probably represent older lineages. There is thus some evidence for migration across Shelikov Strait: Kodiak may have been colonized from the Alaska peninsula mainland, or there may have been more recent interchange between Kodiak and the mainland. The emerging genetic picture indicates that there may be some gene flow provided by females (mtDNA) between Kodiak and the mainland. Given the greater dispersal of males, genetic interchange via males should be greater than that of females; this possibility is currently under investigation. The narrow straits separating Kodiak, Afognak, and other islands in the archipelago are not an effective barrier to gene flow by either sex.

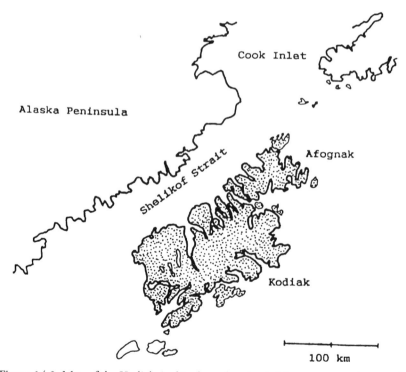

Figure 14.3. Map of the Kodiak Archipelago showing habitat permanently occupied by *Ursus arctos middendorffi* (stippled area). After ADFG (1993).

Unimak Island, adjacent to the Alaska peninsula in the Aleutian chain, is estimated to support 250 (200–300) brown bears (ADFG 1993). This population is almost certainly maintained by immigration from the mainland, which is less than 3 km distant. Resident populations are not found on the other Aleutian islands.

### Prince William Sound: Montague and Hinchinbrook Islands

Both Montague and Hinchinbrook support brown bear populations (Figure 14.4). Population estimates are 55 to 139 for Hinchinbrook, 23 to 58 for Montague, 10 to19 for Hawkins Island, and 11 to 22 for the Kenai Peninsula mainland adjoining western Prince William Sound (Nowlin 1993). No brown bears are considered resident on the islands of western Prince William Sound. Brown bears, as noted, can swim between these islands (Miller and Ballard 1982). It is likely that there is gene flow between Montague and Hinchinbrook, which are about 13 km apart, and it is more likely that migration takes place between Hinchinbrook, Hawkins Island, and the Alaskan mainland near Cordova, which are separated by less than 3 km. With only about 40 resident bears, Montague may occasionally be "rescued" by immigrants crossing the islands to the west from the Kenai Peninsula (less than 10 km apart) or migrating from Hinchinbrook to the east. Kayak Island, less than 3 km off Cape Suckling in the Gulf of Alaska, is estimated to have 3 to 10 brown bears (ADFG 1993).

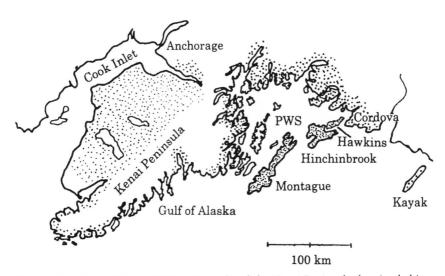

Figure 14.4. Map of Prince William Sound and the Kenai Peninsula showing habitat permanently occupied by brown bears (stippled area). After ADFG (1993).

## The Kenai Peninsula

The Kenai Peninsula in south-central Alaska supports an islandlike population of brown bears (Figure 14.4). It is connected by land to the rest of Alaska by a narrow corridor that is heavily glaciated. Gene flow through this corridor is no doubt greatly restricted. Estimates of population size are around 300 bears, with about 181 (90–270) in the western portion of the peninsula, 96 (50–150) on the eastern edge of the peninsula bordering the Gulf of Alaska and further north (ADFG 1993), and an additional 11 to 22 bordering Prince William Sound (Nowlin 1993). Gene flow probably occurs also into the Kenai through the coast and islands of Prince William Sound. Human population growth and development are proceeding rapidly on the Kenai, and salmon streams are crowded with anglers during salmon runs. Thus the population is probably reduced from historic levels.

## The ABC Islands

Admiralty, Baranof, and Chichagof islands (Figure 14.5) support brown bears at high population densities (ADFG 1993). Admiralty has an area of about 4306 km² and an estimated 1660 bears (1494–1824): a density of 386 bears per 1000 km². Baranof covers about 4159 km² and has about 816 brown bears (719–913): a density of 196 bears per 1000 km². Chichagof has an area of 5445 km² and an estimated 1625 brown bears (1501–1772): a density of 298 bears per 1000 km². Kruzof Island has an area of 518 km² and supports an estimated 127 brown bears (121–133): a density of 245 bears per 1000 km² (ADFG 1993). Each of the larger islands may support viable populations, and each probably functions as its own "mainland," although there is sufficient interchange that they share the same mtDNA lineages (Waits et al. 1996a; Talbot and Shields in press).

It is unlikely that Kruzof has a self-sustaining population; its numbers are probably augmented regularly by immigration from Chichagof over distances of less than 1 km between intermediate islands. It is unlikely that bear populations have gone extinct on any of the main islands since they were first colonized. This island group contains a large number of smaller islands that are presently separated by narrow expanses of water. Brown bears visit these islands and may remain on them for one or more generations, but the islands are not large enough to support a population over time. There are frequent anecdotal reports of bears swimming between Baranof and Chichagof (E. Young, Alaska Department of Fish and Game, personal communication, 1994), and a brown bear was observed 8 km from shore in Chatham Strait where Admiralty and Baranof are separated by 22 km of water (J. Faro, Alaska Department of Fish and Game, personal communication, 1994). Another bear was observed swimming in Freshwater Bay over 3 km from shore by

Titus and Beier (1992), who commented that expanses of water may not be barriers to some bears, especially subadult males. No radio-tagged animals have crossed from the ABC Islands to the mainland (K. Titus, Alaska Department of Fish and Game, as quoted by Talbot and Shields in press).

On the basis of distance alone, there should be significant gene flow between Baranof and Chichagof, which are 2 to 4 km apart in many places. Gene flow between them and Admiralty, about 8 km distant at the closest points, may be somewhat reduced. Recent mtDNA sequence data (Talbot and Shields in press) demonstrate a unique lineage on the ABC Islands that is not found on mainland southeastern Alaska. In addition, none of the mainland types that have been resolved to date are found in ABC brown bears either. Thus, it appears that there is little dispersal of female brown bears between the ABC Islands and the mainland, which can be as close as 4 or 6 km in places. Genetic and fossil data indicate that this is the most ancient lineage

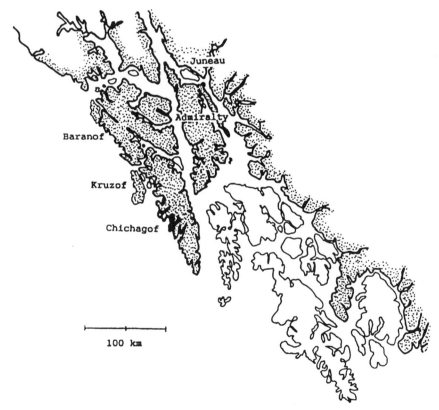

Figure 14.5. Map of the ABC Islands (Admiralty, Baranof, and Chichagof) showing habitat permanently occupied by brown bears (stippled area). After ADFG (1993).

of brown bears in North America and suggest, moreover, that they may have survived in a northwestern coastal refugium during the Pleistocene (Talbot and Shields in press). Further data from both mtDNA and nuclear genes will be necessary to clarify whether males occasionally cross these waters at the present time. In recent times, due to hunting pressure and widespread habitat modification caused by logging and other development, the populations on all three islands have declined but they are considered stable (ADFG 1993).

### Richards Island

Richards Island, on the MacKenzie River delta in Arctic Canada, supports a density of about four bears per 1000 km² (Nagy et al. 1983). It is adjacent to a large mainland population, less than 5 km distant, which ranges throughout the islands of the delta. A 17,318-km² area on the Tuktoyuktuk peninsula and Richards Island supports about 80 bears. Migration between these islands and the mainland is frequent enough that they should be considered a single population. There may be genetic differences between distant areas, but it is not likely to function as a metapopulation. This system is mentioned as an example of an island situation that probably does not act as a barrier to gene flow, as contrasted with Sakhalin and Hokkaido, or the ABC Islands, where genetic differences between island and mainland populations should be more pronounced.

### Are These Metapopulations?

None of the island systems discussed here appears to function as a classic metapopulation (Levins 1969; Gilpin 1987). Each system is connected to some degree to a "mainland" source of colonizing individuals. These mainlands may be large islands (Table 14.1) or continents that support brown bear populations in excess of 1000 individuals and tend to stabilize any true metapopulation dynamics. Nevertheless, the geographic separation affects the system dynamics and these island systems probably function as metapopulations in terms of population genetics (Paetkau et al. 1995) and also in terms of the spatial distribution of individuals (Hassel et al. 1994). With these caveats in mind, and given the more liberal definition of metapopulations exemplified by this volume, we will call these island systems metapopulations.

## Continental Metapopulations

Our discussion of continental populations is based primarily on intensive demographic and genetic studies of a population located in northwestern

TABLE 14.1.

Isolated and semi-isolated island population estimates
for various brown bear populations

| Population | Estimated size | $N_{e1}$[†] | $N_{e2}$[‡] |
|---|---|---|---|
| Sakhalin Island | 1400 | 420 | 574 |
| Kurile Islands | 700 | 210 | 287 |
| Kodiak archipelago | 2732 | 820 | 1120 |
| Admiralty | 1660 | 498 | 680 |
| Baranof | 816 | 245 | 335 |
| Chichagof | 1625 | 488 | 666 |
| Kruzof | 127 | 38 | 52 |
| ABC Island archipelago | 4228 | 1268 | 1733 |
| Kenai Peninsula | 300 | 90 | 123 |
| Hinchinbrook | 97 | 29 | 40 |
| Montague | 40 | 12 | 16 |
| Northern Continental Divide Ecosystem | 681 | 204 | 279 |
| Yellowstone Grizzly Bear Ecosystem | 236 | 71 | 97 |
| Cabinet-Yaak Ecosystem | 17 | 5 | 7 |
| Selkirk Mountains Ecosystem | 30 | 9 | 12 |
| Western Brooks Range[a] | 153 | 46 | 64 |
| Tuktoyuktuk [a] | 80 | 24 | 33 |

*Note:* In calculating $N_e / N$, note that $N$ represents the total number of adults of breeding age in the population. In the field, this is usually difficult to determine accurately and total bears are estimated. Using estimates derived from our data, we determined $N_e / N$ to be about 0.64 (Craighead 1994). When $N$ is used to represent the total bear population (not just those of breeding age), $N_e / N$ is about 0.41.

[†] $N_{e1}$ = using Harris' (1986) estimate of $N_e / N$ = 0.3 and taking $N$ to be the total population size.

[‡] $N_{e2}$ = using estimates from our data of $N_e / N$ = 0.41 when $N$ represents the total population size.

[a] Nonisolated, contiguous arctic populations.

Alaska (Figure 14.6) in the northern foothills of the Western Brooks Range (WBR) around 69°N latitude, 161°W longitude. It encompasses the headwaters of the Colville, Utukok, and Kokolik rivers in an area of 5200 km². Elevation ranges from 400 to 1300 m. The climate is dominated by long severe winters with short cool summers. This is near the northern limits of past and present brown bear range. It is open, treeless arctic tundra and is virtually undisturbed and unhunted. It has probably been inhabited by brown bears since they first arrived from Asia.

The bears inhabiting the area are part of a larger population of grizzly bears found throughout the DeLong Mountains of the Brooks Range. We

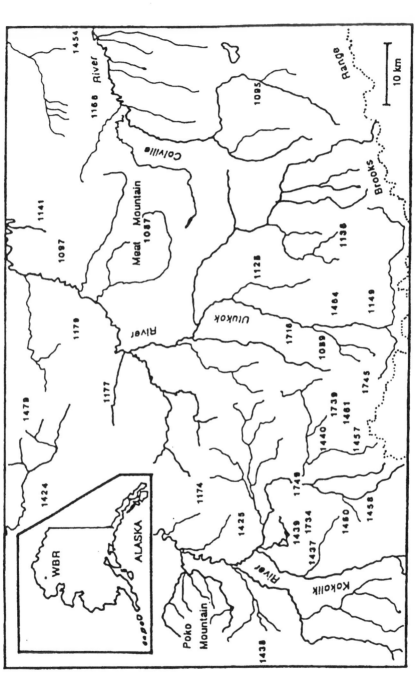

Figure 14.6. Map of the primary study area in the Western Brooks Range, Alaska. Numbers represent the centers of female grizzly home ranges.

assume that the population dynamics and genetics of our study population are representative of most nearctic and palearctic brown bears existing in continuous habitat with no geographic barriers to dispersal. In fact, there are no geographic barriers to bear movement except the Arctic Ocean to the north and the Bering Strait to the west.

The state of knowledge of brown bear genetics is such that we will paint only a general picture of typical population genetics and dynamics based on our work with Paetkau and Strobeck (Paetkau and Strobeck 1994; Paetkau et al. 1995; Craighead et al. 1995). This picture is rapidly being defined, however, by work at the University of Alberta (Paetkau and Strobeck 1996) and the University of Utah (Waits et al. 1996a) in comparing grizzly populations across North America.

The WBR population has been studied continuously since 1977 by the Alaska Department of Fish and Game. During that time 256 individual bears have been captured. Particular emphasis was placed on radio-collaring adult females. Their offspring were subsequently radio-collared just before weaning whenever possible. Adult males were radio-collared whenever they were encountered or as time and available transmitters allowed. Monitoring of radio-collared animals has resulted in detailed knowledge of maternal/offspring relationships for 53 family groups (Reynolds 1978, 1980, 1989, 1991, 1992; Reynolds and Hechtel 1980, 1984). Whole blood or tissue samples were collected from all animals handled since 1986; 152 individuals were used for genetic analysis comprising 30 of the known family groups.

Total DNA was isolated from blood and tissue samples as described by Craighead (1994) and Craighead et al. (1995). Eight DNA microsatellite primer sets were developed by D. Paetkau (Paetkau and Strobeck 1994; Paetkau et al. 1995). These microsatellite loci were amplified using PCR and analyzed in an automatic DNA sequencer as detailed in Paetkau and Strobeck (1994) and Craighead (1994). The biological function of the alleles we examined is not known, but they segregate in a Mendelian fashion and appear to be selectively neutral. Each primer set amplified the two alleles at a single locus, for a total of 16 alleles per individual bear.

### Pedigrees and Paternity

Our study population was estimated in 1992 to consist of 153 bears in a 5200-km$^2$ area: a density of 29.4 bears per 1000 km$^2$ (H. Reynolds, Alaska Department of Fish and Game, personal communication). The 152 bears in our genetic sample included 75 males, 75 females, and 2 bears of unknown sex: 35 males and 45 females were of breeding age (5 years or older). The eight microsatellite primer sets amplified 61 unambiguous alleles in this population. We obtained complete genotypes at eight loci for all of these individual bears (2432 alleles) (Craighead et al. 1995$a$). Mothers of 57 off-spring were known. We were able to identify 12 fathers for 36; the remaining 21 were sired by at least 7 males that had not been sampled. No single male was responsible for more than 11 percent of known offspring, and no males younger than 9 years of age were successful. The fact that three paternal alleles were found at several loci is clear evidence that more than one male could contribute gametes to a litter. Multiple paternity was assigned, or deduced, in one-third of the known litters having two or more cubs (Craighead 1994; Craighead et al. 1995).

Examination of allele frequencies, occurrence of homozygotes, and occurrence of heterozygotes indicates that loci are in Hardy-Weinberg equilibrium. Allele frequency divergence between generations indicates that they are selectively neutral. Mutation or recombination rate was estimated at between 7.1 and 2.2 $\times$ 10$^{-3}$ (Craighead 1994; Craighead et al. 1995).

### Evidence of Population Structure

Our observations and those of other researchers (Pearson 1975, 1977; Glenn 1980; Reynolds 1978; Nagy et al. 1983; Reynolds and Hechtel 1984; Knight et al. 1984; Miller 1984) demonstrate that female offspring tend to remain near their maternal home range after weaning and in some cases establish overlapping home ranges. Mothers and daughters often interact throughout their lives, occasionally foraging and raising young in close proximity to each other. Because adjacent females tend to be more closely related than distant ones, we might expect that genetic subdivision would occur over time; homozygotes would accumulate in local areas, and there would be a noticeable deficiency of heterozygotes (Wahlund effect) when the population is viewed as a panmictic unit.

We examined the data for evidence of a Wahlund effect at a local scale (following the definitions of Hanski and Gilpin 1991) within the study area. Within the WBR population, expected homozygosity (of genotype ii) was calculated as $p_i^2$ for each homozygote at each locus, where $p_i$ is the frequency of the $i$th allele. Average expected heterozygosity ($H$) was calculated as 1 − $\Sigma$ $p_i^2$ and was 0.747 for the WBR sample. Measures of heterozygosity can differ greatly, depending on the molecular techniques used. Allozyme analysis (K. Knudsen, University of Montana, personal communication) and minisatellite probes (Craighead 1994) tend to give lower values of $H$ than we obtained using microsatellite primers.

Observed $H$ in our sample was calculated by enumerating all heterozygotes and homozygotes at each locus from the database. Differences between expected and observed $H$ were examined using chi-square contingency analysis to determine if there was a deficiency of heterozygotes in the population. We found no evidence of population subdivision in the WBR (Craighead 1994).

Because we know that females do not tend to migrate, the apparent panmixia at this scale seems to be due to the wide-ranging movements of the male segment of the population. This interpretation is supported by our paternity analysis: of 57 offspring whose mothers were known, only 36 were fathered by males for which we had genetic samples, despite the fact that we sampled males in the study area intensively over a 6-year period (Craighead et al. 1995a, 1995, 1996). Twenty-one of our known offspring were fathered by males that had not been captured and probably utilized the study area as only a part of their home range or passed through it as transients.

Long movements of male grizzlies are supported by radio-tracking data. Subadult males, in particular, often travel long distances (Glenn 1980; Craighead and Mitchell 1982; Reynolds and Hechtel 1984). One adult male, resident in our study area for several years, was later shot near Barrow, a distance of about 300 km (H. Reynolds, Alaska Department of Fish and Game, personal communication). Another adult male traveled 163 km to the Arctic Ocean coast and then returned (Reynolds and Hechtel 1984). Thus, individual males may occasionally travel hundreds of kilometers and then breed successfully.

With this spatial structure it is difficult to say where one metapopulation ends and another begins. Paetkau et al. (1995) have analyzed gene flow in polar bear metapopulations and found significant differences in allele frequencies between sample areas across the Canadian Arctic. If we use Hanski and Gilpin's (1991) definition of the metapopulation scale—"the scale at which individuals infrequently move from one place (population) to another, typically across habitat types which are not suitable for their feeding and breeding activities, and often with substantial risk of failing to locate another suitable habitat place in which to settle"—the differences in allele frequencies found between distant populations (Paetkau et al. 1995) indicate that a metapopulation structure does exist for polar bears on a large scale: on the order of several hundred kilometers.

A somewhat smaller scale applies to brown bear populations. An analysis of North American brown bear populations (Paetkau and Strobeck 1996) demonstrates a cline in allele frequencies throughout the occupied range. Genetic distance correlates closely with physical distance. Heterozygosity is reduced from about 70 percent in northwestern populations to 55 percent in the Yellowstone ecosystem. The Kodiak population, which is reproductively healthy, has 26 percent heterozygosity, however, so this does not indicate inbreeding depression (Paetkau and Strobeck 1996). Brown bears do not migrate over such great distances as polar bears; consequently, the metapopulation scale for brown bears is expected to be somewhat smaller. Extinction and colonization events do not occur within local populations, but gene flow is restricted and areas of negative growth rate are probably "rescued" by adjacent areas with positive growth rates.

### Size of Local "Populations"

The local population unit in a continental metapopulation may conceivably be defined as the largest or the mean home range size, as the greatest or the mean dispersal distance, or perhaps as the genetic neighborhood size (Wright 1969). Wright defined neighborhood size with the equation

$$N_e = \pi \delta \sigma^2$$

where $N_e$ = genetic effective size of the population
$\pi$ = 3.1416
$\delta$ = number of breeding adults per unit area
$\sigma^2$ = one-way variance of distance between birth and breeding sites

In our Western Brooks Range population we estimated values of $N_e$ = 63.6 using Wright's (1931) original equation for unequal breeding sex ratio and $N_e$ = 116 using Hill's (1972) equation, which incorporates variance in progeny number (Craighead 1994). Using the more conservative estimate of $N_e$ = 63.6 and a density of 15.6 breeding adults (from a total of 29.4 bears) per 1000 km² (H. Reynolds, Alaska Department of Fish and Game, personal communication, 1993), we obtain a value of $\sigma^2$ = 36 km. Accordingly, 99 percent of the individuals in a "population," by this rough estimate, have their offspring within a radius of $3\sigma$, or about 108 km (36,644 km²), around a given point.

An area this size in the Western Brooks Range would support about 570 effective breeding brown bears. There could be roughly five "populations" of this size in the arctic metapopulation between the Bering Strait and the MacKenzie delta. This is a much larger area than that occupied by our study population (which we know to be too small for genetic differentiation) and is also much larger than that of many of the island populations discussed earlier (Table 14.1). This implies that females in island populations have a more restricted pool of male gametes, since the genetic neighborhood is reduced by geographical barriers.

Accordingly, genetic variability is probably lower on isolated islands and on human-fragmented islands of habitat (varying with the length of time since isolation). Island and coastal habitats where salmon are present as a food source, however, maintain densities of brown bears that are about tenfold greater than areas of arctic or interior habitats (ADFG 1993; Servheen 1993).

The genetic neighborhood, or size of a "population," in these habitats will be correspondingly smaller than in the Arctic, since a population of 570 bears (or whatever the true genetic neighborhood encompasses) will be supported in a smaller area.

## Are Continental Populations Metapopulations?

Historically, brown bear populations expanded across Eurasia during the Pliocene/early Pleistocene and across North America during the late Pleistocene/early Holocene. Their ability to utilize a wide variety of habitat types would have resulted in a more or less continuously occupied area. Local populations have maintained relatively high levels of genetic diversity through gene flow resulting from the extensive movements of males, but allelic frequencies differ as distance increases (Paetkau and Strobeck 1996). This structure does not fit the classic metapopulation definitions of Levins (1969) or Gilpin (1987) but may function as a metapopulation in terms of population genetics (Paetkau et al. 1995; Paetkau and Strobeck 1996) or the spatial distribution of individuals (Hassel et al. 1994). Although this situation stretches the definition of a metapopulation, human activities are fragmenting the distribution of continental grizzly bear populations, and in an increasing number of areas bears are being reduced to a metapopulation.

## Human-Fragmented Island Metapopulations

Habitat fragmentation caused by development has been a major force around the world. In some cases it has reached a point where habitat patches have become isolated, and brown bears, to survive, must function as a metapopulation.

### Europe

Brown bears appear to have survived the Pleistocene glaciations in two refugia in Europe, resulting in two distinct mtDNA lineages (Kohn et al. 1995, 1996). Brown bears subsequently expanded throughout Europe and even inhabited the British Isles until the tenth century (Curry-Lindahl 1972; Servheen 1990). Since then, populations in Europe have been severely reduced and habitat has been fragmented by human modification (Figure 14.7). Viewed as a metapopulation, Europe has a large "mainland" population in northwestern Russia and Finland. Other smaller and more isolated populations exist throughout Europe (Table 14.2). Many of these populations in Western Europe are so isolated that they probably cannot function as

Figure 14.7. Map of brown bear distribution in Europe. Stippled areas depict known brown bear populations. After Servheen (1990).

TABLE 14.2.
Estimated brown bear populations in Europe

| Area | Estimated population | Source |
|---|---|---|
| Sweden and Norway | >700 | Swenson (1994), Swenson et al. (1996) |
| Finland | 450 | Pullainen (1989) |
| Poland | 70–75 | Jakubiec and Buchalczyk (1987) |
| Slovakia and Czech Republic | 700 | Rosler (1989) |
| Southwestern Russia and Romania | 6000 | Rosler (1989) |
| Yugoslavia (1989) | 1600–2000 | Isakovic (1970), Huber (1992) |
| Bulgaria | 850 | Spiridonov and Spassov (1992) |
| Albania | Unknown | Servheen (1990) |
| Greece | 90–170[a] | Mertzanis (1989) |
| Italy | 50, 10–16[b] | Boscagli (1987), Zunino (1992) |
| Pyrenees Mountains | 20–30 | Camarra and Parde (1992) |
| Spain | 93–103, 17[b] | Clevenger et al. (1987) |

[a] Total for two populations.

[b] Estimates for each of two populations.

a metapopulation, and some are likely to go extinct in the near future unless they are colonized by translocating bears from other areas. One Italian population is now estimated to contain only two or three bears based on DNA analysis of scat samples (Kohn et al. 1995, 1996). Brown bear populations are expanding in Sweden, however, where male bears have recolonized areas much more rapidly than females (Swenson et al. 1996). The brown bear populations in Eastern Europe—Romania, Greece, Albania, and southern Yugoslavia—occur in forested ecosystems separated by lowland areas of human development, and there is a possibility of a functioning metapopulation if linkages are maintained (Servheen et al. 1996). However, both legal and illegal hunting have increased in Russia, Romania, and Bulgaria since the collapse of the Soviet Union (C. Servheen, U.S. Fish and Wildlife Service, personal communication), and civil war in the former Yugoslavia has reduced bear numbers and further fragmented the habitat (Huber 1994).

### Eurasia

Brown bears are distributed throughout Eurasia (Figure 14.8) with a large "mainland" population in tundra and taiga forests of Russia extending into

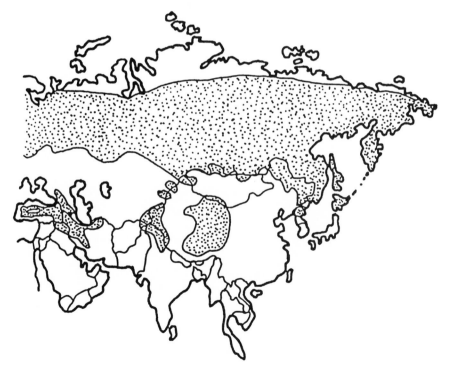

Figure 14.8. Map of brown bear distribution in Eurasia. Stippled areas depict known brown bear populations. After Servheen (1990) and M. Kretchmar, Institute of Biological Problems of the North, Magadan, Russia (personal communication).

neighboring areas of North Korea, Mongolia, and China (Servheen 1990). The distribution and population sizes in these areas are not clearly known. Large areas supporting brown bears are found in central China and in Turkey, and there are smaller population "islands" along the China/Russian border and in parts of northern India, Pakistan, Iran, Iraq, Syria, and possibly Lebanon (Servheen 1990). Habitat in all these areas is declining, and there is little or no protection from hunting. There are barriers to dispersal, such as the Gobi Desert, and intensive human development in lowland areas so that it is not known whether some of these areas may function as a metapopulation.

### North America

Brown bear populations in the lower 48 states have also been greatly reduced in number, and their habitats have been fragmented by human development. The remnant populations form a potential metapopulation with three major large "islands" of bear habitat in the United States and adjacent Canada; these have been designated as the Northern Continental Divide Ecosystem (NCDE), Bitterroot Ecosystem (BE), and Yellowstone Grizzly Bear Ecosystem (YGBE) recovery zones by the U.S. Fish and Wildlife

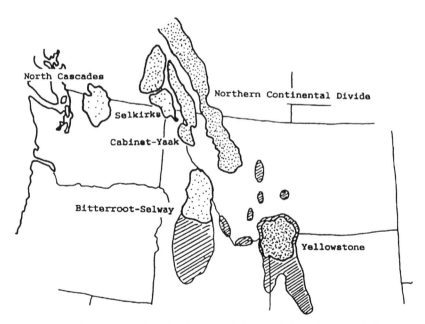

Figure 14.9. Map of brown bear habitat islands in North America. Stippled areas depict recovery zones; cross-hatched areas signify additional habitat. After Servheen (1990).

Service (Servheen 1993) (Figure 14.9). The cores of these areas are protected as national parks (in both the United States and Canada) and wilderness areas. Additional lands around these cores are at present undeveloped and contain brown bear habitat. Smaller islands of habitat exist, largely in Canada but extending over the U.S. border in the Cabinet-Yaak Ecosystem (CYE) and Selkirk Mountains Ecosystem (SE). A small and more isolated island of habitat is located in the North Cascades in Washington and British Columbia. These areas have also been designated as recovery zones. Areas of habitat designated as recovery zones have been calculated as sufficient to support a minimum brown bear population size (recovery goal) deemed large enough to be self-sustaining (Servheen 1993). Between the three large-island recovery zones there are small "stepping stones" of habitat that might be linked as corridors for movement if human development has not created impassable barriers. There is some evidence that bears occasionally move, or attempt to move, along these corridors (Picton 1986).

There is currently no clear consensus on brown bear numbers within the NCDE. Figures range from a Montana Department of Fish, Wildlife, and Parks estimate of 549 to 813 bears (a density of about 22.1 per 1000 km² ) (Dood et. al. 1986) to a U.S. Fish and Wildlife Service estimate of a minimum 306 bears (87 adult females) (Servheen 1993), with a recovery goal of 391. At least 236 bears (67 adult females) are estimated for the YGBE, with a recovery target of 15 females with cubs per year (a density of about 11.2 bears per 1000 km²). Only 15 to 20 bears (a density of about 2.2 bears per 1000 km²) are estimated as a minimum population for the Cabinet-Yaak (within the United States), with a recovery goal of 106 for the area on both sides of the border. A minimum of 25 to 36 bears (a density of about 4.9 bears per 1000 km²) is estimated in the Selkirk Recovery Zone in both the United States and Canada, with a recovery goal of 90 bears on both sides of the border (Servheen 1993). Although the Bitterroot and North Cascades have historically supported brown bear populations, at present there are few, if any, bears residing in these areas.

## Can We Design a Metapopulation?

As noted earlier, self-sustaining populations of brown bears on islands or groups of islands have persisted perhaps for millennia. Island metapopulations known to have persisted in isolation have probably maintained populations in excess of 1000 brown bears. These island populations are contrasted with those of smaller islands in Table 14.1. Yet, smaller "islands" such as Unimak (250), the Kenai Peninsula (300), Kruzof (127), Hinchinbrook (97),

Montague (40), and the Kurile Islands have also persisted for long periods of time adjacent to large "mainland" sources of immigrants.

The success of brown bears on island systems offers hope for the maintenance of a metapopulation of brown bears in the lower 48 states if functional routes for dispersal can be maintained between the core reserve areas. If such routes are closed by human activity, translocation may be required to maintain a functioning metapopulation structure.

## Acknowledgments

We thank Harry Reynolds for his insights and collaboration in studying continental grizzly bear populations, David Paetkau and Curtis Strobeck for their help with the genetic analysis, Michael Gilpin for his assistance in understanding metapopulations, Dale McCullough and two anonymous reviewers for help with the manuscript, and all of the bear geneticists cited in the text for their cooperation and support.

## REFERENCES

Alaska Department of Fish and Game (ADFG). 1993. Estimated brown bear populations in different Alaskan Game Management Units and Subunits. Table 1 in S. M. Abbott, ed., *Brown Bear: Management Report of Survey and Inventory Activities*. Federal Aid in Wildlife Restoration. Juneau: Alaska Department of Fish and Game.

Barnes, V. G., and R. B. Smith. 1996. Brown bear abundance and population trend monitoring on Kodiak Island, Alaska. In *Proceedings of the International Conference on Bear Research and Management* 10:in press.

Boscagli, G. 1987. Brown bear mortality in central Italy from 1970 to 1984. Cited in Servheen (1990).

Camarra, J. J., and J. M. Parde. 1992. The brown bear in France—status and management in 1985. Cited in Servheen (1990).

Churcher, C. S., and A. V. Morgan. 1976. A grizzly bear from the middle Wisconsin of Woodbridge, Ontario, Canada. *Journal of Earth Science* 13:341–347.

Clevenger, A. P., F. J. Purroy, and M. S. DeBuraaga. 1987. Status of the brown bear in the Cantabrian Mountains, Spain. Cited in Servheen (1990).

Cowan, I. M. 1972. The status and conservation of bears (Ursidae) of the world—1970. *Proceedings of the International Conference on Bear Research and Management* 2:343–367.

Craighead, F. L. 1994. Conservation genetics of grizzly bears. Ph.D. thesis, Montana State University, Bozeman.

Craighead, F. L., D. Paetkau, H. V. Reynolds, E. R. Vyse, and C. Strobeck. 1995. Microsatellite DNA analysis of paternity and reproductive success in Arctic grizzly bears. *Journal of Heredity* 86:255–261.

Craighead, F. L., D. Paetkau, H. V. Reynolds, C. Strobeck, and E. R. Vyse. 1996. The use of microsatellite DNA analyses to infer breeding behavior and demographic processes in an Arctic grizzly bear population. *Proceedings of the International Conference on Bear Research and Management* 10:in press.

Craighead, J. J., and J. A. Mitchell. 1982. Grizzly bear (*Ursus arctos*). Pages 515–556 in J. A. Chapman and G. A. Feldhamer, eds., *Wild mammals of North America: Biology, Management, Economics.* Baltimore: Johns Hopkins University Press.

Cronin, M. A., S. C. Amstrup, G. W. Garner, and E. R. Vyse. 1991. Interspecific and intraspecific mitochondrial DNA variation in North American bears (*Ursus*). *Canadian Journal of Zoology* 69:2985–2992.

Curry-Lindahl, K. 1972. The brown bear in Europe: Decline, present distribution, biology, and ecology. *Proceedings of the International Conference on Bear Research and Management* 2:74–80.

Domico, T. 1988. *Bears of the World.* New York: Facts on File.

Dood, A., R. D. Brannon, and R. D. Mace. 1986. Final programmatic environmental impact statement: The grizzly bear in northwestern Montana. Helena: Montana Department of Fish, Wildlife, and Parks.

Dunishenko, Y. M. 1987. Distribution and numbers of the brown bear in Siberia and Far East. Cited in Servheen (1990).

Gilpin, M. E. 1987. Spatial structure and population vulnerability. Pages 125–139 in M. E. Soulé, ed., *Viable Populations for Conservation.* Cambridge: Cambridge University Press.

Glenn, L. P. 1980. Morphometric characteristics of brown bears on the central Alaska Peninsula. *Proceedings of the International Conference on Bear Research and Management* 4:313–319.

Hanai, M. 1985. The status of Japanese black bear and a few problems of its management. Cited in Servheen (1990).

Hanski, I., and M. Gilpin. 1991. Metapopulation dynamics: Brief history and conceptual domain. *Biological Journal of the Linnean Society* 42:3–16.

Harris, R., ed. 1986. Results of the workshop on grizzly bear population genetics. Missoula: U.S. Fish and Wildlife Service, Office of Grizzly Bear Recovery Coordinator.

Hassell, M. P., H. N. Comins, and R. M. May. 1994. Species coexistence and self organizing dynamics. *Nature* 270:290–292.

Hill, W. G. 1972. Effective size of populations with overlapping generations. *Theoretical Population Biology* 3:278-289.

Huber, D. 1992. The brown bear in Yugoslavia. Cited in Servheen (1990).

———. 1994. Bears and bear research in Croatia. *International Bear News* 3:2.

Isakovic, I. 1970. Game management in Yugoslavia. Cited in Servheen (1990).

Jakubiec, Z., and T. Buchalczyk. 1987. The brown bear in Poland: Its history and present numbers. Cited in Servheen 1990.

Knight, R., B. Blanchard, and D. Mattson. 1984. Influences of the Fishing Bridge area on the Yellowstone grizzly bear population. Missoula: U.S. Department of the Interior Interagency Grizzly Bear Study Team Report.

Kohn, M., F. Knauer, A. Stoffela, W. Schroder, and S. Paabo. 1995. Conservation genetics of the European brown bear—a study using excremental PCR of nuclear and mitochondrial sequences. *Molecular Ecology* 4:95–103.

———. 1996. Conservation genetics of the European brown bear. *Proceedings of the International Conference on Bear Research and Management* 10:in press.

Levins, R. 1969. Some genetic and demographic consequences of environmental heterogeneity for biological control. *Bulletin of the Entomological Society of America* 15:237–240.

Matheus, P. 1996. The late Pleistocene paleoecology of short-faced bears (*Arctodus simus*) and brown bears (*Ursus arctos*) in eastern Beringia. *Proceedings of the International Conference on Bear Research and Management* 10:in press.

Mertzanis, G. 1989. Considerations on the situation of the brown bear (*Ursus arctos*) in Mediterranean areas. Cited in Servheen (1990).

Miller, S. 1984. Black bear and brown bear. Big game studies, VI Susitna Hydroelectric Project. 1983. Annual report. Juneau: Alaska Department of Fish and Game.

Miller, S., and W. Ballard. 1982. Density and biomass estimates of an interior Alaskan brown bear, *Ursus arctos*, population. *Canadian Field Naturalist* 96:448–454.

Nagy, J. A., R. H. Russell, A. M. Pearson, M. C. Kingsley, and C. B. Larsen. 1983. A study of grizzly bears on the barren grounds of Tuktoyuktuk Peninsula and Richards Island, Northwest Territories, 1974 to 1978. Edmonton: Canadian Wildlife Service.

Nowlin, R. 1993. Game Management Unit 6. In S. M. Abbott, ed., *Brown Bear: Management Report of Survey and Inventory Activities.* Federal Aid in Wildlife Restoration. Juneau: Alaska Department of Fish and Game.

Paetkau, D., and C. Strobeck. 1994. Microsatellite analysis of genetic variation in black bear populations. *Molecular Ecology* 3:489–495.

———. 1996. Pedigree analysis, diversity, and population structure: bear genetics below the species level using microsatellites. *Proceedings of the International Conference on Bear Research and Management* 10:in press.

Paetkau, D., W. Calvert, I. Stirling, and C. Strobeck. 1995. Microsatellite analysis of population structure in Canadian polar bears. *Molecular Ecology* 4:347–354.

Pearson, A. M. 1975. The northern interior grizzly bear *Ursus arctos* L. Canadian Wildlife Service Report Series, no. 34, Ottawa.

———. 1977. Habitat, management, and the future of Canada's grizzly bears. Pages 33–40 in *Proceedings of the Symposium on Canada's Threatened Species and Habitats.* Ottawa: Canadian Nature Federation.

Picton, H. 1986. A possible link between Yellowstone and Glacier grizzly bear populations. *Proceedings of the International Conference on Bear Research and Management* 6:7–10.

Pullainen, E. 1989. The status of the brown bear in northern Europe. Cited in Servheen (1990).

Revenko, I. A. 1996. Status and distribution of brown bears in Kamchatka, Russian Far East. *Proceedings of the International Conference on Bear Research and Management* 10:in press.

Reynolds, H. V. 1978. Structure, status, reproductive biology, movement, distribution and habitat utilization of a grizzly bear population in NPR-A. Federal Aid in Wildlife Restoration. Final report, 105-C studies work group No. 3. Juneau: Alaska Department of Fish and Game.

———. 1980. North slope grizzly bear studies. Federal Aid in Wildlife Restoration. Final report, W-17-6 and W-17-7. Juneau: Alaska Department of Fish and Game.

———. 1989. Grizzly bear population ecology in the Western Brooks Range, Alaska. Progress report 1988. Fairbanks: Alaska Department of Fish and Game and U.S. Department of the Interior National Park Service.

———. 1991. Progress report: Grizzly bear population ecology in the Western Brooks Range, Alaska. Fairbanks: Alaska Department of Fish and Game.

———. 1992. Grizzly bear population ecology in the Western Brooks Range, Alaska. Progress report 1990 and 1991. Fairbanks: Alaska Department of Fish and Game and U.S. Department of the Interior National Park Service.

Reynolds, H. V., and J. Hechtel. 1980. Structure, status, reproductive biology, movement, distribution, and habitat utilization of a grizzly bear population. Federal Aid in Wildlife Restoration. Progress report, W-17-11. Juneau: Alaska Department of Fish and Game.

————. 1984. Structure, status, reproductive biology, movement, distribution, and habitat utilization of a grizzly bear population. Federal Aid in Wildlife Restoration. Progress report, W-21-1, W-22-1, and W-22-2. Juneau: Alaska Department of Fish and Game.

Rosler, R. 1989. The status of the brown bear in central and eastern Europe. Cited in Servheen (1990).

Servheen, C. 1990. The status and conservation of the bears of the world. Eighth International Conference on Research and Management. Monograph Series, no. 2, Victoria, British Columbia.

————. ed. 1993. Grizzly bear recovery plan. Missoula: U.S. Fish and Wildlife Service.

Servheen, C., P. Sandstrom, and S. Meitz. 1996. Habitat fragmentation factors and linkage considerations for the maintenance of global bear populations. *Proceedings of the International Conference on Bear Research and Management* 10:in press.

Shields, G. F., and T. D. Kocher. 1991. Phylogenetic relationships of North American ursids based on analysis of mitochondrial DNA. *Evolution* 45:218–221.

Spiridonov, G., and N. S. Spassov. 1992. Status of the brown bear in Bulgaria. Cited in Servheen (1990).

Swenson, H. 1994. Sweden and Norway: Historic and present status of the brown bear in Scandinavia. *International Bear News* 3:5.

Swenson, H., F. Sandegren, A. Bjarvall, and P. Wabakken. 1996. Living with success: Research needs when a brown bear population is expanding. *Proceedings of the International Conference on Bear Research and Management* 10:in press.

Talbot, S., and G. F. Shields. In press. Phylogeography of Alaskan brown bears (*Ursus arctos*) and paraphyly among the Ursidae. *Molecular Phylogenetics and Evolution*.

Titus, K., and L. R. Beier. 1992. Population and habitat ecology of brown bears on Admiralty and Chichagof Islands. Research progress report, W-23-4. Juneau: Alaska Department of Fish and Game.

Tsuruga, H., S. Ise, M. Hayashi, T. Mizutani, Y. Takahashi, and H. Kanagawa. 1994a. Application of DNA fingerprinting in the Hokkaido brown bear (*Ursus arctos yesoensis*). *Journal of Veterinary Medical Science* 56:887–890.

Tsuruga, H., T. Mano, M. Yamanaka, and H. Kanagawa. 1994b. Estimate of genetic variations in Hokkaido brown bears (*Ursus arctos yesoensis*) by DNA fingerprinting. *Japanese Journal of Veterinary Research* 42:127–136.

Waits, L., M. Kohn, S. Talbot, G. F. Shields, P. Taberlet, S. Paabo, and R. H. Ward. 1996*a*. Mitochondrial DNA phylogeography of the brown bear (*Ursus arctos*) and implications for management. *Proceedings of the International Conference on Bear Research and Management* 10:in press.

Waits, L., D. Paetkau, C. Strobeck, and R. H. Ward. 1996*b*. A comparison of genetic variability in brown bear populations from Alaska, Canada, and the lower 48 states. *Proceedings of the International Conference on Bear Research and Management* 10:in press.

Wright, S. 1931. Evolution in mammalian populations. *Genetics* 16:97–159.

———. 1969. *Evolution and the Genetics of Populations.* Vol. 2: *The Theory of Gene Frequencies.* Chicago: University of Chicago Press.

Zunino, F. 1992. The brown bear in central Italy—status report 1985. Cited in Servheen (1990).

# 15

# Metapopulation Theory and Mountain Sheep: Implications for Conservation

*Vernon C. Bleich, John D. Wehausen,*
*Rob Roy Ramey II, and Jennifer L. Rechel*

Researchers and managers traditionally have emphasized the need to protect steep, rocky terrain with which mountain sheep (*Ovis canadensis*) characteristically are associated (Bleich et al. 1990*a*). Management plans for this species in southwestern deserts commonly have defined mountain sheep populations on the basis of their geographic location, usually a single mountain range (Bureau of Land Management 1988). While mountain sheep occasionally cross the broad valleys that separate the majority of desert mountain ranges, little thought was previously given to the concept that sheep might regularly move between these desert mountain ranges (Schwartz et al. 1986). Wilson et al. (1980), however, emphasize that any areas used by mountain sheep might be essential for their survival, and recent telemetry investigations (Ough and deVos 1984; Jaeger 1994) suggest that intermountain movements occur frequently, necessitating an expanded concept of mountain sheep populations and habitat.

Andrewartha and Birch (1954) postulated that natural populations occurring over any substantial area will be composed of a number of local populations. Levins (1970) coined the term "metapopulation" to describe such systems of populations, but this concept has been applied only recently to wild populations (as in Shaffer 1985). Schwartz et al. (1986) first characterized the probable relationships of demes of mountain sheep inhabiting isolated mountain ranges of the American Southwest as metapopulations, and they used an area of the eastern Mojave Desert of California as an example. Bleich et al. (1990*a*) discussed another, more southern Mojave Desert "metapopulation" relative to the juxtaposition of traditional mountain sheep habitat and movements by sheep between these islands of rocky terrain; these authors emphasized the need to maintain opportunities for movement between disjunct habitat patches (corridors) to ensure genetic diversity within metapopulations. Ramey (1993) further advocated a management approach that recognizes the metapopulation structure of wild sheep. Subsequently, the U.S. Bureau of Land Management embarked on a landscape-level plan that emphasizes the spatial and geographic relationships of mountain sheep habitat.

Metapopulation concepts recently have become fashionable among conservation biologists (Hanski 1991). Habitat fragmentation has created numerous situations where once-continuous distributions of species have come to resemble metapopulations but may not function as such. Hanski and Gilpin (1991) have cautioned that the dynamics of such fragmented populations generally are poorly understood and must be adequately explored before correct management prescriptions can be developed. Mountain sheep, however, have a naturally fragmented distribution (Bleich et al. 1990*a*) and may meet many of the predictions consistent with the structure of metapopulations. Indeed, Hanski and Gilpin note that "delimitation of local populations is often subjective, unless the environment consists of discrete habitat patches—which is the situation that has prompted metapopulation thinking and to which metapopulation thinking most naturally applies." Mountain sheep occur in such discrete habitat patches.

To date, metapopulation concepts have been applied to mountain sheep only from a genetic standpoint (Schwartz et al. 1986; Ramey 1993). Metapopulation theory is based as much on the spatial relationships of habitat patches (Gilpin 1987), however, as on the demographic consequences of extinctions and colonizations within the metapopulation (Hanski 1991). Applications of the metapopulation concept generally have been limited to species that are known to meet the demographic assumptions of the models (Hanski and Gilpin 1991). Because the dynamics of most mountain sheep populations are not well known, the application of metapopulation theory to the management of these ungulates has not been widely implemented.

The metapopulation model requires a colonization rate adequate to balance local extinctions, but colonization has been considered a rare event in mountain sheep (Geist 1971). In contrast, extinctions of populations are well documented (Wehausen et al. 1987; Torres et al. 1994), which appears consistent with the metapopulation model. Most of these extinctions have been human in cause, however, many from diseases contracted from domestic sheep (Foreyt and Jessup 1982; Goodson 1982; Jessup 1985; Onderka et al. 1988; Foreyt 1989) and may not result in an equilibrium that could have existed prior to such changes. Thus, a fundamental question about the dynamics of mountain sheep metapopulations concerns rates of natural extirpation and colonization. From a conservation standpoint, however, these questions are academic. Mountain sheep metapopulations have been altered considerably and the primary management question is: What strategy will best assure their continued viability?

## Mountain Sheep and the Metapopulation Model

In this chapter we explore how well mountain sheep in California fit the assumptions of a metapopulation model, how anthropogenic changes have altered metapopulation structure, and what the most appropriate management approaches for conservation of this species are. We use topography and historical distributions of mountain sheep in eastern and southern California to delineate probable metapopulations that existed prior to their alteration by human activity. We test the hypothesis that mitochondrial DNA (mtDNA) haplotype frequencies differ among our geographically defined metapopulations, as a validation of their biological reality. Because colonization by females should occur more readily within metapopulations than between them, and because mtDNA is maternally inherited, mtDNA patterns should reflect colonization patterns of females (with the possible rare exceptions of males that might have been sampled outside their natal mountain ranges). We also present direct evidence of colonizations and natural extinctions and discuss substructuring of populations that result in a multitiered structure of mountain sheep metapopulations. Finally, we explore anthropogenic changes to these metapopulations and discuss management strategies that might best enhance the long-term persistence of wild sheep in the deserts of North America.

### Geographic Metapopulations

Mountain sheep once were widespread in California, with populations concentrated in the northeast, the Sierra Nevada, and the Mojave and Sonoran deserts in the southeastern part of the state (Figure 15.1*a*). There were more than 92 historic populations of this species when mountain ranges in desert regions (Figure 15.2) or winter ranges of sheep in the Sierra Nevada are considered to represent geographically distinct populations.

We restricted our analyses to the central and southeastern parts of California (the Sierra Nevada and the deserts) where at least 60 populations occur (Torres et al. 1994). Most of these populations occupy public lands managed by a host of federal agencies (Bureau of Land Management, Forest Service, Department of Defense, National Park Service). Three "subspecies" of mountain sheep (Cowan 1940) have legal status in California (California

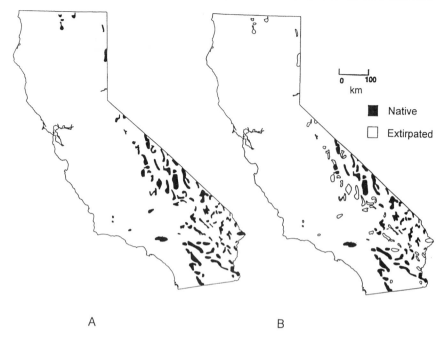

A          B

Figure 15.1. **A.** Historical distribution of mountain sheep in California, circa 1850.
**B.** Extant and extirpated populations of mountain sheep in California, circa 1970.

Fish and Game Commission 1995) despite lack of biological support for one
of these (Ramey 1993; Wehausen and Ramey 1993). In this chapter we treat
mountain sheep in California collectively as *O. canadensis*.

To define the historical distribution of mountain sheep, we relied on data
summarized by Wehausen et al. (1987) and Torres et al. (1994), which included only those geographic areas for which there is reliable evidence for the
prior existence of permanent populations. We also mapped extant populations, including both native or translocated animals. Each area was delineated
on a 1:100,000 USGS topographic map using the contour line that best defined the individual mountain range or the range known to be used by individual populations within large mountain masses (such as the Sierra Nevada).

We used a geographic information system (ARC/INFO, Environmental
Systems Research Institute, Redlands, California) for spatial analyses. We created contour lines at various distances around each mapped population
(buffer distance) and defined potential metapopulations by the connectivity
of those lines. We then considered the relationship between buffer distance
and resultant number of metapopulations in the context of known vagility of
females to define metapopulations for further analysis. Additionally, we used

Figure 15.2. Typical habitat of mountain sheep in the southeastern California deserts is on mountain ranges separated by intervening low-desert expanses. Photo by V. C. Bleich.

the geograpic information system (GIS) to determine changes in metapopulations within historical times prior to and following the construction of fenced, interstate highways in the desert regions of California. We used an extensive data set ($N$ > 10,000 aerial telemetry locations) obtained during 1983–1994 to document movements within and between "geographic" metapopulations. We also used telemetry data and direct observations (Wehausen 1979, 1992; Holl and Bleich 1983; Andrew 1994; Jaeger 1994) that documented the existence of multiple demes of ewes within geographic areas that previously had been thought to contain only single populations.

By 1970, the number of native populations in California had been reduced to 52, with the largest proportion of these occurring in the Mojave and Sonoran deserts (Figure 15.1$b$). Mountain sheep are extinct in northeastern California and in 22 desert mountain ranges, as are 11 populations in the Sierra Nevada. Moreover, this species purportedly disappeared from three areas in the Transverse Ranges, far removed from what is considered typical habitat of mountain sheep. Because no extant populations of this species occur north of Lake Tahoe and management options in northeastern California are limited by the presence of domestic livestock (Northeastern California Bighorn Sheep Interagency Advisory Group 1991), we limit further discussion to the area south of Lake Tahoe.

When we examined the relationship between buffer distance and the resultant number of metapopulations, the number of metapopulations declined rapidly with increasing buffer distance up to 7.5 km; beyond 7.5 km the slope of this relationship decreased markedly (Figure 15.3). Consequently, we chose a 7.5-km buffer distance to define metapopulations, whereby mountain sheep occupying ranges lying less than 15 km from one another are part of the same metapopulation. Using the probable distribution of mountain sheep in 1850, the 7.5-km buffer distance resulted in the definition of 13 metapopulations in southeastern California, of which 8 were insular populations (Figure 15.4*a*). Of the remaining 5 metapopulations, 2 consisted of paired populations and 3 contained from 10 to 50 subpopulations. Despite many extirpations, in 1970 the number of metapopulations remained similar to that in 1850 because additional metapopulations created via fragmentation were balanced by the loss of isolated, small metapopulations (insular populations and population pairs; Figure 15.4*b*). As a result, the mean number of populations composing these remaining metapopulations (including insular populations) declined from 7.7 to 4.0. Since 1970, as a result of the proliferation of the interstate highway system, these metapop-

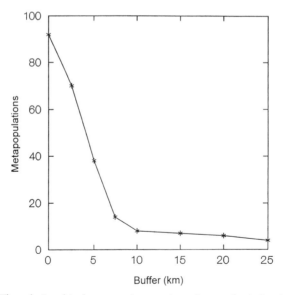

Figure 15.3. The relationship between the number of spatially defined metapopulations declines rapidly with increasing distance from the base of the mountain up to 7.5 km. Beyond that point, the number of metapopulations remains almost constant as distance increases.

ulations have experienced additional fragmentation (Figure 15.4*c*) and the mean number of populations composing each metapopulation has declined to 2.6. This situation has been ameliorated slightly through reestablishment of populations via translocation (Bleich 1990; Bleich et al. 1990*b*), but some of these were not considered in a larger landscape perspective and resulted in the establishment of six additional insular populations (Figure 15.4*c*).

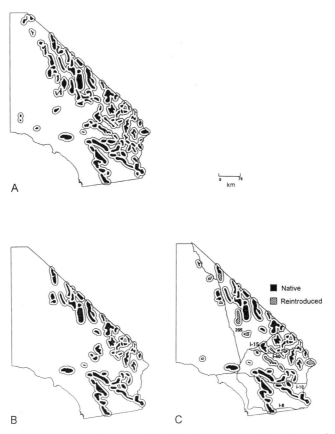

Figure 15.4. **A.** We delineated 13 spatially defined metapopulations in southeastern California, using populations of mountain sheep extant circa 1850. Of these, 8 were insular populations, and 2 were composed of 2 populations each; the remaining 3 consisted of from 10 to 50 populations. **B.** In 1970, the number of spatially defined metapopulations in southeastern California remained similar to that circa 1850 because additional metapopulations created by fragmentation were offset by the loss of isolated, small metapopulations (insular populations and population pairs). **C.** Since 1970, proliferation of the interstate highway system in southeastern California (I-8, I-10, and other routes) has further fragmented mountain sheep metapopulations.

### Haplotype Frequencies

Mitochondrial DNA data were available from 164 mountain sheep inhabiting 18 mountain ranges within our study area (Ramey 1993). We tabulated mtDNA data by metapopulation and haplotype and tested for differences in the relative frequencies of mtDNA haplotypes among metapopulations with a $G$-test (Zar 1984). We eliminated three rare haplotypes that Ramey (1993) detected in only six individuals, due to their small contribution to the data set and because their inclusion would have resulted in less powerful tests. Hence our analyses were based on mtDNA from 158 individuals that occurred in 18 mountain ranges.

Haplotype frequencies (Figure 15.5) differed significantly ($G = 212.3$, df = 20, $P < 0.001$) among six metapopulations defined by the distribution of mountain sheep in 1850 (Figure 15.4$a$). All sheep sampled from the Sierra Nevada had a unique haplotype, and 64.3 percent of those from the White Mountains/Death Valley metapopulation had yet another unique haplotype, both of which strongly influenced this result. Consequently, we eliminated those metapopulations and repeated our analysis. Again, significant differences existed in the haplotype frequencies among the four southern metapopulations ($G = 65.4$, df = 6, $P < 0.001$). Because the San Gabriel Mountains are so isolated from the other southern metapopulations, we further compared the haplotype frequencies among the three desert metapopulations and significant differences persisted ($G = 53.1$, df = 4, $P = 0.001$). Finally, we compared the two largest metapopulations within the desert and significant differences again persisted ($G = 10.6$, df = 2, $P = 0.005$).

### Defining Metapopulations

The probability of a sheep crossing between mountain ranges is a function of the distance between habitat patches (Gilpin 1987) and differs between sexes (Ramey 1993) because males and females follow different strategies to maximize reproductive fitness (Bleich 1993). Schwartz et al. (1986) summarized evidence that mountain sheep can traverse intermountain distances greater than 20 km. Our spatial definition of metapopulations (that is, occupying mountain ranges located less than 15 km from one another) was selected, in part, because of the philopatric tendencies of females, even though some intermountain movements by females have exceeded this distance (Ough and deVos 1984; Jaeger 1994).

### Genetics and Population Substructuring

Schwartz et al. (1986) and Bleich et al. (1990$a$) have documented considerable opportunity for intermountain movements by males and females within metapopulations. The results of our mtDNA analyses are consistent with in-

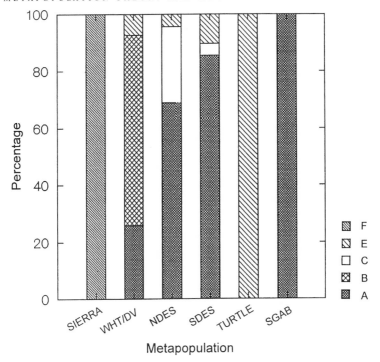

Figure 15.5. The frequency of five mtDNA haplotypes differed among six spatially defined metapopulations that were extant circa 1850 in southeastern California. SIERRA = Sierra Nevada metapopulation ($n$ = 9); WHT/DV = White Mountains–Death Valley metapopulation ($n$ = 27); NDES = Northern Desert metapopulation ($n$ = 45); SDES = Southern Desert metapopulation ($n$ = 48); TURTLE = Turtle Mountains metapopulation ($n$ = 10); SGAB = San Gabriel Mountains metapopulation ($n$ = 19).

frequent movements by females between our defined metapopulations and suggest that our geographic definition reflects biological reality.

The typical view of a mountain sheep population as a group of individuals occupying a particular mountain range is an oversimplification of their demography and genetics (Wehausen 1992; Ramey 1993). Despite documented seasonal movements and dispersal events, most females occupy relatively predictable home ranges that often are restricted to a limited part of a particular mountain range and may have little or no overlap with ranges of other groups of females (Andrew 1994; Jaeger 1994).

Geist (1971) has noted strong fidelity of females to specific winter ranges in the Rocky Mountains. Subsequently, Festa-Bianchet (1986) suggested that females with similar home ranges consist largely of related individuals; such groups, therefore, may represent matrilines. Elsewhere in the Rocky

Mountains, Stevens and Goodson (1993) have described a similar population structure. In the Mojave Desert, evidence for population substructuring has been reported by Wehausen (1992), Cunningham et al. (1993), Jaeger (1994), and V. Bleich (unpublished data). Similar evidence has been found in the Sierra Nevada (Wehausen 1979) and San Gabriel Mountains (Holl and Bleich 1983).

Because of the conservative dispersal behavior of female mountain sheep, nuclear gene flow probably occurs primarily through the movements of males between mountain ranges, which occur much more often than for females (Ramey 1993). If female substructuring reflects matrilines (Avise 1995), however, it should be reflected in geographic patterns of mtDNA haplotypes. The finding that the three isolated "metapopulations" in our data set each exhibited a single haplotype (Figure 15.5) is consistent with this interpretation on a large geographic scale. These three populations were geographically more isolated from adjacent mountain ranges and thus would be expected to receive fewer immigrants. It is possible that each was founded by a lone female, resulting in a single matriline.

That female demes may represent matrilines has important implications for both conservation and metapopulation concepts regarding mountain sheep. If the conservative behavior of female sheep results in infrequent emigrations leading to the founding of subpopulations that are matrilines, then matrilines may be the real operational metapopulation units. Similarly, metapopulation dynamics will be driven by the extirpation and founding of matrilines, rather than by the larger and more complex demographic units previously postulated. If such populations are extirpated, they are unlikely to recover or be reestablished via recruitment of nonindigenous females (Avise 1995) because of the characteristically low dispersal rates of female mountain sheep. Founder events (possibly involving a single female), lineage sorting within populations (that is, matrilineal groups), and low rates of mtDNA gene flow between populations are the most likely factors contributing to the observed patterns of mtDNA variation in southeastern California (Ramey 1993). Low frequency of female dispersal may lead to local fixation of mtDNA haplotypes, even if the population was founded by more than one female having different haplotypes. The much higher rate of intermountain movement by males (Ramey 1993) presumably is important in maintaining heterozygosity of nuclear genes and helps assure potential success of colonizing females.

Substructuring of female populations and differences in male–female dispersal distances (Ough and deVos 1984; Ramey 1993) have important implications for mountain sheep genetics, demography, and management.

Combining data from subpopulations may obscure independent dynamics, as Wehausen (1992) has documented for the Old Woman Mountains; one subpopulation was increasing while the other was decreasing in size. The concept of population substructuring has been incorporated into a recovery and conservation plan for the Sierra Nevada (Sierra Bighorn Interagency Advisory Group 1984). More recently, Stevens and Goodson (1993) have cautioned about similar situations in the Rocky Mountains. Without careful attention to substructuring within populations, managers could operate under the false premise that the overall population was increasing when in fact a segment of it was heading toward extinction.

### Evidence for Natural Extirpations

The phenomenon of natural, independent extirpations of subpopulations within a metapopulation is an important component of metapopulation theory. Losses of mountain sheep populations are well documented (Wehausen et al. 1987; Torres et al. 1994), but the causes of these extirpations are rarely known. Some losses presumably resulted from unregulated market hunting during the latter part of the nineteenth century (Buechner 1960; Wehausen 1985). Other extirpations resulted from diseases contracted from domestic sheep (*Ovis aries*) that seasonally were herded through desert regions and pastured in summer at higher elevations in the Sierra Nevada, White, and Sweetwater mountains (Jones 1950; Wehausen 1988). Experimental mixing of domestic sheep and native mountain sheep has repeatedly resulted in fatal pneumonia among native sheep (Onderka and Wishart 1988; Foreyt 1989; Callan et al. 1991).

The number of extirpations of mountain sheep that have resulted from nonanthropogenic causes is speculative. Predation can significantly reduce local populations (Wehausen 1992) and may have been responsible for some local extinctions. Similarly, extended drought and resultant poor forage conditions and limited surface water (drying up of springs) may have caused the disappearance of several populations in southeastern California (Weaver and Mensch 1971).

### Colonization Events

Colonization is considered a rare event among mountain sheep (Geist 1971), and reliable documentation of such behavior is lacking. Exploratory behavior probably is a necessary precursor to changes in the distribution of females and may occur in varying degrees from intramountain to intermountain movements. Colonizations likely begin with the exploration of unoccupied areas, followed by establishment of seasonal migrations, some of which may become emigrations. A lamb born after its mother's emigration to a new range may be more likely to remain there (and contribute to colonization). A lamb born within its mother's natal range is less likely to subsequently emigrate.

Although mountain sheep inhabiting arctic and alpine ecosystems frequently exhibit migratory behavior (Geist 1971; Geist and Petocz 1977; Wehausen 1980; Holl and Bleich 1983; Festa-Bianchet 1986), few instances of migration have been described in desert ecosystems. Since 1984, we have documented seasonal movements by female sheep between six pairs of mountain ranges in the eastern Mojave Desert. Most of these have been migrations representing seasonal visits to neighboring mountain ranges, and females have given birth in at least five of these seasonally occupied ranges (Table 15.1). In one instance, a yearling female emigrated from her natal range, established a permanent home range, and produced at least one offspring in the neighboring mountain range. Observations of female sheep in mountain ranges far removed from permanent populations ($\bar{x}$ = 51.5 km, $n$ = 4; McQuivey 1978) may represent additional evidence of exploratory and migratory behavior by females. Further use of radiotelemetry is likely to yield additional information on exploratory and seasonal movements and occasional colonization events.

### Implications of Metapopulation Theory for Conservation

There are four probable causes of metapopulation extinction (Hanski 1991). First, if extirpation rates of local populations exceed colonization rates, then extinction of the metapopulation is inevitable. Metapopulation persistence is contingent on the factors affecting extinction and colonization rates of populations within the metapopulation. Gilpin and Soulé (1986) have described four population extinction vortices: demographic, distribution, genetic, and adaptation, each with a potential role in the persistence of local populations. Mountain sheep are large, iteroparous mammals, and populations can persist for considerable periods with little recruitment because of their longevity (Wehausen 1992). They are also physiologically buffered from short-term environmental changes; for instance, they can live year-round in some hot deserts with no surface water (Krausman et al. 1985). Because of these characteristics, mountain sheep will not be as affected by the demographic and distribution extinction vortices as smaller, more $r$-selected species. They may, however, be more susceptible to genetic and adaptation extinction vortices due to lower population densities and longer generation times.

The rate of loss of genetic heterozygosity within a metapopulation is inversely related to the number of occupied habitat patches (Gilpin 1991). It is likely, however, that demographic processes are more important than genetics in the long-term persistence of populations within metapopulations (Lande

TABLE 15.1.

Evidence for migration and colonization of vacant
habitats (emigration) by female mountain sheep in the
Mojave and Sonoran deserts of California

| Type of event and location (source/destination) | Birth occurring | Number of events |
|---|---|---|
| *Migration* | | |
| Old Woman/Ship Mts. | No | 1 |
| Old Woman/Iron Mts. | Yes | 1 |
| Old Dad Peak/Cowhole Mt. | Yes | 3 |
| Kingston/Mesquite Mts. | Yes | 14 |
| Clark/Spring Mts. | Yes | 20 |
| Marble/Bristol Mts. | Yes | 4 |
| *Emigration* | | |
| Marble/Bristol Mts. | Yes | 1 |

1988; Caro and Laurenson 1994). Although Berger's (1990) conclusion that populations of less than 50 sheep will not survive for more than 50 years seems to bode poorly for the persistence of mountain sheep metapopulations, a reanalysis of data from California indicates much lower extinction probabilities (J. Wehausen, unpublished data). Traditional thinking suggests that small habitat patches are less important as sources of colonizers than large patches because the latter support larger populations and thus send out more colonists (Gilpin 1991). This does not mean, however, that smaller patches play no role in the metapopulation process. While smaller populations may receive more genetic variation than they export and thus contribute little to the effective size of metapopulations (Gilpin 1991), they may in some cases be critical stepping-stone populations in a larger colonization process (Gilpin 1987), making them important in the long-term conservation of mountain sheep in the American Southwest (Bleich et al. 1990*a*). Small populations and small patches of suitable habitat should not be undervalued (Krausman and Leopold 1986).

The second cause of metapopulation extinction (Hanski 1991) occurs where only two stable equilibria exist, such that metapopulations below a certain size are destined to extinction while larger ones may persist. Metapopulations above a certain size can become extinct if stochastic processes cause them to drop below that threshold. Such an outcome, which depends on immigration contributing to the dynamics of individual populations, is unlikely to be important to mountain sheep, given their low vagility and concomitant low colonization rates.

Immigration–extinction stochasticity is a third potential cause of meta-population extinction when the number of local populations is small (Hanski 1991). Fragmentation of desert regions of eastern California has led to a decline in the average size of the extant mountain sheep metapopulations, making them more vulnerable to such extinction.

The fourth potential cause of metapopulation extinction involves regional stochasticity (Hanski 1991) because correlated dynamics tend to reduce metapopulation persistence (Hanski 1989). Weaver and Mensch (1971) have implied that a long drought period may have caused the extinction of several populations, and such an outcome would be consistent with the concept of correlated dynamics. However, Wehausen (1992) found differing dynamics of populations within metapopulations as well as between subpopulations within a mountain range, both of which would favor longer metapopulation persistence. Where regionally correlated stochasticity occurs, the low dispersal rates of mountain sheep may bode poorly for the long-term persistence of metapopulations of this species.

## Lessons

Although numerous extirpations have occurred and anthropogenic barriers have fragmented historical metapopulations, our analysis suggests that 19 extant metapopulations currently exist in the Great Basin, Mojave, and Sonoran deserts and in the Sierra Nevada of California. Habitat fragmentation is the most significant threat faced by this species. Unoccupied habitat patches represent an important aspect of mountain sheep metapopulation dynamics because these areas may be the sites of future populations (Nunney and Campbell 1993). Moreover, given their potential importance to evolutionary processes, suboptimal habitats and peripheral populations should not be overlooked (Hoffmann and Parsons 1991; Lesica and Allendorf 1995).

Managers might best consider reestablishing extirpated populations in proximity to occupied habitat, thereby enhancing the probability of dispersing males encountering females. Because female dispersal appears to be rare, nuclear gene flow via male immigration from adjacent occupied patches is more apt to occur than the establishment of additional populations by females dispersing from translocation sites. In essence, the maintenance of existing metapopulations should precede the establishment of new metapopulations.

The California Department of Fish and Game has reestablished six populations that were spatially isolated from the nearest extant population, and one of these is isolated from all other populations by an interstate highway. In retrospect, these translocations probably had limited value to the long-term persistence of mountain sheep as a species unless establishment of a large, continuous population or a metapopulation is possible at each site. But the establishment of isolated populations may be appropriate in certain situations—for example, to avoid catastrophic loss of a rare ecotype. Because population subdivision often prevents the spread of pathogens (Dobson and May 1986), management policy for the ecotype of mountain sheep occurring in the Sierra Nevada has recommended that newly established populations be disjunct from each other (Sierra Bighorn Interagency Advisory Group 1984). In this way, managers have reduced the probability of an epizootic depleting this ecological race. The ultimate success of this approach, however, may depend on establishing adequate metapopulations in each of these isolated locations.

Most metapopulations of mountain sheep are distributed across lands managed by a diversity of agencies, and there are not many examples (such as, Keay et al. 1987; Bleich et al. 1991) of the interagency coordination needed in comprehensive management programs (Bailey 1992). Indeed, interagency competition and bureaucratic inertia frequently have thwarted the preservation of biodiversity on public lands (Grumbine 1990). A landscape-level approach to management of lands (Agee and Johnson 1988) is necessary to ensure that reserves of adequate size are established and protected. Moreover, interagency cooperation is necessary to implement the strategies whereby such reserves are managed for the long-term persistence of large carnivores and ungulates (Salwasser et al. 1987).

Maintaining viable populations of mountain sheep in North America is contingent on protecting habitats for this species, but habitat protection alone will not be sufficient in many cases (Soulé et al. 1979; Belovsky et al. 1994). Public lands, including those harboring occupied and unoccupied habitats of mountain sheep, offer opportunities for the long-term conservation of these specialized ungulates. The apparent conformance of this species to the predictions of metapopulation models dictates that government agencies cooperate to ensure that opportunities for colonization are not precluded by further fragmentation of habitats through thoughtless blocking of movement corridors.

The ramifications of a metapopulation structure for mountain sheep conservation are clear: managers must ensure that anthropogenic extirpations are minimized and that opportunities for natural recolonization by females and the migration of nuclear genes via males are not impeded. For mountain sheep, the future is now. Adequate planning and interagency cooperation will best serve the future of this species. Opportunities to develop proactive, rather than reactive, conservation strategies should not be wasted (Wilcove 1987).


## Acknowledgments

We thank J. Barrette and M. C. Nicholson for suggesting the method used to define geographic metapopulations; A. M. Pauli for assistance with analyses; M. E. Gilpin for conversations that helped shape the direction of this chapter; and R. T. Bowyer, P. R. Krausman, and S. G. Torres for helpful comments on the manuscript. This is a contribution from the California Department of Fish and Game's Mountain Sheep Management Program.

## REFERENCES

Agee, J. K., and D. K. Johnson, eds. 1988. *Ecosystem Management for Parks and Wilderness.* Seattle: University of Washington Press.

Andrew, N. G. 1994. Demography and habitat use of desert-dwelling mountain sheep in the East Chocolate Mountains, Imperial County, California. M.S. thesis, University of Rhode Island, Kingston.

Andrewartha, H. G., and L. C. Birch. 1954. *The Distribution and Abundance of Animals.* Chicago: University of Chicago Press.

Avise, J. C. 1995. Mitochondrial DNA polymorphism and a connection between genetics and demography of relevance to conservation. *Conservation Biology* 9:686–690.

Bailey, J. A. 1992. Managing bighorn habitat from a landscape perspective. *Biennial Symposium of the Northern Wild Sheep and Goat Council* 8:49– 57.

Belovsky, G. E., J. A. Bissonette, R. D. Dueser, T. C. Edwards, Jr., C. M. Luecke, M. E. Ritchie, J. B. Slade, and F. H. Wagner. 1994. Management of small populations: Concepts affecting the recovery of endangered species. *Wildlife Society Bulletin* 22:307–316.

Berger, J. 1990. Persistence of different-sized populations: An empirical assessment of rapid extinctions in bighorn sheep. *Conservation Biology* 4: 91–98.

Bleich, V. C. 1990. Costs of translocating mountain sheep. Pages 67–75 in P. R. Krausman and N. S. Smith, eds., *Managing Wildlife in the Southwest.* Phoenix: Arizona Chapter of The Wildlife Society.

———. 1993. Sexual segregation in desert-dwelling mountain sheep. Ph.D. thesis, University of Alaska, Fairbanks.

Bleich, V. C., J. D. Wehausen, and S. A. Holl. 1990*a*. Desert-dwelling mountain sheep: Conservation implications of a naturally fragmented distribution. *Conservation Biology* 4:383–390.

Bleich, V. C., J. D. Wehausen, K. R. Jones, and R. A. Weaver. 1990*b*. Status of bighorn sheep in California, 1989, and translocations from 1971 through 1989. *Desert Bighorn Council Transactions* 34:24–26.

Bleich, V. C., C. D. Hargis, J. A. Keay, and J. D. Wehausen. 1991. Interagency coordination and the restoration of wildlife populations. Pages 277–284 in J. Edelbrock and S. Carpenter, eds., *Natural Areas and Yosemite: Prospects for the Future.* Denver: USDI National Park Service.

Buechner, H. K. 1960. The bighorn sheep in the United States: Its past, present, and future. *Wildlife Monographs* 4:1–174.

Bureau of Land Management. 1988. Rangewide plan for managing habitat of desert bighorn sheep on public lands. Washington, D.C.: USDI Bureau of Land Management.

Callan, R. J., T. D. Bunch, G. W. Workman, and R. E. Mock. 1991. Development of pneumonia in desert bighorn sheep after exposure to a flock of exotic wild and domestic sheep. *Journal of the American Veterinary Medical Association* 198:1052–1056.

California Fish and Game Commission. 1995. *Fish and Game Code of California.* Longwood, Fla.: Gould Publications.

Caro, T. M., and M. K. Laurenson. 1994. Ecological and genetic factors in conservation: A cautionary tale. *Science* 263:485–486.

Cowan, I. M. 1940. Distribution and variation in the native sheep of North America. *American Midland Naturalist* 24:505–580.

Cunningham, S. C., L. Hanna, and J. Sacco. 1993. Possible effects of the realignment of U.S. Highway 93 on movements of desert bighorns in the Black Canyon area. Pages 83–100 in P. G. Rowlands, C. van Riper III and M. K. Sogge, eds., *Proceedings of the First Biennial Conference on Research in Colorado Plateau National Parks.* Washington, D.C.: USDI National Park Service.

Dobson, A. P., and R. M. May. 1986. Disease and conservation. Pages 345–365 in M. E. Soulé, ed., *Conservation Biology: The Science of Scarcity and Diversity.* Sunderland, Mass.: Sinauer Associates.

Festa-Bianchet, M. 1986. Seasonal dispersion of overlapping mountain sheep ewe groups. *Journal of Wildlife Management* 50:325–330.

Foreyt, W. J. 1989. Fatal *Pasteurella haemolytica* pneumonia in bighorn sheep after direct contact with clinically normal domestic sheep. *American Journal of Veterinary Research* 50:341–344.

Foreyt, W. J., and D. A. Jessup. 1982. Fatal pneumonia of bighorn sheep following association with domestic sheep. *Journal of Wildlife Diseases* 18:163–168.

Geist, V. 1971. *Mountain Sheep: A Study in Behavior and Evolution.* Chicago: University of Chicago Press.

Geist, V., and R. G. Petocz. 1977. Bighorn sheep in winter: Do rams maximize reproductive fitness by spatial and habitat segregation from ewes? *Canadian Journal of Zoology* 55:1802–1810.

Gilpin, M. E. 1987. Spatial structure and population vulnerability. Pages 125–139 in M. E. Soulé, ed., *Viable Populations for Conservation.* Cambridge: Cambridge University Press.

———. 1991. The genetic effective size of a metapopulation. *Biological Journal of the Linnean Society* 42:165–175.

Gilpin, M. E., and M. E. Soulé. 1986. Minimum viable populations: Processes of species extinction. Pages 19–34 in M. E. Soulé, ed., *Conservation Biology: The Science of Scarcity and Diversity.* Sunderland, Mass.: Sinauer Associates.

Goodson, N. J. 1982. Effects of domestic sheep grazing on bighorn sheep populations: A review. *Proceedings of the Biennial Symposium of the Northern Wild Sheep and Goat Council* 3:287–313.

Grumbine, R. E. 1990. Viable populations, reserve size, and federal lands management: A critique. *Conservation Biology* 4:127–134.

Hanski, I. 1989. Metapopulation dynamics: Does it help to have more of the same? *Trends in Ecology and Evolution* 4:113–114.

———. 1991. Single-species metapopulation dynamics: Concepts, models and observations. *Biological Journal of the Linnean Society* 42:17–38.

Hanski, I., and M. Gilpin. 1991. Metapopulation dynamics: Brief history and conceptual domain. *Biological Journal of the Linnean Society* 42:3-16.

Hoffmann, A. A., and P. A. Parsons. 1991. *Evolutionary Genetics and Environmental Stress.* Oxford: Oxford University Press.

Holl, S. A., and V. C. Bleich. 1983. San Gabriel mountain sheep: Biological and management considerations. San Bernardino, Calif.: USDA Forest Service, San Bernardino National Forest.

Jaeger, J. R. 1994. Demography and movements of mountain sheep (*Ovis canadensis nelsoni*) in the Kingston and Clark mountain ranges, California. M.S. thesis, University of Nevada, Las Vegas.

Jessup, D. A. 1985. Diseases of domestic livestock which threaten bighorn sheep populations. *Desert Bighorn Council Transactions* 29:29–33.

Jones, F. L. 1950. A survey of the Sierra Nevada bighorn. M.A. thesis, University of California, Berkeley.

Keay, J. A., J. D. Wehausen, C. D. Hargis, R. A. Weaver, and T. E. Blankinship. 1987. Mountain sheep reintroduction in the central Sierra: A cooperative effort. *Transactions of the Western Section of the Wildlife Society* 23:60–64.

Krausman, P. R., and B. D. Leopold. 1986. The importance of small populations of desert bighorn sheep. *Transactions of the North American Wildlife and Natural Resources Conference* 51:52–61.

Krausman, P. R., S. Torres, L. L. Ordway, J. J. Hervert, and M. Brown. 1985. Diel activity of ewes in the Little Harquahala Mountains, Arizona. *Desert Bighorn Council Transactions* 29:24–26.

Lande, R. 1988. Genetics and demography in biological conservation. *Science* 241:1455–1460.

Lesica, P., and F. W. Allendorf. 1995. When are peripheral populations valuable for conservation? *Conservation Biology* 9:753–760.

Levins, R. 1970. Extinction. Pages 77–107 in M. Gesternhaber, ed., *Some Mathematical Questions in Biology.* Providence, R.I.: American Mathematical Society.

McQuivey, R. P. 1978. The desert bighorn sheep of Nevada. Biological Bulletin 6. Reno: Nevada Department of Wildlife.

Northeastern California Bighorn Sheep Interagency Advisory Group. 1991. California bighorn sheep recovery and conservation guidelines for northeastern California. Alturas, Calif.: Modoc National Forest.

Nunney, L., and K. A. Campbell. 1993. Assessing minimum viable population size: Demography meets population genetics. *Trends in Ecology and Evolution* 8:234–239.

Onderka, D. K., and W. D. Wishart. 1988. Experimental contact transmission of *Pasteurella haemolytica* from clinically normal domestic sheep causing pneumonia in Rocky Mountain bighorn sheep. *Journal of Wildlife Diseases* 24:663–667.

Onderka, D. K., S. A. Rawluk, and W. D. Wishart. 1988. Susceptibility of Rocky Mountain bighorn sheep and domestic sheep to pneumonia induced by bighorn and domestic livestock strains of *Pasteurella haemolytica. Canadian Journal of Veterinary Research* 52:439–444.

Ough, W. D., and J. D. deVos, Jr. 1984. Intermountain travel corridors and their management implications for bighorn sheep. *Desert Bighorn Council Transactions* 28:32–36.

Ramey, R. R., II. 1993. Evolutionary genetics and systematics of North American mountain sheep: Implications for conservation. Ph.D. thesis, Cornell University, Ithaca.

Salwasser, H., C. Schonewald-Cox, and R. Baker. 1987. The role of interagency cooperation in managing for viable populations. Pages 159–173 in
M. E. Soulé, ed., *Viable Populations for Conservation*. Cambridge: Cambridge University Press.

Schwartz, O. A., V. C. Bleich, and S. A. Holl. 1986. Genetics and the conservation of mountain sheep *Ovis canadensis nelsoni*. *Biological Conservation* 37:179–190.

Shaffer, M. L. 1985. The metapopulation and species conservation: The spe-
cial case of the northern spotted owl. Pages 86–99 in R. Gutiérrez and A.
Carey, eds., *Ecology and Management of the Northern Spotted Owl in the
Pacific Northwest*. General Technical Report PNW-185. Portland: USDA
Forest Service.

Sierra Bighorn Interagency Advisory Group. 1984. Sierra Nevada bighorn
sheep recovery and conservation plan. Bishop, Calif.: Inyo National
Forest.

Soulé, M. E., B. A. Wilcox, and C. Holtby. 1979. Benign neglect: A model of
faunal collapse in the game reserves of East Africa. *Biological Conservation* 15:259–272.

Stevens, D. J., and N. J. Goodson. 1993. Assessing effects of removals for
transplanting on a high-elevation bighorn sheep population. *Conservation Biology* 7:908–915.

Torres, S. G., V. C. Bleich, and J. D. Wehausen. 1994. Status of bighorn
sheep in California, 1993. *Desert Bighorn Council Transactions* 38:17–28.

Weaver, R. A., and J. L. Mensch. 1971. Bighorn sheep in northeastern Riverside County. California Department of Fish and Game, Wildlife Management Administrative Report 71-1:1–8.

Wehausen, J. D. 1979. Sierra Nevada bighorn sheep: An analysis of management alternatives. Cooperative Administrative Report. Bishop, Calif.:
Inyo National Forest and Sequoia, Kings Canyon, and Yosemite National Parks.

———. 1980. Sierra Nevada bighorn sheep: History and population
ecology. Ph.D. thesis, University of Michigan, Ann Arbor.

———. 1985. A history of bighorn management in the Sierra Nevada. Pages
99–105 in D. Bradley, ed., *State of the Sierra Symposium*, 1985–86. San
Francisco: Pacific Publishing.

———. 1988. The historical distribution of mountain sheep in the Owens
Valley region. Pages 97–105 in *Mountains to Deserts: Selected Inyo Read-
ings*. Independence, Calif.: Friends of the Eastern California Museum.

———. 1992. Demographic studies of mountain sheep in the Mojave
Desert: Report IV. Unpublished report. Bishop, Calif.: California Department of Fish and Game.

Wehausen, J. D., and R. R. Ramey II. 1993. A morphometric evaluation of the peninsular bighorn subspecies. *Desert Bighorn Council Transactions* 37:1–10.

Wehausen, J. D., V. C. Bleich, and R. A. Weaver. 1987. Mountain sheep in California: A historical perspective on 108 years of full protection. *Western Section of the Wildlife Society Transactions* 23:65–74.

Wilcove, D. S. 1987. From fragmentation to extinction. *Natural Areas Journal* 7:23–29.

Wilson, L. O., J. Blaisdell, G. Welsh, R. Weaver, R. Brigham, W. Kelly, J. Yoakum, M. Hinks, J. Turner, and J. DeForge. 1980. Desert bighorn habitat requirements and management recommendations. *Desert Bighorn Council Transactions* 24:1–7.

Zar, J. H. 1984. *Biostatistical Analysis.* Englewood Cliffs, N.J.: Prentice-Hall.

# 16

# From Bottleneck to Metapopulation: Recovery of the Tule Elk in California

*Dale R. McCullough, Jon K. Fischer, and Jonathan D. Ballou*

> The inland we found to be far different from the shoare, a goodly
> country and fruitful soil, stored with many blessings fit for the use
> of man: infinite was the company of very large and fat deer, which
> we saw by thousands as we supposed in a herd.
> (Francis Drake, July 1579 [Bourne 1653])

For thousands of years the tule elk (*Cervus elaphus nannodes*) thrived on arguably the richest agricultural land in the world: the Central Valley and adjacent coastal valleys and hills of California. Sir Francis Drake was the first European to arrive on this coast and see tule elk. His description of elk occurring in the thousands was echoed frequently in the subsequent centuries as explorers wrote of elk herds darkening the plains and presenting animal spectacles that rivaled those of the East African savanna.

This multitude of elk vanished in the 25 years following the 1849 gold rush under the onslaught of a tidal wave of human immigration and rampant market hunting (McCullough 1969; Phillips 1976; Koch 1987). But the tule elk's fate, in any event, was sealed by its choice of habitat—prime agricultural land "with many blessings fit for the use of man." From perhaps half a million animals, the tule elk population crashed to as few as two (McCullough 1969).

## History

> Mr. A. C. Tibbet . . . [who had] a first-class reputation for veracity . . . formerly a deputy game warden at Bakersfield, knew the country well, and conducted us to the likely places. He says that when the drainage canal was put in from Buena Vista Lake, about 1874 or 1875, there were just two elk left in the whole region, male and female. From these two, Tibbet believes that all of the present

herd (estimated at 500 head) have descended. (Joseph Grinnell field notes, April 1912)

The low in numbers is of interest because of the genetic bottleneck and the loss of genetic diversity. Grinnell's (1912) field notes also include an interview with W. S. Tevis, who maintained that the area of survival was a willow jungle where hundreds of elk could have hidden. This statement was based on conjecture, though, not on a demonstrated knowledge of the elk. On 30 April 1912, Grinnell, Tevis, Tibbet, and two other men took an auto tour of the elk range in the vicinity of the current Tupman Reserve. It was Tibbet, the elderly former game warden, not Tevis, who guided the tour, locating bands of elk and specifying the approximate number in each location. In separate interviews, Tibbet estimated around 500 elk at the time, whereas Tevis and others maintained there were about one-half that number. After counting a minimum of 322 elk in that one day, and hearing Tibbet's accounts of elk in other areas, Grinnell observed that Tibbet's estimate may not have been too high.

Grinnell went to lengths to establish his informants' reliability, and several individuals do not come off well. But it is clear that Grinnell held Tibbet in high regard and was impressed by his knowledge of the elk and other wildlife in the region. Indeed, the life-history information on elk that Tibbet conveyed to Grinnell agreed with research done some 50 years later (McCullough 1969). The survival of only two individuals of the appropriate sexes, although possible, is perhaps too reminiscent of Adam and Eve. Whatever the number, when Van Dyke (1902) visited the protected population some 20 years later in 1895, only 28 elk were present. Any reasonable backward projection gives a minuscule number, and we have arbitrarily used five individuals in our estimate.

Although state laws were passed in 1852 and 1854 to give 6-month closed seasons for elk and in 1873 to grant full protection, these were straws in the wind. We can thank a private citizen, Henry Miller of the Miller and Lux Ranch, for saving the tule elk. Fortunately, he could afford to do so, being owner of the largest ranch in California, covering most of the western half of the San Joaquin Valley. When the small remnant of elk was discovered on his property, he ordered strict protection and put up a $500 reward for reporting violations of this rule. Later in the 1930s, when the state legislature balked at creating a reserve for elk, he provided land for a temporary reserve until the political maneuvering subsided and the Tupman Reserve was established.

Unlike many other species, and despite inbreeding and loss of genetic diversity, the tule elk has always shown a remarkable capacity for population growth if protected from human killing. This was true of the original remnant in the San Joaquin Valley, which increased from the 28 in 1895 to an estimated 100 in 1902 (Van Dyke 1902) and to 145 in 1905 (Merriam 1921), as well as for the most recent establishment of new populations (Figure 16.1). The behavioral plasticity of the tule elk contributes to this success, but it also leads to problems of agricultural damage. Such was the conflict in 1905 when the creation of a tule elk metapopulation by translocation was begun.

## Translocation

Henry Miller turned the herd over to the U.S. Biological Survey in 1904 to be moved to a holding pen in Sequoia National Park. In retrospect, this plan to confine the herd to a pen in an area that was marginal habitat at best, and on the fringe or outside of the original range, seems ludicrous. That it was implemented under the direction of C. Hart Merriam, director of the U.S. Biological Survey and one of the best ecologists of the time, should give us all pause—and reason to contemplate which of our own efforts might seem similarly ridiculous 90 years hence.

In November 1904 a crew of 35 horsemen attempted to round up and herd the tule elk to a pen in Sequoia National Park (Merriam 1921). The elk turned and broke back through the line. Eight elk were roped, but only one calf survived the ordeal to spend a lonely year in Sequoia. In October 1905 another 30 elk were roped and 20 survived to reach Sequoia. This nucleus of 21 elk built up and soon escaped from the pen. In 1914 there were 50 in the pen and some outside (Taylor 1916), and the remainder had escaped by 1920. Predictably, they moved into agricultural areas (Ferguson 1914; Reddington 1922), and by 1926 they were all gone (Grinnell 1933). No descendants were derived from the Sequoia population.

The next set of transplants was conducted by the California Academy of Sciences in 1914 and 1915 (Evermann 1915, 1916). These elk were captured

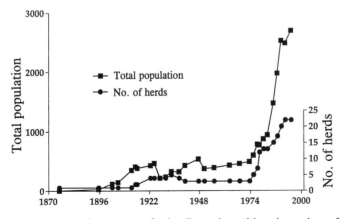

Figure 16.1. Total population size of tule elk in the wild and number of separate herds from 1875 to 1994.

in a large corral in an alfalfa field, a strategy that resulted in substantially lower mortality. In 1914, some 54 elk were transplanted to seven localities, and in 1915, some 92 elk were transplanted to fourteen localities (Table 16.1). Evermann (1916) summarized the project: 146 elk moved to nineteen localities, with 25 mortalities and 3 calves born for a net 124 elk.

Most of these transplants failed. Of the free-roaming populations only the Del Monte Park herd near Monterey, California, was a success, and this population contributed to the extant population via relocation of 21 elk to Cache Creek in 1922 (Table 16.2). Of the zoo animals, only the Griffith Park Zoo in Los Angeles and Roeding Park Zoo in Fresno maintained tule elk over a long time. At least one animal from the Roeding Park Zoo probably contributed genes to the current population at three different sites (see Table 16.2 footnote). The Del Monte Park herd thrived in the gardens and golf courses around Monterey's famous 17 Mile Drive, but it inevitably became a nuisance. One old bull made a habit of treeing golfers and eventually was collected and deposited in the Museum of Vertebrate Zoology at the University of California, Berkeley. A residue of shotgun pellets and small-caliber rifle bullets was found when his skeleton was cleaned. Transplants besides the successful one to Cache Creek were made in the late 1920s and early 1930s with at least 48 head removed, 22 going to the Hearst Ranch at San Simeon. None of these transplants was successful so far as is known. There is no record of their contributing to current populations.

The Yosemite National Park population was established with four elk in 1921 and one in 1922 transplanted from Del Paso Park in Sacramento and nine elk from San Joaquin Valley in 1922. Early mortalities reduced these numbers to six in 1922, but the population increased thereafter until its removal to Owens Valley in 1933 (Moffitt 1934). The elk were maintained in a 28-acre enclosure; they were released for a while, but hazard to visitors led to their reconfinement.

Continuing agricultural damage by the original San Joaquin population and poaching by workers in the developing oil fields of the area resulted in major problems with the wild herd. As early as 1912 Joseph Grinnell had recognized that a reserve was the only feasible solution (Grinnell 1912; Hunter 1913). The problem came to a head in the 1920s following subdivision of the Miller and Lux Ranch to small farms. After a protracted political struggle the Tupman Reserve (currently the Tule Elk State Reserve; Conrad and Waldron 1993) was established in 1932 (Burtch 1934), and most of the remaining wild population was confined to the reserve. The last four wild elk were collected in 1938 (specimens in the Museum of Vertebrate Zoology, University of California, Berkeley). The natural forage of the reserve has never been adequate, and this population has been maintained with alfalfa supplementation (Boyd et al. 1993; Conrad and Waldron 1993).

Following the demise of the San Joaquin wild population, the Owens Valley group grew to become the major tule elk population (Figure 16.2). Owens Valley lies east of the Sierra Nevada, outside the original distribution. In addition to the 26 elk moved from Yosemite in 1933, another 28 elk from the Tupman Reserve were introduced in 1934 (Table 16.2). This population increased in subsequent years and has been estimated by aerial counts every year since 1943 (McCullough 1969).

The Owens Valley herd caused damage to fences and alfalfa fields, and the local ranchers soon advocated control or removal. The following history is from McCullough (1969, 1978). Periodic hunts have occurred since 1943. The first two hunts of tule elk were approved by the California Fish and Game Commission on an ad hoc basis. Continuing pressure from livestock interests led to the adoption in 1952 of a policy to hold the Owens Valley population between 125 and 275 head. The next hunt in 1955 resulted in killing 144 of a population of 301. There was poor behavior by hunters and excessive crippling loss. When another hunt was proposed for 1960 to

Figure 16.2. A tule elk harem group in the Independence herd area of Owens Valley, California. The harem bull is to the right of the cow-calf group. Note the open habitat and dry climate (mean rainfall at this site is about 125 mm per year), characteristic for this subspecies of elk. Note also the shallow, dry evaporation pond that is typical of the flood areas occupied by the bottomland herds in the Owens Valley.

TABLE 16.1.

Relocations of San Joaquin Valley tule elk by the California Academy of Sciences in 1914 and 1915

| Area | Date | Number (male:female) | Fate[a] |
|---|---|---|---|
| 1. Danzinger private property near Los Angeles (unfenced) | 1914 | 6 (3:3) | 2 died, others unknown |
| 2. Doheny private property near Los Angeles (fenced) | 1914 | 10 (6:4) | Unknown |
| 3. Evans private property near Riverside (fenced) | 1914 | 4 (3:1) | 1 died, others unknown |
| 4. Balboa Park, San Diego (fenced) | 1914 | 12 (8:4) | 3 died |
| | 1915 | 22 (3:19) | Maintained for many years; relocated to San Luis Island |
| 5. Enslen Park, Modesto (fenced) | 1914 | 2 (2:0) | Unknown |
| 6. Boulder Creek, Big Basin (unfenced) | 1914 | 10 (5:5) | 2 died, others unknown |
| | 1915 | 4 (0:4) | 3 died, others unknown |
| 7. Del Monte Park, Monterey (unfenced) | 1914 | 10 (5:5) | 1 died, others increased; 21 relocated to Cache Creek; remainder killed by 1940 |
| 8. Moony Grove Park, near Visalia (fenced) | 1915 | 4 (1:3) | Unknown |
| 9. Roeding Park Zoo, Fresno (fenced) | 1915 | 1 (0:1)[b] | Population maintained for many years (see Table 16.2 footnote) |
| 10. Zapp's Park, Fresno (fenced) | 1915 | 1 (1:0) | Unknown |

| | | | |
|---|---|---|---|
| 11. Loinaz private park, Fresno (fenced) | 1915 | 2 (1:1) | Unknown |
| 12. Lisenby private reservation, Madera County (fenced) | 1915 | 3 (1:2) | Unknown |
| 13. Alum Rock Park, San Jose (fenced) | 1915 | 4 (2:2) | 1 female died; others unknown |
| 14. Eden Valley Ranch, Mendocino Co. (fenced) | 1915 | 12 (2:10) | 9 died in snows of first winter; others unknown |
| 15. City Park, Petaluma (fenced) | 1915 | 12 (2:10) | 1 female and 1 male died; others unknown |
| 16. Del Paso Park, Sacramento (fenced) | 1915 | 12 (3:9) | 1 died; others relocated to Yosemite |
| 17. Laveaga Park, Santa Cruz (unfenced) | 1915 | 6 (2:4) | Unknown |
| 18. Dunne Ranch, San Felipe (fenced) | 1915 | 5 (1:4) | Unknown |
| 19. Vancouver Pinnacles, San Benito Co. (unfenced) | 1915 | 4 (1:3) | Unknown |

Source: Evermann (1916).

[a] Most transplants with fates listed as unknown failed in a few years. There is no record of any current elk originating from these sources.

[b] Evermann (1916) notes that this park already had a bull from another source. Because this source is not known, it raises the possibility of interbreeding with another subspecies.

TABLE 16.2.

List of tule elk populations with origins and fates

| Population | Origin | Source (no. individuals) | Fate |
|---|---|---|---|
| San Joaquin | Pristine | Original survivors of bottleneck | Moved to Tupman Reserve in 1932; last wild individuals shot in 1933 |
| Del Monte Park | 1914 | San Joaquin (10) | Survivors killed between 1937 and 1939 |
| Yosemite | 1921 | Del Paso Zoo (3) Tupman Reserve (7) | Moved to Owens Valley in 1933 |
| Cache Creek | 1922 | Del Monte Park (21) | Current population |
| Tupman Reserve | 1932 1933 1981 | San Joaquin (63) San Joaquin (77) Owens Valley (2) | Current population |
| Owens Valley | 1933 1934 1972 | Yosemite (27) Tupman Reserve (28) Tupman Reserve ( 5) | Current population with 6 subpopulations |
| San Luis NWR | 1974 1975 1987 | San Diego Zoo (18) Fresno Zoo (4)[a] Detroit Zoo (5) | Current population |
| Grizzly Island Wildlife Area | 1977 1977 1979 | Tupman Reserve (8) Concord Naval Weapons Stn. (1) Point Reyes (1)[a] | Current population |
| Concord Naval Weapons Station | 1977 1977 | Tupman Reserve (7) Owens Valley (31) | Current population |
| Jawbone Canyon | 1977 | Owens Valley (?) (17) | Extinct (?) |
| Point Reyes Nat. Seashore | 1978 1981 | San Luis NWR (10) Owens Valley (3) | Current population |
| Mt. Hamilton | 1978 1978 1980 1981 | Owens Valley (16) Owens Valley (26) Owens Valley (21) Owens Valley ( 2) | Current population with 4 (?) subpopulations |
| Lake Pillsbury | 1978 1978 1978 1980 | Owens Valley (13) Owens Valley (24) Grizzly Island (1) Owens Valley (58) | Current population |
| Potter Valley | 1978 | Dispersed from Lake Pillsbury (primarily) | Current population |
| Fort Hunter Liggett | 1978 1979 1981 | Tupman Reserve (22) San Luis NWR (2) Owens Valley (26) | Extinct (?) Current population |
| Camp Roberts | 1978 1983 | Tupman Reserve (21) Tupman Reserve (13) | Current population |

*continues*

TABLE 16.2. *Continued*

| Population | Origin | Source (no. individuals) | Fate |
|---|---|---|---|
| Laytonville | 1979 | San Luis Island NWR (20) | Current population |
|  | 1980 | Owens Valley (23) | |
|  | 1982 | Mt. Hamilton (2) | |
| La Panza | 1983 | Tupman Reserve (20) | Current population |
|  | 1985 | Owens Valley (103) | |
|  | 1985 | Potter Valley (16) | |
|  | 1989 | Tupman Reserve (36) | |
| Fremont Peak | 1983 | Tupman Reserve (20) | Current population |
| Bartlett Springs | 1985 | San Luis NWR (59) | Current population |
| So. San Benito | 1985 | Owens Valley (47) | Current population |
|  | 1985 | Owens Valley (16) | |
|  | 1986 | Grizzly Island (57) | |
| Brushy Mt. (Covelo) | 1986 | Grizzly Island (19) | Current population |
|  | 1986 | Tupman Reserve (11) | |
| Elk Creek | 1988 | Grizzly Island (40) | Current population |
| Western Merced | 1990 | Concord (20) | Current population |
|  | 1992 | Grizzly Island (9) | |
| Parkfield | 1990 | Concord (33) | Current population |
|  | 1990 | Grizzly Island (11) | |
|  | 1993 | Tupman Reserve (1) | |
| San Ardo | 1991 | Grizzly Island (13) | Current population |
|  | 1992 | Grizzly Island (20) | |

[a] One bull ("Herb") was moved to San Luis Island from the Fresno Zoo; he was relocated to Point Reyes in 1978 (Ray 1981). In 1979 he showed extreme emaciation (Gogan 1986) and was tranquilized and moved to Grizzly Island, where he lived until 1981, at which time he was at least 11 years old.

remove 150 animals from a population estimated at 285, a great deal of public opposition arose, leading to the formation of the Committee for the Preservation of the Tule Elk. The hunt proposed for 1960 was called off and the policy reviewed. In 1961 a new policy was adopted by the Fish and Game Commission. The herd was to be held between 250 and 300 head. If less than 300 elk were counted in the annual aerial census, no hunt was to be held. If 300 or more were counted, permits would be issued to reduce the number to 250. Hunters would be controlled by warden/guides to reduce crippling loss and eliminate unsportsmanlike hunter conduct. Under this policy, hunts were held in 1961 (40 removed), 1962 (59), and 1969 (78). With the growing public interest in animal protection, the 1969 hunt caused a great uproar. In 1971 the commission raised the number in Owens Valley to 490.

The arena of conflict then changed from the California Fish and Game Commission to the state and federal legislatures. In 1971 the California legislature passed the Behr Bill, which restricted hunting until 2000 tule elk existed in the state, required reintroduction of tule elk to suitable habitat, and established a maximum for the Owens Valley population of 490. In the same year a series of tule elk refuge bills was introduced to Congress. None of these passed, but a joint resolution of Congress that duplicated the language of the Behr Bill and instructed federal agencies to make appropriate lands available for tule elk reintroductions, was passed in 1976. No hunts of tule elk were held from 1970 to 1989. State and federal legislation was instrumental in tule elk restoration in that it temporarily prohibited hunting and mandated that elk be reintroduced to available habitat. Consequently, after a lull from 1934 to 1970, a new flurry of relocations began (Table 16.2).

Attempts to establish a new herd in Owens Valley, the Mount Whitney herd, began in 1971 with the movement of two elk from the adjacent Goodale herd and continued in 1972 with release of five elk from the Tupman Reserve. The San Luis Island herd was established in a fenced enclosure on a U.S. Fish and Wildlife Service refuge near Los Baños in 1974 with 18 elk from the San Diego Zoo. In 1977, seven elk were released on the Concord Naval Weapons Station (also fenced) (Pomeroy 1986) and eight were released on Grizzly Island, a state wildlife area in the Sacramento River delta near Fairfield. The Point Reyes National Seashore herd was established behind a fence in 1978 with ten animals from San Luis Island, supplemented by three males from Owens Valley in 1981 (Ray 1981; Gogan 1986).

Numerous relocations occurred after 1974 (Table 16.2) to establish new populations and augment existing ones (Ray 1981; Phillips 1985; Willison 1985; Gogan 1986; Pomeroy 1986; Gogan and Barrett 1987; Livezey 1993). These efforts resulted in tule elk populations in 22 sites scattered throughout most of the original range (Figure 16.3). Through these relocation efforts, the statewide tule elk population increased from about 500 in 1971 to about 2700 in 1994 (Figure 16.1).

## Metapopulation Aspects

The most obvious metapopulation aspect of the current tule elk population is its distribution into widely scattered herd units (Table 16.2 and Figure 16.3). Some of these are physically restrained from dispersal by fences or other barriers; others are isolated by agricultural and other human developments that render survival during dispersal impossible. Although the tule elk has usually shown a gratifying ability to increase from small numbers, some

| | |
|---|---|
| A. Elk Creek | I. Grizzly Island |
| B. Laytonville | J. Concord |
| C. Brushy Mt. | K. Mt. Hamilton |
| D. Pillsbury | L. San Luis |
| E. Potter Valley | M. Owens Valley |
| F. Bartlett Spring | N. Fremont Peak |
| G. Cache Creek | O. So. San Benito |
| H. Point Reyes | P. Fort Hunter Liggett |
| | Q. Camp Roberts |
| | R. Tupman |
| | S. La Panza |
| | T. W. Merced |
| | U. Parkfield |
| | V. San Ardo |

Figure 16.3. Geographic location of tule elk herds in 1994.

populations have become extinct. These extinctions have been due either to unsuitable habitat (Jawbone Canyon) or to uncontrolled killing (remnants of the wild San Joaquin Valley population, Del Monte Park, first introduction to Fort Hunter Liggett). Still, given the small numbers in some populations, extinctions due to natural environmental stochasticity would be expected. For example, California recently experienced six successive years of drought for the first time in the historical record. This drought contributed to the extinction of the Goodale subpopulation in Owens Valley (Table 16.3).

Despite repeated occurrences of exceedingly small population sizes, and despite all current survivors descending from a few individuals in the

TABLE 16.3.

Aerial count estimate of the Owens Valley tule elk population by year and subpopulation as determined by California Department of Fish and Game

| Year | Bishop | Tinemaha | Goodale | Indepen-dence | Lone Pine | Whitney | Total |
|------|--------|----------|---------|---------------|-----------|---------|-------|
| 1963 | 54  | 40  | 35  | 72  | 46  | 0  | 247 |
| 1964 | 47  | 55  | 42  | 90  | 48  | 0  | 282 |
| 1965 | 29  | 54  | 8   | 41  | 48  | 0  | 180 |
| 1966 | 49  | 59  | 98  | 56  | 28  | 0  | 290 |
| 1967 | 64  | 90  | 28  | 16  | 48  | 0  | 246 |
| 1968 | 66  | 85  | 62  | 72  | 50  | 0  | 335 |
| 1969 | 67  | 77  | 35  | 92  | 59  | 0  | 330 |
| 1970 | 68  | 39  | 81  | 42  | 62  | 0  | 292 |
| 1971 | 80  | 43  | 76  | 40  | 52  | 0  | 291 |
| 1972 | 89  | 50  | 36  | 50  | 50  | 5  | 280 |
| 1973 | 90  | 84  | 46  | 58  | 62  | 0  | 340 |
| 1974 | 93  | 85  | 56  | 61  | 80  | 0  | 375 |
| 1975 | 95  | 84  | 89  | 86  | 67  | 0  | 421 |
| 1976 | 89  | 96  | 89  | 113 | 91  | 0  | 478 |
| 1977 | 133 | 114 | 116 | 111 | 89  | 19 | 582 |
| 1978 | 121 | 107 | 79  | 113 | 96  | 4  | 520 |
| 1979 | 106 | 95  | 39  | 107 | 101 | 39 | 487 |
| 1980 | 138 | 85  | 44  | 97  | 132 | 37 | 533 |
| 1981 | 73  | 118 | 7   | 91  | 136 | 51 | 476 |
| 1982 | 75  | 95  | 23  | 121 | 97  | 35 | 446 |
| 1983 | 90  | 106 | 39  | 72  | 134 | 64 | 505 |
| 1984 | 102 | 110 | 51  | 169 | 113 | 64 | 609 |
| 1985 | 109 | 112 | 49  | 148 | 113 | 69 | 600 |
| 1986 | 119 | 75  | 32  | 93  | 107 | 60 | 486 |
| 1987 | 136 | 71  | 22  | 76  | 163 | 48 | 516 |
| 1988 | 136 | 65  | 17  | 66  | 185 | 50 | 519 |
| 1989 | 119 | 65  | 0   | 49  | 136 | 30 | 399 |
| 1990 | 86  | 66  | 0   | 41  | 127 | 38 | 358 |
| 1991 | 88  | 70  | 0   | 33  | 108 | 42 | 341 |

bottleneck, inbreeding depression has never been observed in tule elk. Either inbreeding depression was not a problem in tule elk originally, or deleterious genes were purged during the bottleneck period. In any event, tule elk have proved repeatedly able to increase rapidly, even as single pairs, and this has greatly reduced their vulnerability to demographic stochasticity. This ability stands in stark contrast to many other threatened mammal species and probably explains why the tule elk has never been a serious candidate for listing under either the California or the federal endangered species acts despite existing in lower numbers than many listed species.

Another characteristic contributing to its success is the tule elk's capacity to move in seeking appropriate food conditions. This accounts for the frequent damage to agricultural fields, which often provide the best food, particularly during the dry summers of the California Mediterranean climate. Tule elk do not starve quietly: during drought periods they move in search of favorable conditions.

It is this dispersal capacity, in combination with its reproductive ability, that makes the tule elk a good colonizer and enables it to exist as a metapopulation species. Following introduction, adult bulls explore broadly if not restricted by fences or barriers. Within a year of introduction of tule elk near the center of Owens Valley, for example, bulls were seen at the north and south ends of the valley, some 230 km apart (Dow 1934; Grinnell 1934). Similar movements have been reported for many subsequent relocations. Even considerable barriers may be overcome. The three bulls introduced from Owens Valley to Point Reyes in 1981 disappeared. Because it is unlikely that they were poached, it is probable that they either got around the fence isolating Tomales Point or swam Tomales Bay, which elk were known to do before extirpation (Evermann 1915). A tragicomic case of dispersal involved one of the bulls relocated to the Santa Monica Mountains in 1914. It wandered out of the area and ended up in Los Angeles, where it died after being chased through the streets and roped by a motorcycle policeman.

Because the herd provides security, dispersal of females usually occurs in groups, whereas males wander alone (McCullough 1969, 1985). Colonization depends on arrival of groups of females with young that have arisen by separation from other groups (McCullough 1985). Some relocated female groups moved from the relocation site to other areas; subpopulations in the Owens Valley, Mount Hamilton, and La Panza and Potter Valley populations were established in this way.

## Natural Metapopulation Behavior

The Owens Valley population shows all the characteristics of a metapopulation. It is divided into six local subpopulations that are largely independent but experience occasional exchange of animals (Table 16.3 and Figure 16.4). As noted earlier, males were seen throughout the valley within a year following the initial introduction, and males annually move between herds. Most of this movement occurs during the breeding season (McCullough 1969), and these individuals usually return to their home herd area after the rut. These forays clearly are a search for mates and usually are not permanent shifts between local populations. These males are usually of a fairly large size class, but they are not the largest bulls that dominate harems. They secure some matings, hence gene flow, by controlling small groups or single females that become separated from the harems, or sometimes by controlling large harems when the rut is largely over.

How female groups separated and dispersed from the original release site in the Tinemaha herd range (Figure 16.4) is not known. But by the middle 1940s subpopulations were present in the four valley floor areas that contained prime habitat. These were marsh and lake flood areas where spring snowmelt from the Sierra Nevada collected when streamcourses overflowed (McCullough 1969, 1978; Figure 16.2). Owens Valley is a desert receiving an average of less than 125 mm of rain annually, with more years below average than above (McCullough 1969, 1978; Fowler 1985). The tule elk in Owens Valley are supported during the dry season by runoff from snowmelt in the Sierra Nevada, as was the San Joaquin population, which lived in the floodplain areas of closed sinks, Tulare and Buena Vista lakes. All herds used adjacent foothill areas during favorable plant growth periods, and most calves were born in these areas in March and April. As these plants dried up, the elk returned to the bottomlands where they remained for most of the year. This pattern is comparable to the original animals in the Central Valley as described by Tibbet (Grinnell 1912) and others (Evermann 1915).

The Goodale herd covered no valley bottom area, occurring entirely on an alluvial fan of the east slope of the Sierra Nevada. The development and history of this herd is fairly well known (Table 16.3). In the early 1950s some elk moved back and forth from the Independence herd and spent the spring in the lower part of the Goodale area (Jones 1954; McCullough 1969). This movement was comparable to use of hill areas by the other herds in spring. In 1958 this movement stopped. Deer pellet-count transects maintained on Goodale (a deer winter range) by the California Department of Fish and Game and U.S. Forest Service recorded no elk pellets before 1962. For the winter of 1962 to 1963, 2.2 elk days per acre were recorded, and this increased to 26 days per acre for 1963 to 1964 (McCullough 1969).

Part of this buildup can be attributed to an 8-ha burn in 1962 that produced an abundance of herbaceous forage heavily used by elk. The Goodale herd continued to swell in subsequent years (Table 16.3), reaching peak numbers in the late 1970s and early 1980s exceeding 100 in 1977. Thereafter, succession of vegetation on the burn to shrubs resulted in a gradual loss of carrying capacity. Numbers censused declined, and apparently a number of females and young joined the Tinemaha herd to the north (Fowler 1985; California Dept. of Fish and Game 1989). Berbach (1991) began intensive

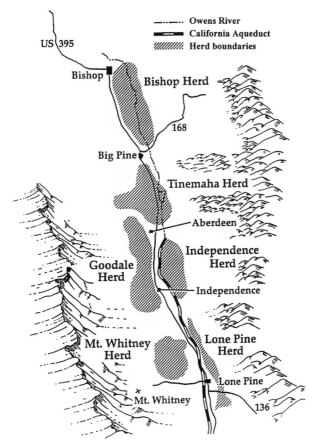

Figure 16.4. Distribution of subpopulations of the Owens Valley herd of tule elk.

studies of this herd in 1987 using radiotelemetry. This study coincided with
California's 6-year drought, and all of the females and young in the Goodale
herd moved out of the area, mostly to the Tinemaha herd to the north that
utilized alfalfa fields in the summer dry period. Only three to five male elk remained in the Goodale herd range in the summer of 1990 (Berbach 1991).
Thus, this local area was colonized by a few individuals in 1960, grew to over
100 in 1977, and declined to extinction in 1990 (Table 16.3), characteristic
of a metapopulation.

The history of the Whitney herd also is known. This area too is on a Sierra
alluvial fan and contains habitat similar to that of the Goodale range. Although tule elk probably would have reached this area naturally with time,
because it was rather isolated and a total of only about 450 tule elk existed

statewide in the 1960s (Table 16.3), it seemed prudent to establish elk by transplant (McCullough 1969).

McCullough (1969) recommended using Goodale herd animals because they were accustomed to this habitat, but the two females moved in 1971 promptly returned to the Goodale area (McCullough 1978). In 1972, five individuals were transplanted from Tupman Reserve. A few sightings of animals in the Whitney area were reported in subsequent years, and 19 were counted in the 1977 aerial census. By the mid-1980s, this herd built up to about 65 animals (Table 16.3) and subsequently declined, apparently due to the extended drought. Effects of drought are ameliorated for the valley bottom herds because tule elk have access to alfalfa fields, and natural forage is better in the marsh and pond bottoms even if dry. The Goodale and Whitney herds had no such recourse and thus were more susceptible to environmental stochasticity.

Other tule elk populations show natural metapopulation characteristics as well. The Mount Hamilton subpopulations were formed by dispersal from the original introduction sites, and a certain amount of exchange of individuals occurs (Phillips 1985). Dispersal is possible between other unrestrained populations (Fort Hunter Liggett, La Panza, Parkfield, and Camp Roberts; Laytonville, Pillsbury, Cache Creek, and Potter Valley). It is likely that as the populations become more established in the Coast and Mount Diablo ranges, exchange of individuals will occur with greater frequency. Some populations, such as Owens Valley and Point Reyes, are large and have extremely low probabilities of extinction. They could function as "mainlands" except that they are isolated by topography (Owens Valley) or barriers (Point Reyes) that restrict natural dispersal to "island" populations. Nevertheless, in the recovery program they can continue to serve as sources to establish populations in unoccupied sites or replenish local populations depleted by environmental stochasticity or human persecution.

## Gene Management

The existing populations of tule elk have undergone several genetic bottlenecks, starting with the reduction of the population in the 1870s to perhaps as few as five individuals. Subsequent bottlenecks occurred each time a new population was founded. These bottlenecks, and the relatively small genetic size of many of the populations, have likely resulted in a significant loss of genetic variation in tule elk.

Information on the origins and the demographic histories of tule elk populations allows us to estimate approximately how much variation has been lost. Genetic variation is lost at a rate inversely proportional to a population's effective size ($N_e$, see Chapter 3 in this volume). The lower the $N_e$, the faster the rate of genetic loss. Effective population size is determined by the breeding structure, sex ratio, and variance in number of offspring produced by males and females. The polygamous breeding structure of elk significantly reduces their effective size. For example, McCullough (1969) estimated that only 17 percent of the adult males participate in breeding, compared to 90 percent of the adult females. Reed et al. (1986) have estimated that the effective population size of elk is 11 percent of the total population size. Using the same formula to estimate effective size as Reed et al. (1986, Harris and Allendorf 1989) and data from life tables and variance in reproductive success of tule elk from McCullough (1969 and unpublished data), we estimate that the effective size of a tule elk population can be as low as 2 to 7 percent of census size. Thus a population of 100 may behave genetically like a population of only 2 to 7 individuals. For the purposes of this chapter and considering Reed et al.'s (1986) results, we optimistically have used a value of 10 percent to represent the ratio of effective size to census size in tule elk.

The loss of genetic variation (heterozygosity) in the tule elk populations was estimated using a simulation model based on the history of the individual populations. Information on numbers of animals founding each population, their source, and the subsequent growth of the population (Tables 16.1 to 16.3) was used as the basis for the model. The model began by assuming an infinite historical population size with heterozygosity at 100 percent. Starting with the bottleneck in the San Joaquin population in the 1870s, the heterozygosity was tracked by simulating transmission of alleles from one generation to the next according to the founding events and effective sizes of the various populations over time. For each generation (6-year period) and for each population we determined the effective size of the population (the minimum number of effective animals in that population during that generation) and the effective number (and source) of animals entering the population. Numbers were expressed in terms of effective population size, which was taken to be 10 percent of census size for populations larger than 30 animals. For smaller populations, the $N_e/N$ was considered to be larger than 0.1. (One breeding pair would have an $N_e/N$ close to 1.0.) Therefore $N_e/N$ for $N$ between 2 and 30 was defined as an exponentially decreasing function of $N$ starting at 1.0 and ending at 0.10. Alleles were transmitted from one generation to the next by randomly selecting alleles from the parental generation according to the rules of Mendelian segregation. Each simulation covered the 20 elk generations from 1875 to 1994 and was repeated 10,000 times. Heterozygosities and allele frequencies were summarized as averages over the simulations.

Figure 16.5 shows the modeled loss of heterozygosity over time as a percentage of the amount present in the original pre-bottleneck San Joaquin
population. The retained heterozygosities of the 22 extant populations are
shown as their average, minimum, and maximum values. Between 60 and 80
percent of the genetic diversity has been lost, mostly during the bottleneck in
the 1870s. An 80 percent loss is roughly equivalent to an average inbreeding
coefficient of 0.80. Considering that the inbreeding coefficient of a full-sib
mating is 0.25, on the average, tule elk are about as genetically inbred as an
individual resulting from three generations of full-sib matings.

While much of this loss of variation and inbreeding occurred relatively
early, primarily while the San Joaquin population recovered from its initial
bottleneck, there has been significant continuing loss due to the population
being subdivided into relatively small, isolated populations. The top line
(Total) in Figure 16.5 is the level of diversity that would have been retained if
the population could have been kept as one large population. Thus, the difference between this line and those of the separate 22 populations represents
the loss of variation due to the subpopulation structure of the metapopulation (Chapter 3 in this volume).

The distribution of estimated levels of retained heterozygosity for the 22
populations is shown in Figure 16.6 and Table 16.4. The effect of demographic history on retention of genetic diversity can be illustrated by examining those populations with the lowest and highest variation. The populations with the lowest retained heterozygosity, Point Reyes and Potato Hill,
were both derived from the San Luis NWR populations and each experienced
five sequential bottlenecks (the original San Joaquin bottleneck, followed by
the Tupman Reserve, San Diego Zoo, San Luis NWR, and, finally, their own
founding events). The populations with the highest expected diversity (San
Benito, Fort Hunter Liggett, La Panza, and Owens Valley) were founded with
individuals from different populations with distinct histories and are relatively large populations, thereby minimizing further loss of the restored
diversity.

As mentioned earlier, the percentage loss of diversity in the model's results
is relative to the levels of heterozygosity present in the population prior to the
1870 bottleneck. If the original levels of diversity were already very low, there
may be extremely low or even no remaining heterozygosity. If this were the
case, then the results of the model would be immaterial: all the variation was
lost and there would be no differences between the populations in the level of
diversity retained. But empirical data show that some genetic diversity is in
fact present, although at low frequencies. Kucera (1991) examined the
genetic variability of tule elk from Tupman Reserve and Owens Valley using
starch-gel electrophoresis. Variation was absent for allozymes from blood, but

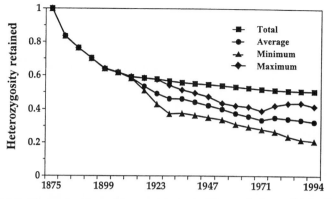

Figure 16.5. Theoretical loss of heterozygosity in 22 populations of tule elk since the original bottleneck in the San Joaquin population in the 1870s. The three bottom lines are the minimum, average, and maximum levels of heterozygosity retained in the 22 populations. The top line is the loss of heterozygosity if the 22 populations had been one large population.

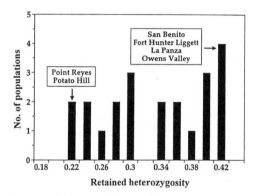

Figure 16.6. Distribution of theoretical levels of retained heterozygosity in 22 tule elk populations.

2 of 30 allozymes from liver and kidney samples were polymorphic. This level of diversity (polymorphism = 5.3 percent) is lower than that found for other elk (Kucera 1991). More recently, B. Lundrigan, K. Ralls, and R. Fleischer conducted DNA fingerprinting analyses of samples taken from four different populations. Their preliminary results show an average pairwise similarity of greater than 90 percent. This is significantly higher than levels found in most outbreeding populations, indicating a high degree of relatedness among tule elk (B. Lundrigan, University of Michigan, personal communication). Both the Kucera (1991) and the Lundrigan et al. study support the model's hypothesis that while tule elk populations have low levels of variability, some

TABLE 16.4.

Estimates of population size and proportion of retained
heterozygosity in 22 populations of tule elk

| Population | Estimated size[a] | Estimated retained heterozygosity[b] |
|---|---|---|
| Brushy Mt. | 30–35 | 0.290 |
| Cache Creek | 325–375 | 0.239 |
| Camp Roberts | 90–100 | 0.279 |
| Concord | 45–55 | 0.365 |
| Elk Creek | 40–50 | 0.269 |
| Fort Hunter Liggett | 200–250 | 0.415 |
| Fremont Peak | 55–65 | 0.257 |
| Grizzly Island | 85–90 | 0.352 |
| La Panza | 550–600 | 0.417 |
| Lake Pillsbury | 20–25 | 0.325 |
| Laytonville | 40–50 | 0.394 |
| Mt. Hamilton | 180–205 | 0.358 |
| Owens Valley | 445–475 | 0.409 |
| Parkfield | 20–30 | 0.390 |
| Point Reyes | 220–240 | 0.209 |
| Potato Hill | 60–65 | 0.216 |
| Potter Valley | 115–120 | 0.321 |
| San Ardo | 20–30 | 0.326 |
| San Benito | 100–125 | 0.417 |
| San Luis NWR | 40–50 | 0.225 |
| Tupman | 30–35 | 0.287 |
| Western Merced | 20–30 | 0.388 |

[a] Based on biannual reports of the California Fish and Game Department
to the California Legislature. The 1994 estimates are preliminary.

[b] Estimated theoretical proportion of original pre-bottleneck heterozygosity retained in the current population.

variation is present. Furthermore, low levels of allozyme and a high degree of
band sharing do not imply lack of variation in other genetic systems. Thus,
while genetic variation is low, prudent management strategies dictate that efforts be made to conserve what variation has been retained.

Numerous studies have illustrated the various deleterious effects of loss of
genetic variation and inbreeding on reproductive fitness, survival, and potential susceptibility to disease (O'Brien et al. 1985; Wildt et al. 1987; O'Brien and
Evermann 1988; Ralls et al. 1988; Brewer et al. 1990). In addition to these direct effects on fitness, loss of genetic variation reduces a population's ability to
adapt through natural selection (Allendorf 1986; Allendorf and Leary 1986).

Furthermore, in small populations the evolutionary process is driven more by genetic drift (the random change in allele frequencies due to errors in sampling small numbers of genes as they are transmitted from one generation to the next) and inbreeding than by natural selection. Deleterious genes are just as likely to become established and fixed as beneficial (adaptive) genes, with the undesirable consequence that the fitness of populations suffers.

Despite these concerns, tule elk appear to suffer little, if any, inbreeding depression, as evidenced by the usually rapid growth of newly founded populations (McCullough 1969, 1978; Gogan and Barrett 1987). It could be argued that the numerous bottlenecks and small population sizes have purged tule elk of deleterious alleles. Yet traits associated with inbreeding in other species (for example, cleft palate) have been observed in tule elk (Gogan and Jessup 1985; Kucera 1991). Moreover, other species with similar inbred histories that also show low levels of variation have shown inbreeding depression when inbred further. This is true of both the cheetah (*Acinonyx jubatus*) and Przewalski's horse (*Equus przewalskii*) when inbred in captivity (Ballou 1989); and the golden lion tamarin (*Leontopithecus rosalia*) shows significant inbreeding depression both in captivity and in the wild (Ralls et al. 1988; Baker and Dietz 1996). Therefore, a history of small population size and low genetic variation does not necessarily assure absence of future inbreeding-related problems as the populations become even more inbred.

Although levels of genetic variation in tule elk are low, management actions should be taken to maintain what variation is present as protection against potential future inbreeding problems and to provide the necessary genetic variability for natural selection. This is particularly important in smaller populations, in which future loss of genetic diversity will be most rapid. In 8 of the 22 tule elk populations (36 percent), the effective size is probably five or less, resulting in a 10 percent loss of genetic diversity per generation.

Genetic management for the retention of genetic diversity within a metapopulation structure consists of translocating animals among populations and maximizing the effective size within populations. Management options to maximize the effective size of individual populations are somewhat limited. While some management actions, such as culling herd sires after a season of successful breeding, might reduce the variance in reproductive success of males (and thus increase $N_e$), a more efficient strategy would be to focus efforts on translocating animals among populations. Translocations can restore heterozygosity to depauperate populations (bring in "new blood") and, if frequent enough, will increase the effective size of the population. Movement of animals among herds is already occurring during the breeding season in several of the larger populations, such as Owens Valley, La Panza, and Mount Hamilton (see the section "Natural Metapopulation Behavior"), and further management actions for these herds are probably not necessary. Planned translocations, however, are needed among isolated populations.

The rate and pattern of managed translocations should be based on the demographic histories of the populations. Receiving animals from the population that was the original source of its founders will not benefit a population as much as receiving animals from a more distantly related population. The demographic histories of the populations are quite complex; populations were founded and supplemented by multiple translocations from different populations at different times (Table 16.2). The simulation model provided a method for developing a hypothesis about the genetic relatedness among populations. Because the simulation model tracked allele frequencies among the different populations, it allowed us to calculate an estimate of Nei's theoretical genetic distances between each pair of populations (Nei 1973). Because of the simplicity of the simulation model, the genetic distances are useful only from a qualitative, rather than a quantitative, perspective. However, using the theoretical genetic distances, we constructed a cluster diagram of the different populations (Figure 16.7). Again it should be stressed that the absolute degree of difference among these populations depends on the absolute amount of genetic diversity remaining in the populations.

The cluster diagram shows six distinct groups of populations. Group 1 is populations founded primarily by transplants from Tupman Reserve; group 3 is those founded primarily by transplants from Owens Valley; group 2 is those hybrids of groups 1 and 3; group 5 is those that came from the original San Joaquin herd via San Diego (and other zoos) and the San Luis NWR; group 4, the Laytonville population, stands out because its founders come from groups 5 and 3. Group 6, Cache Creek, is the most distinct population. It was founded in 1922 and has never been used as a source of transplanted animals.

Metapopulation-management strategies for the translocation of animals should consider this structure. Because populations within groups are more closely related than populations among groups, translocations to benefit restoration and retention of genetic diversity should be between populations from different groups. Genetic theory suggests that on the order of one to five immigrants per generation is needed for effective transfer of genetic variation (Crow and Kimura 1970; Lacy 1987), provided that the translocated animals are effective breeders in their new populations.

We used the simulation model to examine the effects of different management strategies on the retention of genetic diversity for five generations (30 years) into the future. Each population was assumed to remain at its current size, with an effective size of 10 percent of its current size. The following five scenarios were examined:

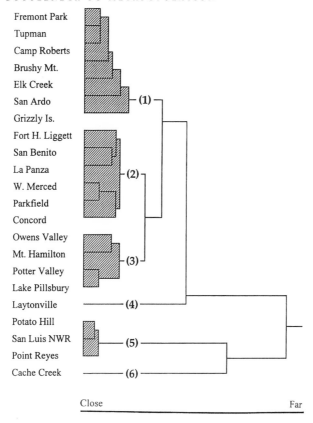

Fremont Park
Tupman
Camp Roberts
Brushy Mt.
Elk Creek
San Ardo — (1)
Grizzly Is.
Fort H. Liggett
San Benito
La Panza — (2)
W. Merced
Parkfield
Concord
Owens Valley
Mt. Hamilton — (3)
Potter Valley
Lake Pillsbury
Laytonville — (4)
Potato Hill
San Luis NWR — (5)
Point Reyes
Cache Creek — (6)

Close — Far

Simulated genetic distance

Figure 16.7. Cluster diagram of simulated genetic distance among tule elk populations. Numbers in parentheses indicate the six groups of populations.

1. No translocation—populations remain in isolation as they currently exist.
2. Each generation one effective migrant is translocated into each population from a different group, and the migrant comes from a population from a different group in subsequent translocations.
3. Two effective migrants are translocated into each population each generation. Selection of the source population is as described in scenario 2.
4. Four effective migrants are translocated into each population each generation. Selection of the source population is as described in scenario 2.
5. Two effective migrants are translocated into each population each generation, but the source is a population within the same group.

Scenario 1 models the loss of variation under no management plan, scenarios 2 through 4 the effect of different migration rates for translocations among groups, and scenario 5 the effect of translocations within groups.

The modeled average level of heterozygosity in the 22 populations over the next 30 years under the five different scenarios is plotted in Figure 16.8. With no management (scenario 1), the population is expected to lose about 10 percent more heterozygosity on the average—in other words, about 25 percent of its existing variation will be lost. The loss will be more rapid in the smaller populations. If transplants occur (two per generation) but population group structure is ignored (scenario 5), further loss of variation is halted but heterozygosity is not restored. This pattern of movement simply tends to increase the effective size of the groups without gaining the benefit of restoring variation. Scenarios 2 to 4 all result in an increase of average genetic diversity within the populations, showing that the higher the transplant rate, the more the variation is restored. But there are diminishing returns: the relative effect of four effective transplants over two is not as great as that of two over one.

Based on these simulations, any movement of animals is likely to have genetic benefit. Therefore, we recommend periodic translocations between populations of different groups and advise that the source of the transplants be rotated over time. We suggest that females be moved because of their greater likelihood of infusing genes into the recipient populations.

Several additional points should be addressed. The first is that about 17 tule elk currently exist in four zoos in North America (ISIS 1994). Although captive breeding programs can serve as effective components of metapopulation-management strategies (Foose et al. 1995), we have not considered them part of the metapopulation-management strategy for tule elk for several reasons. First, there are sufficient animals in wild populations to supply those needed for translocations. Wild-born and experienced animals are almost always preferable to captive-born animals, which are likely to be conditioned and adapted to a captive environment (Griffith et al. 1989; Arnold 1995). Furthermore, the uncertain origin and possible hybridization with other elk subspecies make the captive population a less than ideal source for translocation. Captive breeding, which once played an important role for the preservation of the tule elk, is no longer a necessary component of tule elk conservation.

A second point is the issue of what to do with animals with presumed genetic deformities. It has been recommended that animals with deformities not be used for translocation and, furthermore, that populations where deformed animals have been seen should not be used to supply animals for translocation (Kucera 1991). As the populations become more inbred, however, the occurrence of such animals will likely increase. Curtailing all trans-

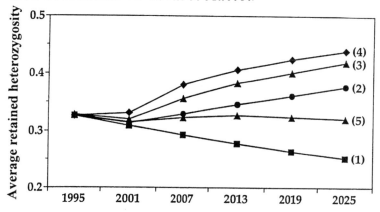

Figure 16.8. Average level of retained heterozygosity in 22 tule elk populations projected five generations into the future for five different management scenarios: (1) no movement of animals among populations; (2) one transplant per generation among population groups; (3) two transplants per generation among groups; (4) four transplants per generation among groups; and (5) two transplants per generation within groups.

plants from populations with observed deformities may be too limiting and ignores the contribution of other genes in the population. While it is prudent not to use deformed individuals for translocation, their populations should not be subsequently excluded from supplying other individuals for transplants.

## Acknowledgments

We wish to express our appreciation to the California Department of Fish and Game for access to their extensive estimates of population sizes for tule elk. For reading the chapter and making many comments for improvement we thank Thomas Kucera, William Lidicker, Jr., James Patton, and Katherine Ralls. Lori Merkle, Margaret Jaeger, and Christina Jordan helped prepare the manuscript.

## REFERENCES

Allendorf, F. W. 1986. Genetic drift and the loss of alleles versus heterozygosity. *Zoo Biology* 5:181–190.

Allendorf, F. W., and R. F. Leary. 1986. Heterozygosity and fitness in natural populations of animals. Pages 57–76 in M. E. Soulé, ed., *Conservation Biology: The Science of Scarcity and Diversity.* Sunderland, Mass.: Sinauer Associates.

Arnold, S. 1995. Monitoring quantitative genetic variation and evolution in captive populations. Pages 295–317 in J. Ballou, M. Gilpin, and T. J. Foose, eds., *Population Management for Survival and Recovery.* New York: Columbia University Press.

Baker, A., and J. Dietz. 1996. Evidence for inbreeding depression and lack of reproductive success in wild golden lion tamarins. Submitted to *Science.*

Ballou, J. D. 1989. Inbreeding and outbreeding depression in the captive propagation of black-footed ferrets. Pages 49–68 in U. S. Seal, E. T. Thorne, M. A. Bogan, and S. H. Anderson, eds., *Conservation Biology and the Black-Footed Ferret.* New Haven: Yale University Press.

Berbach, M. W. 1991. Activity patterns and range relationships of tule elk and mule deer in Owens Valley. Ph.D. thesis, University of California, Berkeley.

Bourne, N. 1653. *Sir Francis Drake Revived.* London.

Boyd, R., E. T. Conrad, and G. P. Waldron. 1993. Elk habitat use and habitat impact at the Tule Elk State Reserve, Tupman, California. Pages 83–91 in R. L. Callas, D. B. Koch, and E. R. Loft, eds., *Proceedings of the Western States and Provinces Elk Workshop.* Sacramento: California Department of Fish and Game.

Brewer, B. A., R. C. Lacy, M. L. Foster, and G. Alaks. 1990. Inbreeding depression in insular and central populations of *Peromyscus* mice. *Journal of Heredity* 81:257–266.

Burtch, L. A. 1934. The Kern County elk refuge. *California Fish and Game* 20:140–147.

California Department of Fish and Game. 1989. Tule elk: Final environmental document. Sacramento: California Department of Fish and Game.

Conrad, E. T., and G. P. Waldron. 1993. Tule elk management in the California state park system: The Tupman Tule Elk Reserve today. Pages 78–82 in R. L. Callas, D. B. Koch, and E. R. Loft, eds., *Proceedings of the Western States and Provinces Elk Workshop.* Sacramento: California Department of Fish and Game.

Crow, J. F., and M. Kimura. 1970. *An Introduction to Population Genetics Theory.* New York: Harper & Row.

Dow, G. W. 1934. More tule elk planted in Owens Valley. *California Fish and Game* 20:288–290.

Evermann, B. W. 1915. An attempt to save California elk. *California Fish and Game* 1:85–96.

————. 1916. The California valley elk. *California Fish and Game* 2:70–77.

Ferguson, A. D. 1914. Fresno Division. General conditions and some important problems. *Biennial Report, California Fish and Game Commission* 23:23–46.

Foose, T. J., L. DeBoer, U. S. Seal, and R. Lande. 1995. Conservation management strategies based on viable populations. Pages 273–294 in J. Ballou, M. Gilpin, and T. J. Foose, eds., *Population Management for Survival and Recovery.* New York: Columbia University Press.

Fowler, G. S. 1985. Tule elk in California: History, current status, and management recommendations. Project report, Contract 83/84-C-698. Department of Forestry and Resource Management, University of California, Berkeley.

Gogan, P. J. P. 1986. Ecology of the tule elk range, Point Reyes National Seashore. Ph.D. thesis, University of California, Berkeley.

Gogan, P. J. P., and R. H. Barrett. 1987. Comparative dynamics of introduced tule elk populations. *Journal of Wildlife Management* 51:20–27.

Gogan, P. J. P., and D. A. Jessup. 1985. Cleft palate in a tule elk calf. *Journal of Wildlife Diseases* 21:463–466.

Griffith, B., J. M. Scott, J. W. Carpenter, and C. Reed. 1989. Translocation as a species conservation tool: Status and strategy. *Science* 245:477–480.

Grinnell, J. 1912. Field notes. Museum of Vertebrate Zoology, University of California, Berkeley.

————. 1933. Review of the recent mammal fauna of California. *University of California Publications in Zoology* 40:71–234.

————. 1934. Field notes. Museum of Vertebrate Zoology, University of California, Berkeley.

Harris, R. B., and F. W. Allendorf. 1989. Genetically effective population size of large mammals: An assessment of estimators. *Conservation Biology* 3:181–191.

Hunter, J. S. 1913. Game conditions in California. *Biennial Report, California Fish and Game Commission* 22:17–25.

ISIS. 1994. *The International Species Information System Database.* Minneapolis: ISIS.

Jones, F. L. 1954. The Inyo-Sierra deer herds. California Fish and Game Federal Aid Project W-41-R. Sacramento: California Department of Fish and Game.

Koch, D. B. 1987. Tule elk management: Problems encountered with a successful wildlife management program. *Transactions of the Western Section of The Wildlife Society* 23:1–3.

Kucera, T. E. 1991. Genetic variability in the tule elk. *California Fish and Game* 77:70–78.

Lacy, R. C. 1987. Loss of genetic diversity from managed populations:

Interacting effects of drift, mutation, immigration, selection, and population subdivision. *Conservation Biology* 1:143–158.

Livezey, K. B. 1993. Tule elk relocated to Brushy Mountain, Mendocino
County, California. *California Fish and Game* 79:131–132.

McCullough, D. R. 1969. The tule elk: Its history, behavior, and ecology.
*University of California Publications in Zoology* 88:1–209.

———. 1978. Case histories: The tule elk (*Cervus canadensis nannodes*).
Pages 173–184 in *Threatened Deer.* Morges, Switz.: International Union
for the Conservation of Nature.

———. 1985. Long range movements of large terrestrial mammals. Pages
444–465 in M. A. Rankin, ed., *Migration: Mechanisms and Adaptive Significance.* Contributions In Marine Science Supplement 27.

Merriam, C. H. 1921. A California elk drive. *Scientific Monthly* 13:465–475.

Moffitt, J. 1934. History of the Yosemite elk herd. *California Fish and Game*
20:37–51.

Nei, M. 1973. The theory and estimation of genetic distance. Pages 45–54 in
N. E. Morton, ed., *Genetic Structure of Populations.* Honolulu: University
of Hawaii Press.

O'Brien, S. J., and J. F. Evermann. 1988. Interactive influence of infectious
diseases and genetic diversity in natural populations. *Trends in Ecology
and Evolution* 3:254–259.

O'Brien, S. J., M. E. Roelke, L. Marker, A. Newman, C. A. Winkler, D.
Meltzer, L. Colly, J. F. Evermann, M. Bush, and D. E. Wildt. 1985.
Genetic basis for species vulnerability in the cheetah. *Science*
227:1428–1434.

Phillips, J. A. 1985. Acclimation of reintroduced tule elk in the Diablo
Range, California. M.A. thesis, San Jose State University, San Jose.

Phillips, W. E. 1976. *The Conservation of the California Tule Elk.* Edmonton:
University of Alberta Press.

Pomeroy, D. R. 1986. Spatial relationships and interspecific behavior of tule
elk and cattle. M.S. thesis, University of California, Berkeley.

Ralls, K., J. D. Ballou, and A. Templeton. 1988. Estimates of lethal equivalents and the cost of inbreeding in mammals. *Conservation Biology*
2:185–193.

Ray, D. T. 1981. Post release activity of tule elk at Point Reyes National
Seashore, California M.S. thesis, University of Michigan, Ann Arbor.

Reddington, P. G. 1922. Introduced elk thrive on Sequoia National Forest.
*California Fish and Game* 8:191.

Reed, J. M., P. D. Doerr, and J. R. Walters. 1986. Determining minimum
population sizes for birds and mammals. *Wildlife Society Bulletin*
14:255–261.

Taylor, W. P. 1916. The conservation of the native fauna. *Scientific Monthly* 3:399–409.

Van Dyke, T. S. 1902. The deer and elk of the Pacific Coast. Pages 167–256 in T. Roosevelt, T. S. Van Dyke, D. G. Elliot and A. J. Stone, eds., *The Deer Family.* New York: Macmillan.

Wildt, D. E., M. Bush, K. L. Goodrowe, C. Packer, A. E. Pusey, J. L. Brown, P. Joslin, and S. J. O'Brien. 1987. Reproductive and genetic consequences of founding isolated lion populations. *Nature* 329:328–331.

Willison, J. M. 1985. The tule elk at Fort Hunter Liggett. M.S. thesis, California Polytechnic State University, San Luis Obispo.

# 17

## Metapopulation Management: What Patch Are We In and Which Corridor Should We Take?

*Dale R. McCullough*

The case histories outlined in the previous chapters should illustrate how ideas of metapopulation dynamics have infused a new intellectual excitement into wildlife management and conservation and brought a sense of hope to the largely gloomy picture of habitat loss and fragmentation. Nevertheless, these gains need to be kept in perspective. The historical, unfragmented landscape was a superior solution to retaining wildlife diversity and abundance, and wherever and whenever possible, unfragmented landscape should be preserved. Saving formerly continuous populations as metapopulations saves but a pale shadow of their original state, though it does offer promise of rescuing species that otherwise would become casualties of human civilization.

One can lament that an enlightened society should recognize and value the beauty, integrity, and essential environmental services of the natural world. But lamentation is not enough. Reality demands that we confront the inexorable expansion of development with its myopic view of what best serves a civilized human community. Even the most optimistic assessment of when we may see this expansionist value system transformed spells doom for a myriad of wild species.

Furthermore, politics and economics being what they are, the advocates and custodians of wildlife are constantly under pressure to adopt the minimum retention of habitat that will save a species. Although this is a poor bargain, pragmatically we must develop our intellectual and professional capability within these constraints because minimal solutions are often the only solutions we can achieve. We have a growing list of endangered species, and we who value wildlife must become proficient at metapopulation management to save as much as possible.

Much basic information is lacking, but in the meantime landscape alteration continues. Wildlife professionals and land use planners have no choice but to do the best job they can with the information available to them. Hopefully this volume will inspire and guide their efforts and illustrate how metapopulation thinking can assist them in making difficult decisions. Perhaps further, it will direct research toward key issues for implementation of metapopulation approaches in real landscapes.

## Weak Links in the Paradigm

We labor under a severe handicap, for we have no history of managing populations as metapopulations. All of the case histories presented in this book are based on recent developments with the exception of the tule elk (Chapter 16), and the tule elk, a human-dependent metapopulation, although a good example of rescue and recovery, is hardly a standard to embrace for other wild species managed as metapopulations. Our goal has to be to establish landscapes in which populations can function themselves by natural means. The cost of saving species through captive breeding and reestablishment in the wild is enormous and, moreover, feasible for only a few charismatic species (such as pandas and condors) or economically valuable ones (such as salmon). Even zoos, whose recent justification has depended heavily on captive breeding, have recognized that saving species in the wild is the only viable long-term solution (Norton et al. 1995; Sunquist 1995).

Given that the ideas of metapopulation management are being developed simultaneously with their implementation, we cannot say, with any certainty, what our ultimate success will be. We dearly hope for many successes, but some failures seem inevitable. We do not have much experience to guide us, and risk assessment is a young discipline. Indeed, the risk posed to other natural environmental values by our use of products with high economic value (fossil fuels, nuclear energy, whaling, marine fisheries) has been systematically underestimated. Habitat fragmentation is being driven by similar economic forces, and it seems likely that habitat plans to preserve wildlife will be pushed to the very edge of viability.

A major technical problem, one that has hardly been touched, is the derivation of landscape designs that will serve the interests of a diverse mix of species in an intact wildlife community. We have reached a stage where we are able to design landscapes that function for the needs of one or a few featured species. But the ultimate goal of conservation is to preserve total biodiversity.

It is clear that dispersal behavior varies enormously across taxa: one size (and spacing) of habitat patches does not fit all (Chapter 4). Large habitat areas widely spaced, as for the northern spotted owl, may serve the needs of the owl while simultaneously retaining continuous populations of small, low-vagility species within habitat patches. But what about the intermediate-sized species for which the spacing of patches exceeds dispersal capacity while the individual patches are too small to maintain viable continuous populations? How do we optimize a system of patches across this array of species? Surely a system that favors the entire mix will be suboptimal from the perspective of many of the individual species. Such an analysis is handicapped by our deficient information about dispersal of most species (Chapter 5). Will corridors funnel animals from patch to patch, or will they become sinks that benefit mainly the predators of the matrix? Can species successfully cross the matrix in the absence of corridors? Which species can maintain continuous populations in the size of patches established, and which are reduced to metapopulations?

The list of questions about the ability of landscape-level habitat-management plans to support mixed-species communities goes on and on. Metapopulation management is clearly a work in progress. Even the case histories presented in this book—although they are among the best-developed applications of metapopulation thinking—are incomplete. Some have strength in demography, others in genetics, and some a bit of both. Most are weak in dispersal evidence. None can claim to document how dispersers relate to the matrix during dispersal.

Similarly, we are in the early stages of integrating traditional processes of community structuring into metapopulation thinking. We know that competition and predation are important forces that influence all populations, including metapopulations. Clearly, the influence of competitors and predators favored by the matrix must be incorporated in metapopulation management—not only their impact on dispersers in the matrix but also their intrusion into the habitat patches themselves (edge effects).

Furthermore, questions remain about the predator and competitor relationships of species naturally coexisting within the patches. It seems likely that scale effects shift ecological balances as the landscape is fragmented into smaller patches. Consequently, equilibria present in the former large habitat areas may be impossible to maintain in the fragmented landscape. The recent review of the competitive displacement of other small rodents by kangaroo rats (*Dipodomys* spp.) (Valone and Brown 1995) illustrates the problem. If the habitat patches are designed for the benefit of kangaroo rats (Chapter 10), the success of that metapopulation could come at the expense of other co-occurring rodent species. Favoring one species at the expense of others may be justified by the endangered status of the first species, but application of metapopulation approaches to protect mixed-species communities may inadvertently result in some species being reduced to disequilibrium by natural ecological processes such as competition and predation.

Another issue that has received insufficient attention is the demographic consequence of sex ratio in persistence of subpopulations of a metapopulation (McCullough 1996). Although sex ratio figures prominently in genetic models of effective population size (Chapter 3), it is less often treated in demographic considerations, perhaps because of the slow transition of metapopulation models from presence/absence for patch occupancy to more realistic models that take into account carrying capacity and other local variables that are not simply correlated with patch size. Obviously, the extinction probability of local populations in real patches is determined by the sex in shorter supply. With vertebrates, this is usually consistently one sex or the other. The four female cougars without a male in an isolated habitat patch (Chapter 13) constitute a case in point, although in that case local extinction was averted by the fortuitous (but timely) dispersal of males into the habitat patch. Sex-ratio effects are encompassed in the larger concept of demographic stochasticity, however, so they have not been ignored entirely. Similarly, simulation models may incorporate sex ratio in their predictions. Nevertheless, the lack of a formal model equivalent to effective population size for genetics has resulted in demographic stochasticity in small populations being considered more commonly with reference to the total numbers than with the numbers of the less abundant sex.

## The Research Agenda

To support metapopulation management the most immediate need is for information on the dispersal behavior of species in a habitat mosaic. Dispersal is the keystone. All landscape designs are predicated on certain assumptions about how animals will move about the landscape, and if these assumptions are incorrect, the design will fail.

Research on dispersal has lagged in comparison with other aspects of life history because of the inherent difficulties in its study. Dispersal usually occurs as a pulse at one season and is short in duration. Dispersal often involves young, previously sequestered animals that are difficult to assess and in any event have low survivorship. These characteristics make it difficult to obtain an adequate sample size of dispersers. Many studies start out with good samples of marked individuals but fail when too few reach dispersal age; then only a portion of the survivors may disperse, leaving a few scattered instances on which to base conclusions.

Finally, dispersal movements often exceed the spatial scale of study areas.

Long-distance dispersals may occur quite irregularly, but they may be sufficient to support metapopulation structure if patch extinctions are relatively rare. Because dispersal may occur over a brief time and involve movements of considerable distance, direct observation is usually inadequate. Visual markings require initial capture, and, while they may work reasonably well for conspicuous birds, they are problematic for secretive mammals, reptiles, and amphibians.

Radiotelemetry has contributed much to what we know about dispersal behavior, and it will continue to be a key tool. It is relatively expensive, however, and establishing an animal's pathways, stopovers, and avoidances requires considerable human effort to keep up with its continuing movements. Nevertheless, the results are indispensable for sound landscape design. To obtain adequate sample sizes, such research will be most useful and cost-effective if it is directed at those classes of animals most likely to disperse. Exceptionally long-distance movements will continue to be difficult to detect, of course, but landscape planning based on the frequency distributions of dispersal distance from sample and study area sizes that do not detect and assess the effects of these rare events will contain a desirable element of conservatism.

Another major research question is how multiple species constituting a community can be incorporated into landscape designs. The "umbrella species" approach has considerable utility, and most single-species metapopulation plans will benefit an array of other species. But little detailed analysis has been done to support such expectations of mutual benefit and, particularly, to determine what species are not favored or are strongly disfavored.

Overall, modeling has proceeded well ahead of field biology with regard to metapopulation dynamics. However, the multispecies problem is one area in which further modeling could make a major contribution. Such modeling is difficult because even simple communities result in models that are extraordinarily complex. Nevertheless, modeling a few species of markedly different niche types (viability in various patch sizes, different dispersal abilities, and so forth) would tell us much about the likelihood of certain landscape patterns to support an array of species. Furthermore, such models might reveal conflict points that cannot be simultaneously resolved across the species array by adjusting size and spacing of habitat patches. Model results would give valuable early warning about problems to be assessed subsequently by biological research.

Landscape planners are grappling with these issues, but of necessity they are currently drawing lines on maps that will determine the future spatial configuration of the wildlife habitats to be retained. Such decisions are being made on the basis of the biological data we have and the best-informed expert opinion on how it will all fit together. Any thoughtful biologist, however, will be troubled by the unpredictable outcome of decisions that will have long-lasting or even permanent consequences.

In the meantime, field biologists and species-group specialists must contribute their best judgments about how the species mix will respond to various landscape designs. This is a stopgap measure. Only careful research results will give solid footing to metapopulation planning. Research priorities must be reordered, therefore, to address the key issues for metapopulation management.

## REFERENCES

McCullough, D. R. 1996. Spatially structured populations and harvest theory. *Journal of Wildlife Management* 61:1–9.

Norton, B. G., M. Hutchins, E. F. Stevens, and T. Maple. 1995. *Ethics on the Ark: Zoos, Animal Welfare, and Wildlife Conservation.* Washington, D.C.: Smithsonian Institution Press.

Sunquist, F. 1995. End of the ark? *International Wildlife* 25(6):23–29.

Valone, T. J., and J. H. Brown. 1995. Effects of competition, colonization, and extinction on rodent species diversity. *Science* 267:880–883.

# About the Contributors

**JONATHAN D. BALLOU** is population manager at the Smithsonian Institution's National Zoological Park in Washington, D.C. His research interests focus on genetic problems of small populations, in particular the effects of inbreeding and outbreeding in captive populations. He has developed genetic-management strategies for maintaining genetic variation in both captive and wild populations and serves as population genetics adviser for endangered species breeding programs such as the golden lion tamarin, black-footed ferret, Przewalski's horse, and Florida panther. Ballou is a member of the Conservation Breeding Specialist Group (CBSG) and the Reintroduction Specialist Group of the International Union for the Conservation of Nature. He recently edited a book on population management (*Population Management for Survival and Recovery*) with M. Gilpin and T. Foose.

**PAUL BEIER** is assistant professor of wildlife ecology at Northern Arizona University's School of Forestry. His research addresses management and conservation problems including population viability and juvenile dispersal of cougars, the relative importance of prey abundance and forest structure to selection of foraging habitat by goshawks, and the influence of the patch size and pattern of small aspen stands on avian communities in conifer-dominated forests. Beier is an active member of The Wildlife Society and the Society for Conservation Biology.

**MARIE-ODILE BEUDELS** is a member of the scientific staff of the Royal Institute for Natural Sciences of Belgium. A historian specializing in the compilation and analysis of archives, she coordinates the Monk Seal Register, a computer database designed to gather all information on the species in a standardized way.

**VERNON C. BLEICH** is a senior wildlife biologist with the California Department of Fish and Game and has worked extensively in the Sonoran, Mojave, and Great Basin deserts of eastern California. He played a prominent role in developing an extensive conservation and restoration program for mountain sheep in California and has received several regional and national awards in recognition of his efforts. He has published more than 50 professional papers or book chapters and is keenly interested in the application of contemporary ecological thought to problems in wildlife conservation. Dr. Bleich is an affiliate faculty member at the University of Alaska Fairbanks and an adjunct faculty member at the University of Rhode Island.

F. LANCE CRAIGHEAD is adjunct professor at Montana State University, director of the Craighead Environmental Research Institute, and vice president of the Northern Rockies Conservation Cooperative. He has been involved in various aspects of grizzly bear research since his high school days in Yellowstone National Park, culminating in his current focus on conservation genetics and the use of GIS technology to model animal movements and gene flow across landscapes. His doctoral research determined paternity, identified cases of multiple paternity, and, using microsatellite DNA analysis, derived estimates of male reproductive success in arctic grizzly bears. As scientific coordinator for American Wildlands in Bozeman, Montana, he is involved in efforts to delineate linkage corridors for animal movements in the northern Rocky Mountains.

JON K. FISCHER is a wildlife biologist with the California Department of Fish and Game. He has been coordinator of the department's elk management program since 1991 and has been involved in reintroducing tule elk to suitable historic habitat and monitoring the status of established herds.

JOHN W. FITZPATRICK is director of the Cornell Laboratory of Ornithology, Ithaca, New York, and was executive director of the Archbold Biological Station, Lake Placid, Florida. He has published over 70 articles, books, and monographs on ecology, demography, biogeography, and social evolution of birds. His research emphasizes conservation biology in Florida and the New World tropics. Currently, his focus is on the ecology and conservation of endangered species whose life histories evolved in close association with natural fire. He serves on the national boards of The Nature Conservancy and National Audubon Society. His books include *The Florida Scrub Jay: Demography of a Cooperative-Breeding Bird* (1984 with G. E. Woolfenden) and *Neotropical Birds: Ecology and Conservation* (1995 with D. F. Stotz, T. A. Parker III, and D. K. Moskovits).

MICHAEL GILPIN is professor of conservation biology at the University of California, San Diego. His early research emphasized evolution, interspecies competition theory, and island biogeography theory. A conservation biologist since 1982, he has pioneered viability modeling and applications of meta-population theory to conservation biological analysis. He has consulted with numerous federal and international agencies on reserve designs and recovery plans. He serves on the editorial board of *Journal of Theoretical Biology* and *Journal of Restoration Ecology*.

R. J. GUTIÉRREZ is a professor of wildlife management, Department of Wildlife, Humboldt State University, Arcata, California. His research em-phasizes habitat ecology, game bird ecology, and endangered species conser-vation. He has worked extensively with all three spotted owl subspecies and has studied their home range, habitat, morphology, population dynamics, and genetics. He serves on the board of trustees of The Nature Conservancy, the Tall Timbers Research Station, the Northern Spotted Owl Recovery Team, and the California Spotted Owl Technical Assessment Team. He is an associate editor of *Wildlife Biology*.

SUSAN HARRISON is a professor of environmental studies at the Univer-sity of California, Davis. Her research focuses on the effects of habitat geom-etry and dispersal on population dynamics and community structure. She has conducted experimental studies on the population and metapopulation dy-namics of herbivorous insects and is currently comparing spatial patterns of plant and insect diversity in patchy versus continuous habitats. She has written numerous reviews on metapopulation theory and its application to conservation.

JOHN HARWOOD is a fellow of St. Edmund's College at Cambridge Uni-versity and has been the head of the United Kingdom Sea Mammal Research Unit since 1978. Much of his research has been concerned with the effects of variation between individuals on the population dynamics of large verte-brates. In particular, he has developed mathematical models of the dynamics of small, and often endangered, populations where individual variation can have significant effects. He serves on the editorial boards of *Conservation Biology*, *Marine Mammal Science*, and *Mathematics Applied in Medicine and Biology*.

PHILIP W. HEDRICK is the Ullman Professor of conservation biology at Arizona State University. His research focuses primarily on conservation and evolutionary genetics from the theoretical, experimental, and molecular per-spectives. He is on the recovery teams for the Mexican wolf and the winter-run chinook salmon and has evaluated the genetic impact of the introduction of Texas cougars into the Florida panther population. His books include *Genetics of Populations*, *Population Biology*, and *Genetics* (with R. Weaver). He is the past president and secretary of the Society of American Naturalists and has served on the editorial boards of *Conservation Biology*, *Evolution*, *Journal of Theoretical Biology*, and *Genetica*.

WALTER D. KOENIG is associate research biologist at the Museum of Vertebrate Zoology and associate adjunct professor in the Department of Integrative Biology, University of California, Berkeley. His interests are primarily in the fields of behavioral ecology and population biology. He has published extensively on avian social behavior, including *Population Ecology of the Cooperatively Breeding Acorn Woodpecker* (with R. L. Mumme) and *Cooperative Breeding in Birds* (with P. B. Stacey), and has also studied dragonflies, oaks, and California tiger salamanders.

WILLIAM Z. LIDICKER, JR., is a professor of integrative biology at the University of California, Berkeley. He has published over 100 articles, books, and monographs, mainly on mammalian ecology, social biology, and systematics. In recent years his research has emphasized population dynamics, spatial structure, dispersal, and landscape ecology. Two recent edited books are *Animal Dispersal: Small Mammals as a Model* (1992 with N. C. Stenseth) and *Landscape Approaches in Mammalian Ecology and Conservation* (1995).

THOMAS R. LOUGHLIN is a wildlife biologist at the National Marine Mammal Laboratory, Alaska Fisheries Science Center, in Seattle, Washington. He heads a program that addresses interactions between Alaskan marine mammals and ecosystem variability related to natural and human causes. Presently his research group is trying to determine the causes of population declines in Alaskan Steller sea lions, northern fur seals, and harbor seals. He has published over 100 articles on marine mammal biology and management. His most recent book is *Marine Mammals and the Exxon Valdez* (1994).

DALE R. McCULLOUGH is a professor of wildlife biology and management in the Department of Environmental Science, Policy, and Management, as well as research conservationist in the Museum of Vertebrate Zoology, at the University of California, Berkeley. He has participated in the recovery program of the tule elk for over 30 years. His research interests focus on the ecology and behavior of large mammals, mainly ungulates and carnivores, and include a long-term study of the George Reserve deer herd in Michigan. In his work he has tried to integrate basic ecology, and more recently conservation biology, with applied management. He has over 90 publications, including the books *The Tule Elk*, *The George Reserve Deer Herd*, edited (with Reg Barrett) *Wildlife 2001: Populations*, and is a coauthor of *Wildlife Policies in the U.S. National Parks*.

**KEVIN S. MCKELVEY** is a research forester for the U.S. Forest Service, Pacific Southwest Station. His research has focused on species conservation in fragmented forested ecosystems. He has developed a spatially explicit model of spotted owl population dynamics that has been used by the Bureau of Land Management, the Forest Service, and the State of Oregon to compare the relative efficacy of competing land management plans. His recent research has focused on evaluation of disturbance dynamics in fire-prone ecosystems, specifically in the Sierra Nevada. He has served as a member of the California Spotted Owl Technical Assessment Team and is currently a special consultant to the Sierra Nevada Ecosystem Project.

**RICHARD L. MERRICK** is a zoologist and the leader of the Steller Sea Lion Task for the National Marine Mammal Laboratory. His research for the past 10 years has dealt with the foraging ecology of the Alaskan population of the Steller sea lion and the potential link between this topic and the decline of the species.

**BARRY R. NOON** is a research ecologist at the U.S. Forest Service's Redwood Sciences Laboratory in Arcata, California. In 1995 he served as chief scientist for the National Biological Service in the Department of the Interior in Washington, D.C. For 6 years he was on the faculty of the Wildlife Department at Humboldt State University. His interests include population ecology of terrestrial vertebrates, conservation biology and planning, and application of science to the resolution of natural resource problems. Recently he has concentrated on spotted owl conservation. He has published over 70 papers and book chapters and is an associate editor of *Journal of Wildlife Management*.

**BILL PRANTY** is a research assistant at Archbold Biological Station, Lake Placid, Florida. He has worked mostly with Florida scrub jays and was recently a participant in a statewide survey. He has published several papers on Florida birds and is currently state compiler for the Florida Ornithological Society's field observations committee. He was acting state coordinator for the Florida breeding bird atlas project and is a coauthor of the atlas. He is the author of *A Birder's Guide to Florida* (1996).

MARY V. PRICE is professor of biology at the University of California, Riverside. Her research considers diverse problems in the evolutionary ecology of animals and plants in arid and montane ecosystems. She has explored factors that determine the diversity and structure of communities of seed-eating desert rodents; how mutualistic interactions between plants and pollinators mold the evolution of flower morphology, mating patterns, and dynamics of plant populations; effects of global warming on montane plant communities; and demography of small mammals in patchy environments. She has served on the editorial boards of *Ecology* and *Evolution.*

ROB ROY RAMEY II is a postdoctoral research associate at the Department of Evolutionary, Population, and Organismic Biology, University of Colorado, Boulder. His research focuses on the application of molecular population genetics and phylogenetics to problems in wildlife conservation and parasitology. He seeks to bridge the gap between pure and applied science by designing research projects that answer interesting theoretical questions while providing management solutions to pressing environmental problems.

CHRIS RAY is a Ph.D. candidate at the University of California, Davis. She studies the effects of spatial population structure and dispersal behavior on long-term population dynamics in the field, in the laboratory, and in the computer.

JENNIFER L. RECHEL is a Ph.D. student at the University of California, Riverside, in the Department of Earth Sciences. Her research focuses on quantifying movement patterns of large mammals in heterogeneous landscapes, with emphasis on how the spatial variation of distance in physical space affects large mammal distributions and advances metapopulation theory. She also conducts research on the spatial processes affecting inter- and intraspecific competition of mule deer and mountain sheep. She has published articles on fire ecology and geographic information systems (GIS).

PER SJÖGREN-GULVE is lecturer and docent in Conservation Biology at Uppsala University in Sweden. His research focuses on viability analyses of single populations and metapopulations using demographic models and occupancy models. The genetic properties of metapopulations in relation to population structure are another focus of interest. The pool frog has been closely studied from these points of view, and other taxa are presently being investigated. Both theoretical and applied aspects of this research are emphasized.

HELEN STANLEY is a research fellow at the Institute of Zoology, London. As a member of the Conservation Genetics Group, she uses molecular techniques to investigate questions of speciation and patterns of genetic variability in terrestrial and aquatic species. Recent investigations have included studies of population structure in selected phocid seal species, as well as the evolutionary relationships among pinnipeds and the South American camelids. As part of the latter project, she is using historical specimens from 7000 B.C. to the sixteenth century. A basic aim of her research is to use genetics to help identify management units and the ranking of conservation priorities.

BRADLEY M. STITH is a Ph.D. student at the University of Florida's Department of Wildlife Ecology and Conservation in Gainesville. His research emphasizes the integration of GIS and remote sensing techniques with spatially explicit population models, using the Florida scrub jay as a case study. His empirical work includes radio-tracking dispersing jays, censusing jays for the statewide survey, and comparing population dynamics of jays in differing habitat quality and in response to habitat restoration. He is a member of the Scientific Advisory Group for the Brevard County, Florida, Scrub Habitat Conservation Plan and is a participant in the committee to develop guidelines for a statewide habitat conservation plan for the Florida scrub jay.

CHARLES VANDERLINDEN is a member of the scientific staff of the Royal Institute for Natural Sciences of Belgium. He is the database manager responsible for the development of the Monk Seal Register.

ERNEST R. VYSE is a professor of biology at Montana State University. He is interested in genetic diversity within and between wildlife populations. Currently, he is using molecular genetics to study mating systems, paternity, dispersal, and migration in wildlife populations. He is acting head of the Biology Department.

JOHN D. WEHAUSEN is a research associate with the University of California's White Mountain Research Station. For more than two decades he has been developing long-term data sets concerning comparative demography of bighorn sheep populations in California. These populations, ranging from the high mountains of the Sierra Nevada and White Mountains to a variety of mountains in the Mojave Desert, exhibit similar diversity in the factors influencing demography. He has also been active in the conservation of bighorn sheep and has published articles on a variety of aspects of this species.

JOHN A. WIENS is professor of ecology in the Department of Biology and the Graduate Degree Program in Ecology at Colorado State University. His research has focused on how landscape structure influences ecological processes, especially movement behavior, and on the organization of communities of birds and ants in semiarid environments. Throughout this work he has emphasized the importance of spatial and temporal variation. In another arena, he has investigated how oil spills affect seabird populations and communities and how studies to document these effects should be designed. He is editor of the *Cambridge Studies in Ecology* series, a member of the editorial boards of *Landscape Ecology, Current Ornithology*, and *Acta Zoologica Academiae Scientiarum Hungaricae*, and has served as editor of *The Auk*. He is a recipient of the Elliott Coues Award of the American Ornithologists' Union and author of a two-volume work: *The Ecology of Bird Communities* (1992).

GLEN E. WOOLFENDEN is Distinguished Research Professor in the Department of Biology at the University of South Florida. He has published over 100 articles, chapters, and books, almost entirely in the field of ornithology. During the past 25 years he has conducted an intensive, long-term study of the behavior and ecology of the Florida scrub jay, a federally threatened species. In 1984 he and coworker John Fitzpatrick published *The Florida Scrub Jay: Demography of a Cooperative-Breeding Bird*, for which they received the Brewster Medal from the American Ornithologists' Union.

ANNE E. YORK is a statistician at the National Marine Mammal Laboratory in the Alaska Fisheries Science Center. Her main research interests are census methods for marine mammal populations and the application of life-history theories to pinniped population dynamics. She has published research papers on northern fur seals, Steller sea lions, harbor porpoise, and fin whales. At present she is working on the comparative population dynamics of all the fur seal species.

# Index